电力安全技术
——安全实践篇（中英双语）

主　编　欧阳仁乐　赵尧麟
副主编　杨　龙　　陈道强　　姜聿涵
　　　　钟运来　　孙睿哲

西南交通大学出版社
·成　都·

图书在版编目（CIP）数据

电力安全技术. 安全实践篇：汉文、英文 / 欧阳仁乐，赵尧麟主编. -- 成都：西南交通大学出版社，2023.12
ISBN 978-7-5643-9595-7

Ⅰ. ①电… Ⅱ. ①欧… ②赵… Ⅲ. ①电力安全 – 教材 – 汉、英 Ⅳ. ①TM7

中国国家版本馆 CIP 数据核字（2023）第 231415 号

Dianli Anquan Jishu — Anquan Shijian Pian（Zhong-Ying Shuangyu）
电力安全技术——安全实践篇（中英双语）

主　编／欧阳仁乐　赵尧麟	责任编辑／张文越
	封面设计／GT 工作室

西南交通大学出版社出版发行
（四川省成都市金牛区二环路北一段 111 号西南交通大学创新大厦 21 楼　610031）
营销部电话　028-87600564　　028-87600533
网址　http://www.xnjdcbs.com
印刷　四川玖艺呈现印刷有限公司

成品尺寸　　185 mm×260 mm
印张　　24.25　　字数　　713 千
版次　　2023 年 12 月第 1 版　　印次　　2023 年 12 月第 1 次

书号　　ISBN 978-7-5643-9595-7
定价　　98.00 元

课件咨询电话：028-81435775
图书如有印装质量问题　本社负责退换
版权所有　盗版必究　举报电话：028-87600562

前 言

电能既是一种经济、实用、清洁且容易控制和转换的能源形态，又是电力部门向电力用户提供由发、供、用三方共同保证质量的一种特殊产品。电能的利用是第二次工业革命的主要标志，从此人类社会进入电气时代。电能被广泛应用在动力、照明、化学、纺织、通信、广播等各个领域，是科学技术发展、人民经济飞跃的主要动力，由此可见电能在我们的生活中起到重大的作用。

然而由于电能的不可视性，人们不易辨识到电的危险，电气事故总是猝不及防，各种安全隐患和事故也如影随形，威胁着人类的健康乃至生命。

如何消除电气安全隐患、防范电气安全事故早已成为一个具有普遍和重要意义的长久课题。这也造就了安全工程领域中的一个重要分支——电力安全技术。电力安全技术是指在电力系统的运行过程中，对其进行安全管理和控制的一系列技术手段。随着现代社会对能源需求的不断增加和使用以及电气设备的不断增多，电力安全问题成为了一个备受关注的话题，因此，电力安全技术的研究和应用显得尤为重要。

根据国家教育事业发展"十四五"规划，突出职业教育类型特色，深入推进改革创新，优化结构布局，推行"1+X"证书制度，深化产教融合等要求，本书充分体现了"十四五"规划的职业教育目标，将电力安全工程知识与电网检修、运行等实际内容结合，构成理论、实践一体化、模块化教学。本书理论知识侧重于电气设备安全运行、电气设备安全检修；电力线路安全运检；现场风险辨识与现场急救；防火与防爆等。

本书由欧阳仁乐和赵尧麟担任主编，杨龙、陈道强、姜聿涵、钟运来、孙睿哲担任副主编。其中欧阳仁乐编写模块五、赵尧麟编写模块六、杨龙编写模块七、陈道强、姜聿涵共同编写模块八、钟运来、孙睿哲共同编写模块九。编写过程中作者参考了大量相关书籍，谨向这些书籍作者致谢，在编写过程中得到编写人员单位和评审人员的大力支持和帮助，在此一并表示由衷的感谢。由于编写时间仓促，经验不足，水平有限，虽经反复修改，仍难免有不妥和不足之处，恳请读者批评指正。

目 录

模块五　电气设备安全运行 ··· 001
Module V　Security operation of electrical equipment ······································ 002

　　任务一　正确实施倒闸操作 ··· 003
　　Task I　Correctly performing switching operations ······································ 010
　　任务二　巡视电气设备 ··· 021
　　Task II　Routine inspections of electrical equipment ··································· 027
　　任务三　高压开关柜停送电操作实训 ·· 036
　　Task III　Training in power interruption and supply operations of HV switch cabinets ······ 039
　　任务四　变电站设备巡视检查实训 ··· 042
　　Task IV　Training in routine inspections of substation equipment ················ 047
　　任务五　10 kV 柱上变压器的停送电操作实训 ··· 055
　　Task V　Training on Power-on/off Operation of 10 kV Pole-mounted Transformers ······ 059

模块六　电气设备安全检修 ··· 065
Module VI　Maintenance of electrical equipment ·· 066

　　任务一　认识主要电气设备检修安全技术 ·· 067
　　Task I　Know the safety technology for maintaining the main electrical equipment ······ 081
　　任务二　认识电气试验安全技术 ·· 100
　　Task II　Learn safety technologies for electric testing ································ 103
　　任务三　高压开关柜故障判断及处理实训 ·· 107
　　Task III　HV switch cabinet fault identification and practical training on fault handling ······ 110
　　任务四　变压器绝缘电阻测试实训 ··· 113
　　Task IV　Practical training on measurement of transformer insulation resistance ······ 116
　　任务五　变压器分接开关的调整 ·· 120
　　Task V　Adjustment of Transformer Tap Changer ······································ 123

模块七　电力线路安全运检 ··· 127
Module VII　Safe operation and inspection of electric power lines ···················· 128

　　任务一　电力线路运行与检修 ·· 129
　　Task I　Electric power line operation and maintenance ······························ 142

  任务二 电力线路带电作业 159
  Task II Live-wire work on electric power line 167
  任务三 电力电缆绝缘测试实训 178
  Task III Practical training on power cable insulation test 184
  任务四 导线在绝缘子子上的绑扎操作实训 191
  Task IV Training on Tying Conductors on Insulators 195

模块八 现场风险辨识与现场急救 199

Module VIII Risk identification and first aid on the site 200

  任务一 作业现场风险辨识及职业病 201
  Task I On-site risk identification and occupational disease 213
  任务二 触电急救 229
  Task II First aid for electric shock sufferers 241
  任务三 紧急救护常识及应用 256
  Task III First aid knowledge and application 267
  任务四 排除作业现场安全风险实训 280
  Task IV Training in eliminating safety risks on the job site 284
  任务五 触电急救实训 288
  Task V Training in first aid for electric shock 293
  任务六 伤口包扎实训 299
  Task VI Training in wound dressing 302

模块九 防火与防爆 305

Module IX Fire and explosion protection 306

  任务一 认识火灾和爆炸 307
  Task I Learn about fire and explosion 313
  任务二 消防设施与消防器材的选择与使用 321
  Task II Selection and use of fire-fighting equipment and facilities 330
  任务三 防火、灭火、逃生与自救 342
  Task III Fire prevention, extinguishing, escape and self-rescue 356
  任务四 灭火器选择与检查实训 374
  Task IV Training in the selection and inspection of fire extinguishers 378

参考文献 382

模块五 电气设备安全运行

电气运行是指在电力生产过程中，为保证发供电设备正常运行，以及事故情况下的紧急处理，发电厂、变电站运行值班人员对发供电设备进行的监视、控制、操作和调整。电气运行安全技术是指运行值班人员在进行上述工作时，为保证人身安全和设备安全，应遵循的技术规范、使用的技术装备，以及采用的技术措施。本模块主要讲述倒闸操作安全技术，以及设备巡视检查安全要求等相关内容。

学习目标：
（1）能说出倒闸操作的概念和作用、倒闸操作的安全技术规范。
（2）能安全进行开关类设备倒闸操作。
（3）能说出电气设备巡视检查的规定和方法。
（4）能说出各种电气设备巡视检查的注意事项。

Module V Security operation of electrical equipment

By electrical operation, we mean in power generation, the on-duty operators of a power plant and substation monitor, control, operate, and adjust power generation and power supply equipment to ensure their normal operation and respond to emergencies if any. Electrical operation security technique refers to the technical specifications followed, technical equipment used, and technical measures adopted by the on-duty operators to ensure personal safety and equipment safety when performing the above work. This chapter mainly describes the security technology of switching operations and the safety requirements for equipment routine inspection.

Learning objectives:

(1) Be able to tell the concept, function, and security technical specification of the switching operations.

(2) Be able to perform switching operations of switching equipment.

(3) Be able to tell the regulations and methods for routine inspections of electrical equipment.

(4) Be able to tell the precautions for routine inspections of all kinds of electrical equipment.

任务一　正确实施倒闸操作

一、倒闸操作概述

倒闸操作是指需使电气设备改变运行状态或使电力系统改变运行方式时，对开关电器的拉合、操作回路的拉合、控制及动力电源的拉合、继电保护装置和自动装置的投退及切换，以及临时接地线的装拆等操作。

发电厂、变电站是电力系统的主要环节，因负荷变化、需改变电气设备或系统运行方式，经常要进行倒闸操作。发电厂、变电站里的电气设备比较多，电气接线也比较复杂，在倒闸操作中不仅有一次回路的操作，也有二次回路的操作。因此可以说，倒闸操作不仅是电气运行的一项重要工作，也是一项经常性、比较复杂的工作。在进行倒闸操作时，一旦发生误操作，不仅会影响供电或损坏设备，还可能危及操作人员的人身安全和电网的安全运行。因此，正确进行倒闸操作是保证发电厂、变电站安全运行的一个重要内容。

倒闸操作应根据值班调度控制（简称调控）人员或运营维护（简称运维）负责人的指令，受令人复诵无误后执行。发布指令应准确、清晰，使用规范的双重名称。发令人和受令人应先互报单位和姓名，发布指令的全过程（包括对方复诵指令）和听取指令报告时应录音，并做好记录。操作人员（包括监护人）应了解操作目的和操作顺序。对指令有疑问时应向发令人询问清楚无误后执行。发令人、受令人、操作人员（包括监护人）均应具备相应资质。

倒闸操作可以通过就地操作、远控操作、程序操作完成。远控操作、程序操作的设备应满足有关技术条件。

二、倒闸操作的分类

1. 监护操作

监护操作顾名思义就是指有人监护的操作。操作时，其中一人（对设备较为熟悉者）作监护。特别重要和复杂的倒闸操作，由熟练的运维人员操作，运维负责人监护。

2. 单人操作

单人操作是指由一人完成的操作。

（1）单人值班的变电站或发电厂升压站操作时，运维人员根据发令人用电话传达的操作指令填用操作票，且复诵无误。

（2）若有可靠的确认和自动记录手段，调控人员可实行单人操作。

（3）实行单人操作的设备、项目及人员需经设备运维管理单位或调控中心批准，人员应通过专项考核。

3. 检修人员操作

检修人员操作即由检修人员完成的操作。

（1）经设备运维管理单位考试合格并批准的本单位检修人员，可进行 220 kV 及以下的电气设备由热备用至检修，或由检修至热备用的监护操作。监护人应是同一单位的检修人员或设备运维人员。

（2）检修人员进行操作的接、发令程序及安全要求应由设备管理单位审定，并报相关部门和调控中心备案。

三、倒闸操作票填写规定

（1）除事故处理、拉合开关的单一的操作、拉开或拆除全站唯一的一组接地刀闸或接地线外，其他操作均应填写倒闸操作票。其中，事故处理指事故发生后的紧急处理，不包括事故处理后的善后操作。上述不用操作票的操作必须事先在值班记录簿上做好记录，持值班记录簿或事故调度命令票，两人一起去操作。

（2）手工填写的操作票必须用统一印刷装订的操作票，操作票用钢笔或圆珠笔填写，票面整洁、字迹工整清楚，有个别字错了可立即在错字上打上"×"，接连书写，不能涂改。但"拉""合"等关键字及设备调度命名、编号和操作时间等关键信息不得修改。

（3）操作执行完毕，在操作票最后一步下面一行左边顶格盖"已执行"章；若最后一步正好位于操作票的最后一行，则在该操作步骤项目栏右边顶格加盖"已执行"章。

（4）在操作票执行过程中因故中断操作，应在已操作完的步骤下边一行顶格居左加盖"已执行"章，并在备注栏内注明中断原因。若此操作票还有几页未执行，则应在未执行的各页操作任务栏右下角加盖"未执行"章。

（5）若一份调度命令票不是一次性全部执行，则应严格在调度员的分项命令对应的每段操作的最末一项盖"暂停"章，模拟预演和正式操作均不得超过盖有此印章的操作项，严禁越项操作。

（6）操作票作废应在操作任务栏内右下角加盖"作废"章，在作废操作票备注栏内注明作废原因；调控通知作废的任务票应在操作任务栏内右下角加盖"作废"章，并在备注栏内注明作废时间、通知作废的调控人员姓名和受令人姓名。若作废操作票含有多页，则应在各页操作任务栏内右下角均加盖"作废"章，在作废操作票首页备注栏内注明作废原因，自第二张作废页开始可只在备注栏中注明"作废原因同上页"。

（7）用计算机打印操作票的站不允许用一般计算机制票，必须使用有防误操作系统的专家系统制票。

（8）操作票以变电站为单位填写，一个变电站对应一本操作票，按值移交，每月由专人进行整理收存。计算机打印的操作票按月装订成册，封面应有当月操作票执行张数、合格张数、合格率的统计。操作票至少保存两年。

（9）"编号"栏：包含变电站电压等级和名称，按月连续编号，如"500 kV 尖山变电站 202006001"，表示 500 kV 尖山变电站 2020 年 6 月第 1 张操作票。

（10）操作中应注意设备的动作、指示、声音情况，正确方可继续操作；在倒闸操作过程中，若有疑问，应立即停止操作，并报告调度员，待询问清楚后，方可继续操作；雷雨时，禁止倒闸操作。

（11）操作票在执行中运行人员不得擅自颠倒顺序、增减步骤、跳步、隔步。

（12）为使操作更顺畅、减少操作时的往复跑路次数，在保证安全的前提下，填票时可考虑采用便利的操作步骤。

四、变电站倒闸操作程序

变电站倒闸操作一般程序如图 5-1 所示。

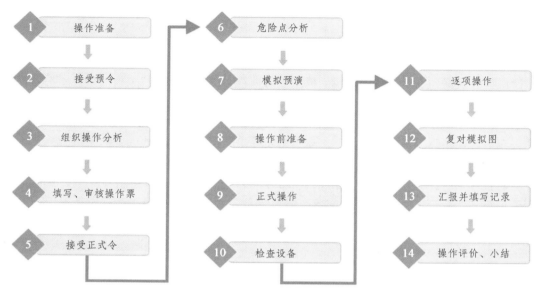

图 5-1　变电站倒闸操作一般程序

1. 操作准备

复杂操作前由站长或值长组织全体当值人员做好如下准备：

（1）明确操作任务和停电范围，并做好分工。

（2）拟订操作顺序，确定装挂地线部位、组数及应设的遮栏、标示牌；明确工作现场邻近带电部位，并制定出相应安全措施。

（3）考虑保护和自动装置相应变化及应断开的交、直流电源和防止电压互感器、所用变二次反高压的措施。

（4）分析操作过程中可能出现的问题和应采取的措施。

（5）与调度联系后写出操作票草稿，由全体人员讨论通过，站长或值长审核批准。

（6）预定的一般操作应按上述要求进行准备。

（7）设备检修后，操作前应认真检查设备状况及一、二次设备的分合位置与工作前是否相符。

2. 调度下达操作命令（操作计划），正值运行人员接受操作命令（操作计划）

接受调度命令由正值运行人员进行。在接受调度操作命令（操作计划）时应打开录音机，首先通报单位、岗位和姓名，如"这里是××220 kV 变电站，我是当值值班负责人（正班）××"，并问清下令人姓名、时间；在填写调度命令时，应在接电话的同时在相关记录上认真填写齐全，字迹清楚、整洁；在填写完毕后，必须予以复诵无误；然后由值班负责人（正班）确定此操作的操作人和监护人。

如果认为该命令不正确，应向调度员报告，由调度员决定原调度命令是否执行。但当执行该项命令将威胁人身、设备安全或直接造成停电事故时，必须拒绝执行，并将拒绝执行命令的理由报告调度员和本单位领导。

3. 操作分析

监护人应根据操作任务，结合实际运行方式、设备缺陷等情况，与操作人分析该操作的关键点和注意事项，重点要防止误操作。

4. 操作人查对图板填写操作票，值班负责人（正班）审票

操作人应根据调度操作命令，到模拟图板前核对无误后（无模拟图板的站可核对监控机中主接线的实时运行情况），由操作人填写操作票；填写时思想要集中，做到顺序不乱、项目不漏、字迹清晰；有计算机防误系统的，必须在"实时开票"栏里进行，禁止直接提取"典型操作票"。操作人在填完操作票后，应进行自审，确认无误后，操作人签名，然后交予操作监护人。

当操作监护人接到操作人交来的操作票后应根据调度命令的要求，结合当时的实际运行方式，对操作票仔细审核，当发现错误后立即指出，并交操作人重新填写；当操作监护人不是当值值班负责人时，操作票还应交予当值值班负责人审核。在操作票审核无误后，监护人和当值值班负责人分别在票中相应栏签名。

5. 调度正式发布操作命令，值班负责人（正班）接受操作任务

当调度正式发布操作命令时，必须由当值值班负责人（正班）接受，在复诵无误后，立即将发布时间、人员记录在相关记录中。

6. 监护人和操作人在系统模拟图上预演

在操作实际设备前，监护人和操作人必须先在模拟图板上进行预演（无模拟图板的可在五防查找计算机上进行预演）；预演要求监护人按操作票的顺序唱票，操作人复诵后在模拟图板上操作（只模拟一次设备）；监护人持操作票，逐项下令，要求声音洪亮。

在图板上进行核对性预演，重点是检查操作票有无错误，只有当监护人、操作人均认为演习正确无误时，方获得通过。

7. 操作前准备

在操作开始前，监护人、操作人应确定操作路线，以及对所需的安全用具进行检查（如绝缘靴、绝缘手套检查试验日期、完好性，验电器检查试验日期、声光信号和完好性），并准备好本次操作所需要的钥匙（工器具室、高压室及高压开关柜门钥匙等）。

8. 正式操作

（1）穿戴要求：按《国家电网公司电力安全工作规程》的规定执行。

（2）操作人持安全用具，监护人持钥匙、票，一起走到设备面前，监护人应核实操作任务与设备是否对应。

（3）监护人按操作票顺序逐项下令，要求声音洪亮。例如：

——监护人下令："拉开××开关！"

——操作人手指设备双重编号牌，响亮复诵："拉开××开关！"

——监护人确认无误后，下令："对，执行！"

——操作人在完成该项操作后，对监护人报告："操作完毕！"或"已执行。"

监护人和操作人共同检查操作质量，确认后，在操作票此项操作执行列打"√"，然后进行下一项操作。

（4）操作中发生疑问时，应立即停止操作向值班调度员或值班负责人报告，弄清问题后，再进行操作。不准擅自更改操作票，不准随意解除闭锁装置。

9. 检查设备，向调度汇报操作任务完成

在全部操作结束后，操作人和监护人再次检查设备情况，如该发出的光字信号是否发出，该亮（灭）的信号灯是否亮（灭），电压表、电流表该有（无）指示的是否有（无）等；当监护人不是当值值班负责人时，应向值班负责人汇报此张操作票已完成，值班负责人（正班）审查确认后即可向调度汇报此操作任务完成。

10. 做好记录，在操作票上盖章并保存

在向调度汇报完毕后，值班负责人（正班）将时间、情况记录在相关记录中，并在操作票最后一项下面左边加盖"已执行"章后将其保存，操作人将钥匙和安全用具放回原处。

11. 小结本次操作

由监护人负责小结本次操作情况，对操作过程进行评价，指出不足，提出注意事项，便于下次工作开展。

五、开关设备操作安全技术

1. 高压断路器操作安全技术

高压断路器是高压系统中最重要的开关电器和保护电器，设有灭弧装置，在额定条件下可拉工作电流、过负荷电流和短路电流，也是倒闸操作的主要操作对象之一。为正确使用和操作断路器，电气运行人员应遵循以下安全技术原则：

（1）熟悉所使用断路器的技术性能。熟悉其技术性能，如灭弧性能和操作性能，是保证正确操作断路器的条件之一。了解设备性能，才能检查到位，保证操作的正确性。

（2）了解所操作断路器当前健康状况。了解断路器当前健康状况，是为了防止在操作过程中出现意外事故，如发生断路器爆炸事故。所操作的断路器是否处于健康状态，或是存在缺陷，这些缺陷对操作断路器是否有影响，电气运行人员应该做到心中有数，避免蛮干。如：SF_6断路器是否存在严重漏气现象；真空断路的真空度是否严重下降；液压操动机构的压力是否正常，是否存在严重漏油现象。这些对断路器操作有重大影响的设备缺陷，运行人员尤其要及时了解，在消除缺陷后才能进行操作。

2. 高压断路器操作要领

（1）不允许带电手动操作合闸。

（2）断路器合闸前，必须投入相关继电保护装置和自动装置，以便在故障设备上或带接地线合闸时，断路器能迅速动作跳闸，避免越级跳闸扩大事故影响范围。

（3）了解当前运行方式对断路器操作的影响，如在某运行方式下合上断路器时其最大短路电流是否大于断路器的开断电流，在某运行方式下操作断路器是否会引起谐振过电压……操作断路器时应避开这些可能导致危险的运行方式。

（4）用控制开关进行断路器合、分闸时，应动作迅速，待指示灯亮后才松手返回；也应注意不要用力太猛，以免损坏控制开关。

（5）断路器操作完成后，应检查相关仪表和信号指示，避免非全相合、分闸，确保动作的正确性。

（6）在误合、分闸可能造成人身伤亡事故或设备事故的情况下，如断路器检修、断路器存在严重缺陷不能分闸、继电保护装置故障、倒母线操作、二次回路有人作业、拉开与断路器并联的旁路开关等，应断开断路器的操作电源。

（7）现场有作业时，对有"遥控"操作的断路器，应将控制方式切换到"就地"控制，断开操作电源，并在操作把手上悬挂"禁止合闸，有人工作"标示牌。

3. 隔离开关操作安全技术

隔离开关在高压系统中主要用于隔离电源，使检修设备与带电设备间有一明显断开点，以保证检修安全。隔离开关还可用于倒母线操作，以及拉合有限制的小电流电路。110 kV GW4-126 型隔离开关如图 5-2 所示。

图 5-2　GW4-126 型隔离开关

由于隔离开关没有专门的灭弧装置，不能开断负荷电流，其安全操作基本技术原则是等电位操作，严禁带负荷拉合隔离开关。

当用隔离开关配合高压断路器做停、送电操作时，应先检查断路器的位置必须在"断开"状态，还应按一定操作顺序进行。停电拉闸操作时，按"断路器→负荷侧隔离开关→电源侧隔离开关"的顺序依次操作；送电合闸操作时应按相反顺序进行，即按"电源侧隔离开关→负荷侧隔离开关→断路器"的顺序依次合闸。

在双母线倒母线操作时，应先合上母联断路器，使双母线等电位，然后按"先合后拉"顺序操作隔离开关，即合上另一组母线的母线隔离开关后，才能拉开原在运行的母线隔离开关。

对带有接地刀闸的隔离开关，应注意主刀闸与接地刀闸的联锁要求。

4. 隔离开关操作要领

（1）在手动合隔离开关时，应动作迅速果断，一合到底，即使出现弧光，也不能中途停顿，更不能将已合闸或将合闸的隔离开关拉回。因为带负荷拉隔离开关会引起更大弧光，使设备损坏更严重，甚至造成支柱绝缘子爆炸和电弧灼伤操作人员。在隔离开关合到底时，也不要用力过猛，以免造成冲击折断支柱绝缘子。

（2）在手动拉开隔离开关时，应分两步进行。第一步是先缓慢拉开动触头，形成一微小间隙，观察是否出现异常弧光，若正常则可进行第二步，迅速将动触头全部拉开；若发现有异常弧光，应立即将动触头重新合上，停止操作，待查明原因后再操作。

（3）对分相操作的隔离开关，一般先拉开中间相，然后再拉开两个边相，有风时应先拉下风相，后拉上风相；合闸操作顺序刚好相反，先合两个边相，最后合中间相。

（4）对远方操作的隔离开关，不应在带电压情况下就地手动操作；当操作失灵时，应查明原因，只有在确定操作正确时，才允许解锁手动操作。

（5）手动操作隔离开关时，应戴绝缘手套，穿绝缘靴。

（6）隔离开关经操作后，特别是远方操作后，必须进行位置检查，确认隔离开关操作到位，即全拉开或全合上，以及位置指示器指示正确；若未操作到位，则要手动操作到位，并检查设备是否存在缺陷。

六、倒闸操作安全注意事项

发电厂、变电站倒闸操作是一项比较复杂的工作，应坚持操作之前"三对照"（对照操作任务和运行方式填写操作票、对照模拟图审查操作票并进行预演、对照设备名称和编号无误后再操作），操作之中"三禁止"（禁止监护人直接操作设备、禁止有疑问盲目操作、禁止边操作边做其他无关事情），操作之后"三检查"（检查操作质量、检查运行方式、检查设备状况）。为保证倒闸操作安全进行，还应注意以下有关事项：

（1）倒闸操作前，必须了解系统当前的运行方式，继电保护、自动装置运行情况，并考虑操作中及操作后电源与负荷的合理分布。

（2）设备送电前，严格执行工作票终结制度，收回工作票，拆除有关临时安全措施（拉开接地刀闸，拆除临时接地线，恢复固定遮栏及常设标示牌等），测量绝缘电阻，对送电设备进行全面检查。

（3）倒闸操作前，应考虑有关继电保护装置、自动装置的投运和整定值调整，防止无保护运行和保护不正确动作。

（4）应注意主设备与所属二次设备的投退顺序，如备用电源自动投入装置、自动重合闸装置、自动励磁调节装置应在主设备投运后投入，在主设备退出运行前退出。

（5）在倒闸操作过程中，应注意分析仪表和信号指示，防止有设备过负荷跳闸或出现其他异常情况。若有疑问，应停止操作重新检查核实，不得擅自改动操作票。

（6）在断路器操作前应检查直流操作电源电压是否正常，若不正常应检查消除后，才能进行操作。

（7）正确对防误装置进行解锁、加锁。

（8）正确使用合格的安全工器具，如验电器、绝缘棒、绝缘手套、绝缘靴等。

Task Ⅰ Correctly performing switching operations

Ⅰ. Overview

Switching operations refer to switching on and off of switching devices, operating circuits, control and power supplies, putting relay protection devices and automatic devices into and out of operation, as well as mounting and removal of temporary grounding wires, etc. when electrical equipment changes the operating state or the electric power system changes the operation mode.

As power plant or substation is the main link of an electric power system, switching operations are usually required when the load changes or when the electrical equipment or system changes the operation mode. There are a lot of electrical equipments in a power plant or a substation, and the electrical wiring is also complex. Switching operations not only concern the operations of primary circuits but also the operations of secondary circuits. Therefore, it can be said that switching operations are not only an important work of electrical operation but also regular and complex work. During a switching operation, any misoperation will not only affect the power supply or damage the equipment, but will also endanger the operator and the safe operation of the power grid. So, correct switching operations are important to ensure the safe operation of power plants and substations.

Any switching operation shall be conducted according to the command of the on-duty dispatching and control officer or the person in charge of operation and maintenance only after the command receiver correctly repeats the command. The command given shall be correct, and clear and use normative double names. Both the command issuer and command receiver shall inform each other of the organization they belong to and their names. The whole process of giving a command (including the command issuer repeating the command) and receiving the command shall be taped and recorded. The operator (including the supervisor) shall know the purpose and sequence of the operation concerned. If there is any doubt about the command, the command issuer shall be asked before executing. The command issuer, the command receiver, and the operator (including the supervisor) should have the corresponding qualifications.

Switching operations can be done locally, remotely, or by using a program. Remotely controlled and program-controlled shall meet relevant technical specifications.

Ⅱ. Classification of switching operations

1. Supervised operations

Supervised operations, as the name suggests, are the operations supervised by someone. A person who is more familiar with the equipment shall be the supervisor of the operation. Switching operations that are especially important and complex shall be carried out by skilled O&M personnel, and supervised by the person in charge of operation and maintenance.

2. One-man operations

One-man operations refer to operations carried out by one person.

(1) In terms of operating the step-up substation attended by one person of a substation or power plant, the O&M personnel shall fill in the operation ticket according to the operation command given by the command issuer on the phone, and correctly repeat the command.

(2) If there is any reliable means of confirmation and automatic recording, the dispatching and control personnel can perform the one-man operation.

(3) The equipment, items, and personnel associated with the one-man operation shall be approved by the equipment O&M management organization or the dispatching and control center, and the personnel concerned shall pass the special appraisal.

3. Maintainer operations

Maintainer operations are operations carried out by maintainers.

(1) The maintainers of the organization, who have passed the examination of and approved by the O&M management organization, can supervise the shift of the electrical equipment of 220 kV or below from hot stand-by duty to overhaul or vice versa. The supervisor should be the maintainer or equipment O&M personnel of the same unit.

(2) The command receiving and issuing procedures and safety requirements for maintainers shall be reviewed and approved by the equipment management organization and submitted to the relevant departments and dispatching and control center for filing.

III. Requirements for filling switching operation tickets

(1) Switching operation tickets shall be completed for operations other than accident response, switching on and off switches, and switching off or removing the only set of grounding knife-switches or grounding wires of the whole station. Accident response, among others, refers to the emergency handling after the accident occurs, excluding the follow-up operations after the accident handling. The above operations requiring no operation tickets must be recorded in the duty logbook in advance and shall be carried out by two people who take the duty logbook or accident dispatching order sheet together with them.

(2) The operation tickets filled in by hand must be ones that are uniformly printed and bound. The operation tickets must be completed with a fountain pen or a ballpoint pen. The face of the ticket should be neat and the handwriting legible. The wrongly written characters, if any, shall be crossed out with a "×" mark. The operation tickets shall be filled in continuously without covering mistakes. Keywords like "opening" and "closing", equipment scheduling name, number, and operation time cannot be modified.

(3) After the operation is completed, set the "Executed" stamp flush left and stamp it on the line below the last step of the operation ticket. If the last step is on the last line of the operation ticket, set the "Executed" stamp flush right and stamp it in the column of this step.

(4) If the operation is interrupted for any reason during the execution of the operation ticket,

set the "Executed" stamp flush left and stamp it on the line below the completed step, and the reason for the interruption should be indicated in the remarks column. If there are several pages of this operation ticket that have not been executed, stamp "Not executed" in the lower right corner of the operation task column of each page not executed.

(5) If a dispatching order sheet is not executed in one go, stamp "Suspended" on the last item of each operation corresponding to the dispatcher's sub-command. Neither simulation rehearsal nor formal operations shall go beyond the operation item with this stamp, and it is forbidden to proceed beyond the stamped operation item.

(6) An operation ticket to be canceled shall be stamped with "Canceled" in the bottom right corner of the operation task column, and the reason for cancellation shall be indicated in the remarks column of the canceled operation ticket. When notifying an operation ticket is canceled, the dispatch and control personnel shall stamp "Canceled" in the bottom right corner of the operation task column and indicate the cancellation time, the notifying dispatch and control personnel's name, and the command receiver's name in the remarks column. If the canceled operation ticket has multiple pages, the bottom right corner of the operation task column of each page shall be stamped with "Canceled", and the reason for cancellation shall be indicated in the remarks column of the first page of the canceled operation ticket. From the second scrapped page, indicated in the remarks column "The reason for cancellation is the same as the previous page" will suffice.

(7) Stations that use microcomputers to print operation tickets are not allowed to use general microcomputers to issue operation tickets and must use an expert system with an anti-misoperation system to issue tickets.

(8) The operation tickets shall be filled in by substations, and each operation ticket book corresponds to a substation. The operation tickets shall be handed over to whoever works on shift and organized and kept by a specially assigned person. The operation tickets printed by computer are bound into a book form by month, and the cover shall bear the number of operation tickets executed in the month, the number of qualified operation tickets, and the qualified rate. The operation tickets shall be kept for at least two years.

(9) "Number" column: including the voltage class and name of a substation. The operation tickets shall be numbered consecutively by month. For instance, "500 kV Jianshan Substation 202006001" indicates the first operation ticket in June for the 500 kV Jianshan Substation.

(10) When it comes to operations, pay attention to the actions, indications, and sound of the equipment, and proceed only if the above is proper. In switching operations, stop immediately in case of any doubts, report to the dispatcher, and proceed after getting a clear answer. Switching operations are forbidden in a thunderstorm.

(11) The operator shall not reverse the order, add or delete any steps, or skip a step when executing an operation ticket.

(12) To make the operations smoother and reduce the number of round trips during the operations, one can consider taking convenient operating steps when filling a ticket as long as it's safe.

Ⅳ. Switching operation procedures for substations

The general switching operation procedures for substations are as shown in Fig. 5-1.

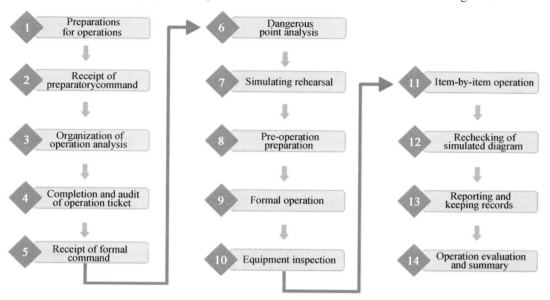

Fig. 5-1 General switching operation procedures for substations

1. Preparations for operations

Before complex operations, the substation head or shift chief-operator will organize all the personnel on duty to make the following preparations:

(1) Specify the scope of an operation task and power interruption and then divide the work.

(2) Propose an operation sequence, determine where to hang the grounding wires, the number of groups of grounding wires, and the barriers and sign boards that should be set up. Identify the live parts near the work site and come up with corresponding safety measures.

(3) Consider the corresponding changes in protective and automatic devices, AC and DC power supplies that should be disconnected, and measures to prevent the secondary reverse high voltage of voltage transformers and substation transformers.

(4) Analyze the possible problems arising from the operations and the measures to be taken.

(5) Contact the dispatcher and draft an operation ticket, which shall be discussed and adopted unanimously and audited and approved by the substation head or shift chief-operator.

(6) Scheduled general operations shall be prepared in accordance with the above requirements.

(7) After equipment maintenance and before operation, carefully check the condition of the equipment and the opening and closing positions of the primary and secondary equipment.

2. The dispatcher issues an operation command (operation plan), and the on-duty operator receives the operation command (operation plan)

The dispatcher's commands will be received by the on-duty operator. Turn on the recorder when receiving a dispatching operation command (operation plan). First, give the organization, position, and name, such as "This is the ××220 kV substation, and I am the on-duty person in charge (the person on regular shift), ××". Then, ask for the name and time of the command issuer. When filling in the dispatching order, carefully fill in relevant records while answering the phone, and the handwriting shall be clear and neat. After completion, it must be repeated correctly. Then, the on-duty person in charge (or the person on regular shift) shall determine the operator and supervisor of the operation.

Report it to the dispatcher if the command is incorrect, who will decide whether to execute the original command. However, if the execution of the command will threaten personal and equipment safety or directly cause power failure, refuse to execute the command and report the reason for the refusal to the dispatcher and the leader of the organization.

3. Operation analysis

The supervisor should analyze the key points and precautions of the operation with the operator according to the operation task, the actual operation mode, equipment defects, etc. The focus shall be put on preventing misoperation.

4. The operator shall check the diagram board and complete the operation ticket, and the on-duty person in charge (or the person on regular shift) shall review the operation ticket

The operator shall, according to the dispatching operation command, check if everything that is right in front of the simulated diagram board (check the operational condition in real time on the monitor if the substation has no simulated diagram board) and complete the operation ticket. When filling the operation ticket, stay concentrated, do not disrupt the sequence, do not omit any items and the handwriting shall be legible. If there is a microcomputer anti-misoperation system, the operation ticket must be completed in the "real-time ticket issuing" column, and it is prohibited to extract a "typical operation ticket" directly. After filling in the operation ticket, the operator shall check it on their own, and if not spotting any error, the operator shall sign it and then hand it over to the operation supervisor.

When receiving the operation ticket handed in by the operator, the operation supervisor shall carefully audit it according to the dispatching command and the actual operation mode at that time, point out the errors immediately after finding them, and give it to the operator for re-filling. If the operation supervisor isn't the on-duty person in charge, the operation ticket shall be handed to the on-duty person in charge for audit. After confirming that the operation ticket is correct during the audit, the supervisor and the on-duty person in charge shall sign their names in the corresponding column of the ticket.

5. The dispatcher officially issues an operation command, and the on-duty person in charge (or the person on regular shift) accepts the operation task

When the dispatcher officially issues an operation command, the on-duty person in charge (or the person on regular shift) must accept it, repeat it and record the issuing time and personnel in the relevant records immediately.

6. The supervisor and the operator have a rehearsal on the simulated diagram of the system

Before operating the actual equipment, the supervisor and the operator must first have a rehearsal on the simulated diagram board (without the simulated diagram board, have a rehearsal on the five-prevention computer). The rehearsal requires the supervisor to read the operation ticket in sequence, and the operator will operate on the simulated diagram board after repeating it (only the primary equipment will be simulated). The supervisor shall give commands one by one according to the operation ticket loudly.

Have a checking rehearsal on the diagram board, focusing on checking if the operation ticket is wrong. Only if both the supervisor and the operator think that the rehearsal is correctly performed, the rehearsal is passed.

7. Pre-operation preparation

Before the operation, the supervisor and the operator shall determine the operation route, and check the required safety appliances (such as insulating boots, insulating gloves, check and test date, intactness, electroscope check and test date, sound and light signal and intactness), and prepare the keys required for this operation (keys to the door of tools and instruments room, HV room and HV switch cabinet, etc.).

8. Formal operation

(1) Wearing requirements: Abide by the provisions of the *Electric Power Safety Working Regulations of State Grid Corporation of China*.

(2) The operator with the safety appliances and the supervisor with the keys and the ticket shall walk to the equipment together. The supervisor shall verify if the operation task and the equipment are consistent.

(3) The supervisor shall give commands one by one according to the order of the operation ticket loudly. For example:

- The supervisor gives the command, "Turn on the ×× switch!"
- The operator point to the double numbering plate of the equipment and repeat loudly, "Turn on the x x switch! "
- After confirming that it was correct, the supervisor gives the command, "Yes, execute! "
- The operator shall report to the supervisor after executing this operation, "Operation done! " or "Executed."

The supervisor and the operator shall jointly check the operation quality. After confirming, they shall mark " √ " in the operation execution column of the operation ticket, and then proceed to the next operation.

(4) In case of any doubt in the operation, stop immediately and report to the on-duty dispatcher or the on-duty person in charge. After figuring this out, proceed with the operation. Do not change the operation ticket without permission, and do not unlock the locked device at will.

9. Check the equipment and inform the dispatcher that the operation task is completed

After the operation is finished completely, the operator and the supervisor shall re-check the equipment. For example, check if the illuminated word signal that should be given is given, check if the signal lamp that should be turned on (off) is on (off), and check if the voltmeter and ammeter that should have (no) indications have indications. When the supervisor is no longer the on-duty person in charge, they should inform the on-duty person in charge that the operation ticket has been completed. The on-duty person in charge (or the person on regular shift), after reviewing and confirming, can inform the dispatcher that the operation task has been completed.

10. Keep records, and stamp and keep the operation ticket

After reporting to the dispatcher, the on-duty person in charge (or the person on regular shift) shall record the time and situation in the relevant records, stamp "Executed" on the left below the last item of the operation ticket, and properly keep it. The operator shall return the keys and safety appliances to their original places.

11. Summarize this operation

The supervisor shall be responsible for summarizing and evaluating this operation, pointing out shortcomings, and putting forward precautions to facilitate the work next time.

V. Operation safety technology of switchgear

1. Operation safety technology of HV circuit breakers

HV circuit breakers are the most important switching devices and protective devices in the HV system. An HV circuit breaker usually comes with an arc extinguishing device and can be used to switch on or off working current, overload current, and short circuit current. It is one of the main devices where switching operation is performed. To correctly use and operate circuit breakers, electrical operators should follow the following safety technology principles.

(1) Be familiar with the technical performance of the circuit breakers used, such as the arc extinguishing performance and operating performance, which is one of the conditions to correctly operate the circuit breakers. Only by knowing the equipment's performance can the operator properly check and use it.

(2) Know the current health status of the circuit breakers operated. To know the current health status of a circuit breaker is to prevent any accidents during operation, such as an explosion of the circuit breaker. The electrical operator should know if the circuit breaker operated is in good health or defective and if such defects could affect its operation to avoid being foolhardy. For example, the operator should check if the SF_6 circuit breaker severely leaks, if the vacuum degree of the vacuum circuit is greatly decreased, if the pressure of the hydraulic operating mechanism is normal, and if there is a serious oil leak. The operator should know these equipment defects that have a significant impact on the operation of the circuit breakers in particular in time, and can operate the circuit breakers only after eliminating the defects.

2. Essential points of operating circuit breakers

(1) It's not allowed to close the circuit breaker manually when it's live.

(2) Before the circuit breaker is closed, the relevant relay protection device and automatic device must be put into operation, so that the circuit breaker can quickly trip and avoid accident escalation due to override tripping in the event of closing of the faulty equipment or the grounding wire.

(3) Understand the impact of the current operation mode on the operation of the circuit breaker. For example, if the circuit breaker is closed in a certain operation mode, you should know if the maximum short circuit current is greater than the breaking current of the circuit breaker; and when operating the circuit breaker in a certain operation mode, you should know if it will cause resonance overvoltage. These dangerous operation modes should be avoided when operating the circuit breaker.

(4) When using a control switch to close and open the circuit breaker, do it quickly and let go after the indicator light goes on, but be careful not to exert too much force to avoid damage to the control switch.

(5) After operating the circuit breaker, check the relevant instruments and signal indications to avoid incomplete closing and opening. This is to ensure that the circuit breaker operates correctly.

(6) Cut off the operational power supply to the circuit breaker when mistakenly closing or opening may cause injuries and deaths or equipment accidents, for example, when the circuit breaker is being overhauled, the circuit breaker has serious defects and cannot be opened, the relay protection device is faulty, changeover busbar operation is underway, someone is working on the secondary circuit, or the bypass switch that is connected in parallel with the circuit breaker is opened, etc.

(7) When someone is working on the site, switch the "remotely controlled" circuit breaker to the "local" control mode, cut off the operational power supply, and hang a sign board reading "No switch closing! Working!" on the operating handle.

3. Operation safety technology of isolating switch

An isolating switch is mainly used to isolate the power supply in an HV system so that there is a clear cut-off point between the equipment under maintenance and the live equipment to ensure the safety of maintenance. An isolating switch can also be used to conduct the changeover busbar operation and switch on and off a limited low-current circuit. A 110 kV GW4-126 isolating switch is as shown in Fig. 5-2.

Because the isolating switch does not have a special arc extinguishing device, the load current cannot be cut off. So, the basic technical principle for safe operation is equipotential operation, and it is strictly forbidden to open and close the isolating switch with load.

When the isolating switch is used in conjunction with the HV circuit breaker to interrupt or supply power, first check if the circuit breaker is in the "off" state. And a certain operating sequence shall also be followed. When it comes to power interruption by opening, the circuit breaker, the isolating switch on the load side, and then the isolating switch on the mains side shall be opened sequentially. For power supply by closing, the isolating switch on the mains side, the isolating switch on the load side, and then the circuit breaker shall be closed sequentially.

Fig. 5-2　GW4-126 isolating switch

When it comes to the changeover duplicate busbar operation, the bus tie circuit breaker shall be closed first to make the double bus equipotential, and then the isolation switches shall be operated in order of "closing before opening", which means that the isolating switch of another group of busbars shall be closed before opening the busbar isolating switch that is operating originally.

For the isolating switch with the grounding knife-switch, attention shall be paid to the interlock requirements for the main knife switch and the grounding knife-switch.

Essential points of operating isolating switches:

(1) When manually closing an isolating switch, close it quickly and completely. Do not stop halfway even if there is an arc, let alone pull back the closed or opened isolating switch. This is because pulling the isolating switch with load will cause a greater arc, more severe equipment damage, and even an explosion of the post insulator and arc burns of the operator. To close the isolating switch completely, do not exert too much force to avoid impacting and breaking the post insulator.

(2) When manually opening the isolating switch, do it in two steps. The first step is to slowly pull open the moving contact terminals till there is a small clearance. Observe if there is an abnormal arc. If everything is normal, proceed to the second step, and quickly pull open the moving contact terminals completely. If there is an abnormal arc, close the moving contact terminals

immediately and stop. Find out the cause before proceeding.

(3) For the isolating switch that is operated by phase, the intermediate phase is generally pulled open before the two side phases. When it's windy, the downwind phase shall be pulled open before the windward phase. The closing sequence is just the opposite of the opening sequence. The two side phases shall be closed before the intermediate phase.

(4) The remotely operated isolating switch with voltage shall not be operated manually on the spot. When it can not be operated remotely, find out the reason. Only when the operation is correct, manual operation can be allowed.

(5) When manually operating the isolating switch, wear insulating gloves and insulating boots.

(6) After operating, especially remotely operating the isolating switch, check the position and confirm that the isolating switch is pulled to an appropriate position, that is, fully opened or fully closed. And check if the position indicator indicates correctly. If the isolating switch is not pulled to an appropriate position, manually operate it till it is in the appropriate position and check if the equipment is defective.

VI. Safety precautions for switching operation

Switching operations of a power plant and substation are complicated, and you should adhere to the "three against" principles before operation (fill in the operation ticket against the operation task and operation mode, review the operation ticket against the simulated diagram and have a rehearsal, and check against the equipment name and number before operation); "three prohibitions" in operation (prohibit the supervisor from directly operating the equipment, prohibit operating blindly with doubts, and prohibit doing other unrelated things while operating); "three checks" after operation (check the operation quality, check the operation mode, and check the equipment status). To ensure the safety of switching operations, the following matters should also be paid attention to:

(1) Before starting a switching operation, it is necessary to know the current operation mode of the system, and the operation of the relay protection device and automatic device, and consider how the power supply and load can be reasonably distributed during and after the operation.

(2) Before supplying power to the equipment, strictly implement the work ticket termination system, recover the work ticket, remove the relevant temporary safety measures (open the grounding knife-switch, remove the temporary grounding wire, restore the fixed barriers and permanent sign boards, etc.), measure the insulation resistance, and comprehensively inspect the equipment to be energized.

(3) Before the switching operation, consider putting the relay protection device and automatic device into operation and adjusting their set values to prevent unprotected operation and incorrect protection.

(4) Pay attention to the sequence in which the main equipment and the secondary equipment are put into or out of operation. For example, the automatic throw-in equipment of emergency

power supply, automatic circuit recloser, and exciting regulator shall be put into operation after the main equipment is put into operation and shall be put out of operation before the main equipment is put out of operation.

(5) When it comes to a switching operation, be sure to analyze the instrument and signal indications to prevent the equipment from overload tripping or other anomalies. If you have any doubts, stop and check it again. Do not change the operation ticket without authorization.

(6) Before operating the circuit breaker, check if the voltage of the DC operational power supply is normal. If the voltage is abnormal, check and eliminate it before proceeding.

(7) Unlock and lock the anti-misoperation device correctly.

(8) Properly use qualified safety tools and instruments, such as electroscopes, insulating rods, insulating gloves, and insulating boots.

任务二 巡视电气设备

一、高压设备巡视一般要求

高压设备巡视是一项专业性强且须慎重对待的经常性工作,是电气运行岗位经常性的,同时也是很重要的一项专业工作。处于运行状态的电气设备,在高电压作用下,其性能和状态的变化主要依靠设备的保护、监视装置、表计等显示,而对于故障和异常萌芽初期的外部现象,则主要依靠值班人员定期的和特殊的检查巡视来发现。因此,设备巡视质量的好坏、全面与否,与人员的运行经验、工作责任心和巡视方法直接相关。它要求电气值班员在当值期间必须对全部电气设备按规定的巡视路线(防止漏巡、减少重复巡视)、各设备的巡视项目认真予以检查和巡视。

根据《国家电网公司电力安全工作规程(变电部分)》第 5.2 条,高压设备巡视的一般要求如下:

1. 对巡视高压设备人员的基本要求

(1)允许单独巡视高压设备的人员应经单位批准并张榜公布。

(2)巡视高压设备时,不得进行其他工作,不得移开或越过遮栏。

(3)巡视配电装置,进入配电室时应随手关门,离开时必须将门锁好。

单独巡视设备时无人监护,发生意外无人及时相助,所以,从安全的愿望出发,应由生产运行部门领导对需要单独巡视设备的值班员是否具备独立工作能力和运行实际经验予以审查。需要单独巡视设备的非值班人员(主要是专业技术人员和生产管理人员),应经常深入实际进行目的性工作,对其是否熟悉设备和《电力安全工作规程》、是否具备单独工作能力及经验进行审查和把握。被批准的单独巡视设备的各类人员名单应书面公布。单独巡视设备时,要严守规章制度,不得从事维修工作,不移越遮栏和做与巡视无关的工作。

2. 对配电室门锁钥匙的基本规定

(1)配电室的门锁钥匙至少要有 3 把,由运行值班人员负责保管,按值移交。一把专供运行值班人员使用,其他可以借给经批准的单独巡视高压设备的人员和经批准的工作负责人使用,但应登记签名,巡视或当日工作结束后交还。

(2)各隔离开关的钥匙不得相互通用。

3. 对巡视环境的要求

(1)雷雨天气,需要巡视室外高压设备时,应穿绝缘靴,且不准靠近避雷器和避雷针。

雷雨天气可能出现大气过电压,阴雨天气容易使设备绝缘性能降低,绝缘脏污处容易发生对地闪络。雷电产生的过电压会使出线避雷器和母线避雷器放电,很大的接地电流流过接地点向周围呈半球形扩散,所产生的高电位也是按照一定的规律降低的。这样,在该接地网引入线和接地点附近,人体步入一定的范围内,两腿之间就存在着电位差,通常称为跨步电压。为防止该跨步电压对运行人员造成伤害,雷雨天气巡视设备时应穿绝缘靴。

阀型避雷器放电时,若雷电流过大或不能切断工频续流就会爆炸。避雷针落雷时,泄雷通道周围存在扩散电压，强大的雷电磁场不仅会在周围设备上产生感应过电压；而且，假如该接地体接地电阻不合格，还可能使地表及设备外壳和架构的电位升得很高，反过来对设备放电形成反击。所以，巡视有关设备时，值班人员与避雷装置必须保持规定的安全距离。通常，避雷、接地装置与道路或建筑物的出入口等处的水平距离应大于 3 m。

（2）地震、台风、洪水、泥石流等灾害发生时，禁止巡视灾害现场。灾害发生后，如需要对设备进行巡视，应制定必要的安全措施，得到设备运维管理单位批准，并至少两人一组，巡视人员应与派出部门之间保持通信联络。

① 适时地对电气设备进行特殊巡视检查，掌握和监视好运行设备存在的缺陷，弄明白缺陷发展的内部条件及与自然气候的联系。对新发现的缺陷，应当做出正确的分析，对重要和紧急的一类缺陷，应立即报告当值调度员和有关领导，加强监视。同时，积极开展与其有关的事故预想，按规定进行必要的检查试验，采取限制缺陷发展的措施，做好记录。

② 出现大风天气时，对电气设备引线、连接线、母线弛度大的地方，应注意观察其摆动情况，防止距离不够而放电；对设备区内易被风吹起的杂物应予以清除，防止吹落在设备带电部分或搭挂在导线上。对于线路断路器，特别是针对架设质量不高、跳闸较多的线路断路器可能出现的问题进行预想，根据现场情况做好检查和试验，以便一旦出现故障时，准确、迅速地予以查处。

③ 在雷雨天气进行岗位值班时，应从思想上做好处理应急事故的精神准备，做好防止过电压保护装置异常而引起的事故处理的预想。对安全检查、巡视用的装备器具提前准备好。严肃值班纪律，认真监督。雷雨过后，检查设备瓷套管、绝缘子有无破损裂纹、闪络放电痕迹，各避雷器运行是否正常、绝缘性能是否完好，引线、连接线是否烧伤、断股，重点检查脏污设备绝缘性能，抄录雷电记录器动作次数及泄漏电流值。

④ 下雪时，是观察负荷设备接头、连接处有无发热的好机会。值班人员应结合具体运行状况对设备进行仔细巡视。发现冰棱柱等缺陷时，按现场运行规定制订方案，进行消除或上报有关部门进行处理。出现雨雾天气时，重点监察绝缘性能存在缺陷和薄弱环节的设备，如油污的断路器套管、电压互感器、瓷质有损伤的绝缘设备的运行情况。根据需要采取超前管理措施，保证设备安全连续运行。

⑤ 火灾、地震、台风、洪水等灾害发生时，如要对设备进行巡视，应得到设备运行管理单位有关领导的批准，巡视人员应与派出部门之间保持通信联络。在重大灾难发生后，是否需要对设备进行巡视，有关领导应慎重考虑。一旦决定巡视，则必须配备必要的和可靠的通信联络手段，保证巡视人员在可控之中。

（3）高压设备发生接地时，室内人员应距离故障点 4 m 以外，室外人员应距离故障点 8 m 以外。进入上述范围人员应穿绝缘靴，接触设备的外壳和构架时，应戴绝缘手套。

当电气设备发生接地时，故障电流通过接地体或接地点向大地作半球形分散，在距接地体或接地点越近的地方半球形的球面越小，越远的地方球面越大。由于电阻与导电面积成反比，所以距接地体或接地点越近的地方电阻越大，越远的地方电阻越小。实测表明，在距接地体或接地点 20 m 以外的地方，电阻已接近于零，因而该处的电位也接近于零。这个等于零的电位，称为电气上的"地"。

电气设备接地部分（如接地外壳、接地线、接地体等）与大地零电位之间的电位差，称为接地时的对地电压。

高压设备发生接地,室内不得接近故障点 4 m 以内,室外不得接近故障点 8 m 以内,主要是防止接地电流扩散形成的电压。当室内电气设备发生接地时,地面虽然窄小,但比较干燥,电流如能经过 4 m 距离的扩散,则其外围基本上没有多高的电位。对于室外电气设备,其地面空间裕度大,但地表潮湿,室外大多都是高电压等级的设备,因此,安全防护禁止进入的半径为 8 m。由于工作需要而进入该范围之内的人员,应穿绝缘靴,以防跨步电压触电;接触设备外壳和架构时,由于从接触点到地面垂直距离与人所站立处的水平距离之间也存在着危险的电位差,人体接触这两部分,就会承受接触电压,所以,应戴绝缘手套将电压隔离。

二、高压设备巡视分类

变电站的设备巡视检查,分为例行巡视、全面巡视、专业巡视、熄灯巡视和特殊巡视。

(1)例行巡视是指对站内设备及设施外观、异常声响、设备渗漏、监控系统、二次装置及辅助设施异常告警、消防安防系统完好性、变电站运行环境、缺陷和隐患跟踪检查等方面的常规性巡查,具体巡视项目按照现场运行通用规程和专用规程执行。配置机器人巡检系统的变电站,机器人可巡视的设备可由机器人巡视代替人工例行巡视。

(2)全面巡视是指在例行巡视项目的基础上,对站内设备开启箱门检查,记录设备运行数据;检查设备污秽情况,检查防火、防小动物、防误闭锁等有无漏洞;检查接地引下线是否完好;检查变电站设备厂房等,对这些方面进行的详细巡查。全面巡视和例行巡视可一并进行。需要解除防误闭锁装置才能进行巡视的,巡视周期由各运维单位根据变电站运行环境及设备情况在现场运行专用规程中明确。

(3)熄灯巡视指夜间熄灯开展的巡视,重点检查设备有无电晕、放电,接头有无过热现象。

(4)专业巡视指为深入掌握设备状态,由运维、检修、设备状态评价人员联合开展的对设备的集中巡查和检测。

(5)特殊巡视指因设备运行环境、方式变化而开展的巡视。遇有以下情况,应进行特殊巡视:大风后;雷雨后;冰雪、冰雹后和雾霾过程中;新设备投入运行后;设备经过检修、改造或长期停运重新投入系统运行后;设备缺陷有发展时;设备发生过负载或负载剧增、超温、发热、系统冲击、跳闸等异常情况;法定节假日、上级通知有重要保供电任务时;电网供电可靠性下降或存在发生较大电网事故(事件)风险时段。

各类巡视完成后应填写巡视记录,其中全面巡视应持标准作业卡巡视,并逐项填写巡视结果。

三、高压设备巡视基本方法

1. 用手触试

用手触试电气设备来判断缺陷和故障虽然是一种必不可少的方法,包括在整个巡视过程中经常会用到手;但首先需要强调的是:必须分清可触可摸的界线和部位,明确禁止用手触试的部位,如旋转电机的中性点,变压器、消弧线圈的中性点及其接地装置应视为带电设备,严禁触摸。

(1)对于一次设备,用手触试检查之前,应当首先考虑安全方面的有关问题。如对带电运行设备的外壳和其他装置,需要触试检查温度时,先要检查其接地确属良好,同时还应站好位置，注意保持与设备带电部位的安全距离。

（2）对于二次设备的检查，如感试装置继电器等元件是否发热，对于非金属外壳的可以直接用手摸，对于金属外壳且接地确实良好的，也可以用手触试检查。

2. 用眼看

用双目来测视设备看得见的部位，通过观察它们的外表变化来发现异常现象，是巡视检查最基本的方法，例如，标色设备漆色的变化，裸金属色泽、充油设备油色等的变化，充油设备的渗漏，设备绝缘的破损裂纹、污秽，旋转设备的磨损，化学设备的腐蚀，等。

3. 用鼻子嗅

人的嗅觉器官可以说是人的一个向导，对于某些气味（如绝缘烧损的焦糊味）的反应，比用某些自动仪器的反应要灵敏得多。人的嗅觉功能虽然因人而异，但对电气设备有机绝缘材料过热所产生的气味，正常人都是可以辨别的。电气值班人员在巡视中，一旦嗅到绝缘烧损的焦糊味，应立即寻找发热元件的部位，并判别其严重程度，如是否冒烟、变色及有无其他异音异状，从而对症查处。

4. 用耳听

带电运行的电气设备，不论是静止的还是旋转的，在交流电压作用下，有很多都能发出表明其运行状态特征的声音。变压器正常运行时，平稳、均匀、低沉的"嗡嗡"声是我们所熟悉的。它的产生是由于交变磁场反复作用振动的结果。电气值班员随着经验和知识的积累，只要熟练地掌握了这些设备正常运行时的声音情况，那么，遇有异常时，用耳朵或者借助于听音器械（如听音棒等），就能通过它们高低、节奏、声色的变化、杂音的强弱来判断电气设备的运行状态。

5. 用仪器检测

上述用眼、耳、鼻巡视检查的方法，因为设备初期发热不伴随有音、色、味的异常，因此，仅靠人的感官判断是比较困难的和不可靠的。为了及早发现设备的过热，有条件的地方应尽可能地利用先进的检测仪器，测量运行中设备的温度或有关参数。

发电厂和变电站的电气设备建设投运时，已装置有各种检测仪器，它们都是固定安装在具体位置或盘面上的，交接班时均按规定的周期进行测试检查。电气设备绝缘故障大多是在带电状况时由于过热老化而引起的，只有在带电运行时才会出现，因此带电进行检测能更真实地反映设备状况。巡视设备所用的便携式检测仪器，主要是红外线测温仪。它利用一种灵敏度较高的热敏感应辐射元件，检测由被测物发射出来的红外线，是一种实用的带电测温仪表。

四、变电站设备巡视注意事项

1. 变压器的巡视检查注意事项

（1）巡视检查人员除应遵守《国家电网公司电力安全工作规程》中有关巡视检查的安全规定外，还应注意现场对着装和工器具使用的特别要求，如在接地网接地电阻不符合要求的情况下进行室外变压器巡视检查时按规定穿着绝缘靴，保持与中性点接地线有足够的安全距离，巡视室内变压器时不得单独移开或跨越遮栏进入隔离围栏内，等。

（2）在特殊情况下，如大风、大雾、大雪、雷雨后和气温突变等天气，应对变压器进行特殊巡视检查，如有无积物、积雪，套管绝缘子有无严重电晕、闪络放电现象，避雷器动作次数，套管有无放电痕迹、破损，等。

（3）变压器过负荷运行期间，要加强对变压器的巡视检查，重点检查变压器运转声音、冷却器运行情况、变压器上层油温等。

（4）气体继电器动作后，应对变压器进行外部检查，并抽取气样进行分析。

（5）夜间巡视时，应注意引线接头、线夹处有无过热、发红及严重放电现象。

（6）发现变压器有异常现象，要按运行规程及时汇报处理。如发现油轻微渗漏，要汇报并记录设备缺陷，可采取一些临时措施，待检修时再作彻底处理；若漏油严重造成油面过低，则要立即汇报，准备停运；如发现变压器内部有放电声，应立即停运检查。

2. 高压断路器的巡视检查注意事项

（1）巡视检查人员除应遵守《国家电网公司电力安全工作规程》中有关巡视检查的安全规定外，还要遵守现场运行规程的安全规定，如在巡视检查时不得随意打开间隔门，对气体绝缘开关设备（GIS）配电装置巡视检查不得单人进行。进入高压配电室应先开启通风机并不小于 15 min；在报警状态下不得进入 SF_6 高压配电室，如确需要进入则要戴防毒面具、手套，穿防护衣。

（2）在气温低的冬天，要按运行规程开启加热器，对 SF_6 断路器进行加热，防止气体黏度增大而导致开断能力下降。

（3）在断路器故障跳闸、强送电后，高温、高负载期间以及气温突变时，应进行特殊巡视检查。

（4）遥视检查中发现断路器有异常现象时，按影响程度作加强监视或停用检查处理。如发现引线接头处示温片指示接头过热，可加强监视，或采用减负荷的降温方式进行处理；若发现真空断路器泄漏、SF_6 断路器气压过低或液压操动机构严重漏油，应立即拉开控制回路电源，报告调度后经倒闸操作退出运行，作停用检查或修理。

3. 隔离开关的巡视检查注意事项

（1）巡视检查中发现隔离开关有异常现象时，按影响程度作加强监视或停用检查处理。如发现连线接头处示温片指示接头过热，可加强监视，或采用转移负荷的方式进行处理；如发现绝缘子网络或操动机构异常，应报告调度后经倒闸操作退出运行，做停用清扫或检修处理。

（2）对隔离开关的防误操作安全装置缺陷应作为设备严重缺陷对待。

4. 互感器的巡视检查注意事项

巡视检查中发现互感器有异常现象时，按具体情况作加强监视，或停用检查处理。在处理中应注意：

（1）在对母线电压互感器作紧急停用处理时，如发现电压互感器冒烟着火，严禁用隔离开关拉开故障的电压互感器，应使用断路器断开。

（2）在处理电流互感器二次回路开路时，值班人员应穿绝缘鞋、戴绝缘手套和使用绝缘工具。

5. 电容器的巡视检查注意事项

巡视检查中发现电容器有异常现象时，按具体情况作加强监视和改善运行条件，或停用检查处理。

（1）发现电容器外壳膨胀变形，应采用强力通风以降低电容器温度；如发生群体变形应及时停用检查。

（2）发现电容器渗漏，应加强监视，并减轻电容器负载和降低周围环境温度，但不宜长期运行；若渗漏严重，则应立即汇报并做停用检查处理。

（3）因过电压造成电容器跳闸，应对所有设备进行特殊巡视检查；若未发现问题，也要在 15 min 后才可试合闸。

Task II Routine inspections of electrical equipment

I. General requirements for routine inspections of HV equipment

Routine inspections of HV equipment are highly professional and regular work that must be handled with care. They are carried out by people in the electrical operation position periodically and are also very important professional work. When electrical equipment is operating, its changes in performance and status under high voltage are shown by its protection devices, monitoring devices, instruments, meters, etc. In addition, the external phenomena in the early stage of a fault or anomaly are mainly spotted by operators on duty in their regular and special checks and routine inspections. Therefore, whether an equipment inspection is well performed or whether it is comprehensive is directly related to people's operating experience, work responsibility, and ways of the routine inspection. Therefore, the electrical operators on duty shall carefully check and inspect all electrical equipment according to the prescribed inspection route (to avoid any omissions and repetitions in the routine inspection) and the specified routine inspection items of each piece of equipment while on duty.

According to Article 5.2 of the *Electric Power Safety Working Regulations (Power Transformation) of State Grid Corporation of China*:

1. Basic requirements for people routinely inspecting HV equipment

(1) People who are allowed to routinely inspect HV equipment alone shall be approved by the organization and their names shall be posted.

(2) When routinely inspecting HV equipment, do not perform other work, and do not move away or cross the barriers.

(3) When routinely inspecting power distribution units, close the door when entering the power distribution room; and close the door when leaving.

When routinely inspecting HV equipment alone, if there is no one supervising, no one will help you timely in case of an accident. Therefore, for the sake of safety, the leaders of the production and operation department shall examine if the operators on duty who need to routinely inspect the equipment alone have the ability to work on their own and have practical operating experience. The non-duty personnel who need to routinely inspect the equipment separately (mainly professional technicians and production management personnel) often perform actual purposeful work in depth. So, they shall be examined to see if they are familiar with the equipment and the *Electric Power Safety Working Regulations* and if they have the ability and experience to work alone. The list of all kinds of people approved to routinely inspect equipment alone shall be posted in writing. When inspecting equipment alone, strictly abide by the rules and regulations, do not perform maintenance work, and do not cross the barriers to do something that is irrelevant to the routine inspection.

2. Basic provisions for the key of the door lock of the power distribution room

(1) There should be at least 3 keys to the door lock of the power distribution room, which should be kept by the on-duty operator and handed over to whoever works on shift. One key shall be used by the on-duty operator, and others can be lent to the people who are approved to routinely inspect HV equipment alone and the approved person in charge of work, but registration and signature are required, and the keys shall be returned after routine inspection or work is completed.

(2) The key of an isolation switch shall not be used for another isolating switch.

3. Requirements for routine inspection environment

(1) When it's necessary to have a routine inspection of outdoor HV equipment in a thunderstorm, wear insulating boots and do not approach lightning arresters and lightning rods.

Atmospheric overvoltage may occur when a thunderstorm comes. On cloudy and rainy days, the insulation of equipment may be degraded and flashover to ground is likely to be seen where the insulation is dirty. The overvoltage generated by lightning cause the lightning arresters of the outgoing line and the busbar to discharge, a large grounding current flows through the grounding point and spread into a hemispherical shape, and a high potential generated is also reduced according to a certain rule. In this way, near the inlead and grounding point of the grounding grid, when a person steps into a certain range, there will be a potential difference between his/her legs, which is usually known as step voltage. To prevent harm from step voltage, operators should wear insulating boots when routinely inspecting equipment against a thunderstorm.

When a valve type lightning arrester is discharging, the lightning arrester will explode if the lightning current is too large or the power-flow current cannot be cut off. When the lightning rod captures lightning, there is a diffusion voltage around the lightning release channel, and the strong lightning magnetic field will not only generate induced overvoltage on the surrounding equipment, but also, if the grounding resistance of the grounding body is not qualified, raise the potential of the earth's surface and equipment housing and architecture to a very high level, and discharge toward the equipment and impact the equipment. Therefore, when routinely inspecting the relevant equipment, the operator on duty shall maintain the prescribed safe distance with lightning protection devices. Generally, the horizontal distance between a lightning protection device/ grounding device and an entrance/exit of a road or structure shall be greater than 3 m.

(2) In the event of an earthquake, typhoon, flood, mudslide, and other disasters, routine inspection of the affected site is forbidden. After a disaster occurs, if it is needed to have a routine inspection of the equipment, formulate necessary safety measures, and obtain approval from the equipment O&M management organization. A group of at least two people is required. And the inspectors shall keep in touch with the dispatching department.

① Timely conduct special routine inspection of electrical equipment, grasp and monitor the defects of operating equipment, and understand the internal conditions of defect developments and the relations to climate and weather. For any new defects, analyze them correctly. For important and urgent defects, immediately report them to the on-duty dispatcher and relevant leaders, and strengthen monitoring. With the above said, actively premeditate any accidents related to such defects, carry out necessary inspections and tests according to the regulations, take measures to limit the development of defects, and keep records.

② Where the leads, connecting lines, and busbars of electrical equipment go slack greatly when it's blowing a gale, be sure to observe their swing to prevent discharging due to an insufficient distance. Any debris that is easily blown up by the wind in the equipment area should be removed to prevent falling onto a live part of the equipment or hanging on a conductor. For circuit breakers of a circuit, especially for the circuit breakers of a circuit that is poorly erected and trips many times, premeditate any problems that the circuit breakers could have, and inspect and test them according to the site conditions. Once any fault appears, deal with it accurately and quickly.

③ When working on duty in a thunderstorm, be mentally prepared to tackle any emergencies. Premeditate the ways to prevent any accidents arising from an anomaly of the overvoltage protection device. Prepare equipment for safety checks and routine inspections in advance. Strictly implement the duty disciplines and carefully supervise how they are implemented. After a thunderstorm, check if the porcelain bushings and insulators of the equipment are damaged, cracked, or have any flashover and discharge marks, if the lightning arresters working properly, if the insulation is intact, and if the leads and connecting lines are burned or broken. And more importantly, check the insulation of dirty equipment, and write down the operation times and leakage current value of the lightning recorder.

④ Snowy days provide a good opportunity to observe if the joints and connections of the load equipment are heated. Operators on duty shall have a routine inspection of the equipment carefully according to the specific operation conditions. When spotting any icicles and other defects, come up with a solution according to the on-site operation regulations, and eliminate or report such defects to the relevant departments for tackling. When it's rainy or foggy, focus on monitoring the operation of equipment with defective insulation and weak links, such as the operation of circuit breakers with oil-stained bushings, voltage transformers, and equipment with damaged porcelain insulation. Take management measures beforehand as required to ensure the safe and continuous operation of the equipment.

⑤ In the event of a fire, earthquake, typhoon, flood, and other disasters, if it is needed to have a routine inspection of the equipment, obtain approval from the leaders of the equipment O&M management organization, and the inspectors shall keep in touch with the dispatching department. This is a new article of the *Electric Power Safety Working Regulations of State Grid Corporation of China*. After a major disaster, whether it's necessary to have a routine inspection of the equipment shall be deliberated by the leaders concerned. Once the routine inspection decision is made, necessary and reliable means of communication must be provided to ensure that the inspectors are under control.

(3) When the HV equipment is grounded, people standing indoors shall maintain a minimum distance of 4 m from the fault point while people standing outdoors shall be at least 8 meters away

from the fault point. Personnel entering the above ranges shall wear insulating boots, and those who contact the equipment housing shall wear insulating gloves.

When the electrical equipment is grounded, the fault current diffuses hemispherically to the earth via the grounding body or ground point. The closer the fault current is to the grounding body or ground point, the smaller the spherical surface of the hemisphere is; the farther the fault current is, the larger the spherical surface is. Because the resistance is inversely proportional to the conducting area, a place that is closer to the grounding body or ground point has a greater resistance while a farther place has a smaller resistance. The measurements show that the resistance is close to zero at a place 20 m away from the grounding body or ground point, so the potential there is close to zero. This zero potential is electrically called the "ground".

The potential difference between the grounding part of the electrical equipment (such as the grounding housing, grounding wire, and grounding body) and the zero potential of the ground is called the voltage to ground at the time of grounding.

When the HV equipment is grounded, people standing indoors shall maintain a minimum distance of 4 meters from the fault point while people standing outdoors shall be at least 8 meters away from the fault point. The main purpose is to prevent voltage formed due to grounding current diffusion. When indoor electrical equipment is grounded, although the floor is narrow and small, it is relatively dry, and if there is a 4 m range in which the current can diffuse, there is basically no high potential outside that range. For outdoor electrical equipment, the ground space margin is large, but the surface is wet, and most outdoor equipment is of high voltage class. So, the radius of inaccessible safety protection is 8 m. People who enter this range for work should wear insulating boots to prevent electric shock caused by a step voltage. When touching the equipment housing and architecture, there is also a dangerous potential difference between the contact point and the ground, and the contact point and the person's standing point, and the human body that touch these two parts will be subjected to a contact voltage. Therefore, wear insulating gloves to isolate the voltage.

II. Classification of routine inspections of HV equipment

Routine inspections of substation equipment are divided into regular routine inspection, comprehensive routine inspection, professional routine inspection, lights-out routine inspection, and special routine inspection.

(1) Regular routine inspections are periodical inspections, such as checking the appearance of the equipment and facilities in the substation, abnormal sound, equipment leakage, malfunction alarms of the monitoring system, secondary devices and auxiliary facilities, intactness of the firefighting and security systems, the operating environment of the substation, and tracking and inspecting defects and hidden dangers. The specific routine inspection items shall be subject to general procedures and special procedures for on-site operation. In the substation equipped with a robot patrol inspection system, the robot can perform the regular routine inspection in place of people as long as the equipment permits.

(2) Comprehensive routine inspections are developed on the basis of regular routine inspection items, including open-door inspection of substation equipment, equipment operation data recording, presence of any dirt on equipment, presence of any vulnerabilities in fire prevention, small animal prevention, and misoperation-preventive locking, intactness of grounding down conductor, substation equipment and powerhouse checks, and other detailed inspections. Comprehensive routine inspections and regular routine inspections can be done at the same time. If it is necessary to remove the misoperation-preventive locking device to conduct routine inspections, the routine inspection period shall be specified in the special procedures for on-site operation by the O&M organizations according to the operating environment and equipment conditions of the substation.

(3) Lights-out routine inspections are conducted at night when lights are out, focusing on checking if the equipment has corona or is discharging, and if the joint is overheated.

(4) Professional routine inspections are the centralized inspections and testing of equipment jointly carried out by O&M, maintenance, and equipment status evaluation personnel to have a deep understanding of the equipment status.

(5) Special routine inspections are carried out due to the changes in the operating environment and mode of the equipment. Special routine inspections are required: ① after a gale; ② after a thunderstorm; ③ after snowing and hailing, and when it's smoggy; ④ after the new equipment is put into operation; ⑤ after the equipment is put back into the system after maintenance, upgrading, or long-term shutdown; ⑥ when the equipment's defect develops; ⑦ in the event of overload or spiking load, overtemperature, heating, system impact, trips, and other anomalies of the equipment; ⑧ when it's statutory holidays or when the superior organization notifies an important task of guaranteeing power supply; ⑨ when there is a decrease in the reliability of power supply by the power grid or there is a higher risk of power grid accidents (events).

(6) After all kinds of routine inspections, keep records of them. The comprehensive routine inspections, among others, shall be conducted with a standard job card, and the routine inspection results shall be written down item by item.

Ⅲ. Basic methods for routine inspections of HV equipment

1. Touch

Although hand-touching electrical equipment is an essential way to tell if it is defective or faulty, and hands are frequently used throughout routine inspections, it has to be emphasized first that you must distinguish the boundaries and parts that can be touched from parts that can not be touched by hand as expressly stipulated. For example, the neutral point of the rotating motor, transformer, and the neutral point and grounding device of the arc suppression coil should be regarded as live equipment, and it's strictly prohibited to touch.

(1) To touch primary equipment, first you should consider safety-related issues. When it comes to touching the housing of equipment that operates on electricity and other devices to check the temperature, first, you should make sure that the grounding is good, stand at a proper place, and

maintain a safe distance from the live part of the equipment.

(2) When it comes to checking secondary equipment, such as touching a device's relay or other elements to see if they are heated, you can directly touch the housing if it's not metallic; if it's a metallic housing that's well grounded, you can touch it with your hands.

2. Seeing

See visible parts of the equipment with your eyes. Identifying any anomalies from their appearance changes is a basic way of routine inspection. These changes include the change of paint color of standard color equipment, change of color of bare metal, change of oil color of oil filled equipment, leakage of oil filled equipment, damage, cracks and dirt of equipment insulation, pollution, abrasion of rotating equipment, corrosion of chemical equipment.

3. Sniffing

Our olfactory organ is a guide for us, so to speak. And it responds to some odors (such as the smell of burnt insulation) much more sensitively than some automatic instruments. Although people's olfactory function varies from person to person, average people can tell the smell of overheated organic insulation of electrical equipment. Once smelling burnt insulation during a routine inspection, the electrical operators on duty shall immediately find the heating element and judge how severe it is. For example, check if it's smoking, discolored, produces any abnormal sounds, or has other anomalies to investigate and address the problem according to the signs.

4. Hearing

A lot of electrical equipment that operates on electricity, whether stationary or rotating, under the action of AC voltage, can produce sounds that tell you their operating state characteristics. We are familiar with the smooth, uniform, and low "buzz" sound of a transformer when it's working properly. The sound is a result of repeated vibration caused by an alternating magnetic field. As experience and knowledge grow, as long as the electrical operators on duty can judge the operating state of electrical equipment in the case of an anomaly by ear or using a listening apparatus (such as a listening rod) by changes in the pitch, rhythm, and tone, as well as the loudness of noise as long as they have skillfully grasped the sound of such equipment during normal operation.

5. Testing by instrumentation

Although sometimes routine inspections are conducted by eye, ear, and nose as above mentioned, there is nothing unusual about the sound, color, and smell of the equipment in the initial stage of being heated, so it's difficult and unreliable to make a judgment by our sense organs alone. In order to detect if the equipment is overheated as early as possible, advanced testing instruments shall be used as far as possible to measure the temperature or related parameters of the equipment in operation.

When the electrical equipment of a power plant and substation is constructed and put into operation, a variety of testing instruments have been installed, which are fixed in specific locations or on panels. At the time of shift change, these instruments can be used to test and check the equipment according to the specified period. Electrical equipment insulation faults are mostly caused by overheating and aging when operating on electricity and only occur when operating on electricity. Therefore, testing the equipment when it's operating on electricity can reflect more real conditions of the equipment. Portable testing instruments are used for routine inspections, infrared thermodetectors in most cases, which is a practical live temperature instrument that uses a heat-sensitive radiating element of high sensitivity to detect the infrared rays emitted by the measured object.

IV. Precautions for routine inspections of substation equipment

1. Precautions for routine inspections of transformers

(1) Routine inspectors shall comply with the safety provisions of the *Electric Power Safety Working Regulations of State Grid Corporation of China* on routine inspections and the on-site special requirements for clothing, tools and instruments. For example, when the grounding resistance of the grounding grid does not meet the requirements, and you have to have a routine inspection of outdoor transformers, you should wear insulating boots as required, and maintain a sufficient safe distance from the grounding wire of the neutral point. When routinely inspecting indoor transformers, you are not allowed to just move or cross the barriers to get into the isolating fence.

(2) In special circumstances, such as gales, thick fog, heavy snow, thunderstorms, and sudden temperature changes, special routine inspections shall be carried out on the transformers, to see, for example, if there is accumulation of debris or snow, if the bushing insulator has any serious corona, flashover or discharge phenomena, how many times does the lightning arrester get actuated, and if the bushing has any discharge marks and damage.

(3) When the transformers are operating with overload, it is necessary to strengthen the routine inspections of the transformers, focusing on inspecting the sound of transformer operation, the operation of the coolers, the upper oil temperature of the transformers, etc.

(4) When the gas relay is actuated, the transformer shall be visually inspected and gas samples shall be taken for analysis.

(5) When it comes to routine inspections at night, please note that the lead connectors and wire clamps shall not be overheated, become red, or have serious discharge phenomena.

(6) If a transformer is abnormal, report it in time according to the operation procedures. If a slight oil leakage is found, report and record the defect of the equipment, take some temporary measures, and then thoroughly tackle the defect at the time of maintenance. If the oil level drops too low due to serious oil leakage, report it immediately and prepare for the shutdown. If a discharge sound is heard inside the transformer, it shall be shut down immediately for inspection.

2. Precautions for routine inspections of HV circuit breakers

(1) Routine inspectors shall comply with the safety provisions of the *Electric Power Safety Working Regulations of State Grid Corporation of China* on routine inspections and the safety

requirements of on-site operation procedures. For example, do not open the compartment door at will during a routine inspection, and do not carry out a routine inspection of the GIS power distribution unit alone, when entering the SF_6 HV power distribution room, turn on the ventilator and keep it working for not less than 15 min. Do not enter SF_6 HV power distribution room in the case of an alarm, and wear a gas mask, gloves, and protective suits if you have to.

(2) In the cold winter, turn on the heater according to the operation procedures to heat the SF_6 circuit breaker, to prevent the breaking capacity from decreasing by greater gas viscosity.

(3) After a trip of the circuit breaker caused by a fault, or forced power supply, in a period of high temperature or high load, and when the temperature suddenly changes, conduct a special routine inspection.

(4) When a circuit breaker is found abnormal during a remote inspection, strengthen monitoring or shut it down for inspection according to the impact. If the temperature indicating plate at the lead connector indicates that the connector is overheated, strengthen monitoring or reduce the load to cool it. If the vacuum circuit breaker leaks, the pressure of the SF_6 circuit breaker is too low, or the hydraulic operating mechanism is severely leaking oil, cut off the power supply of the control circuit immediately, report to the dispatcher, conduct switching operations to put it out of operation, shut it down, and check or repair it.

3. Precautions for routine inspections of isolating switches

(1) When an isolating switch is found abnormal during a routine inspection, strengthen monitoring or shut it down for inspection according to the impact. If the temperature indicating plate at the connector of the connecting line indicates that the connector is overheated, strengthen monitoring or transfer the load. If the insulator network or operating mechanism is found abnormal, report to the dispatcher, put it out of operation conduct switching operations, shut it down, clean or maintain it.

(2) Defects of the anti-misoperation device of an isolating switch shall be treated as major defects of equipment.

4. Precautions for routine inspections of mutual inductors

When a mutual inductor is found abnormal during a routine inspection, strengthen monitoring or shut it down for inspection as appropriate. When dealing with it, do the following:

(1) When a busbar voltage transformer is shut down for emergency management, if the voltage transformer is smoking and on fire, it is strictly prohibited to open the isolating switch to cut off the faulty voltage transformer, and use the circuit breaker instead.

(2) When dealing with the open-circuit secondary circuit of the current transformer, the operator on duty should wear insulating shoes, insulating gloves and use insulating tools.

5. Precautions for routine inspections of capacitors

When a capacitor is found abnormal during a routine inspection, strengthen monitoring and improve the operating conditions, or shut it down for inspection as appropriate.

(1) If the capacitor housing has an expansive deformation, ventilate it forcibly to cool the

capacitor. In case a lot of capacitor capacitors have deformed, they shall be shut down for inspection in time.

(2) If a capacitor leaks, strengthen monitoring, reduce the capacitor load, and lower ambient temperature, but it is not suitable for the capacitor to operate for a long time. If it's a severe leak, report it immediately and shut it down for inspection.

(3) If the capacitor trips due to overvoltage, carry out a special routine inspection of all equipment; if no problem is found, closing can be tried after 15 min.

任务三　高压开关柜停送电操作实训

一、作业任务

3人一小组，一人扮演工作负责人，两人扮演班组成员，以"110 kV 智能变电站"为工作环境，按照《国家电网公司电力安全工作规程（变电部分）》中的要求，完成高压开关柜停送电操作。

二、引用标准及文件

（1）《国家电网公司电力安全工作规程（变电部分）》。
（2）《国家电网公司十八项电网重大反事故措施》。

三、作业条件

应在良好的天气操作；作业人员精神状态良好，熟悉工作中的安全措施、技术措施以及现场工作危险点。

四、作业前准备

1. 现场操作的基本要求及条件

勘察现场设备情况，查阅相关技术资料，包括历史数据及相关规程。

2. 工器具及材料选择

工位1个、标准化作业卡1套、中性笔1支、安全帽3顶、绝缘鞋1双、绝缘手套1双、录音笔1支、已执行印章1枚、操作工具巡检钥匙各1套、红背心1件。

3. 危险点及预防措施

（1）高压触电。
预防措施：正确进行模拟预演，做好个人防护。
（2）使用不合格的安全工器具。
预防措施：操作设备前，应正确选取安全工器具并检查合格。

4. 作业人员分工

现场工作负责人（监护人）：×××
现场作业人员：×××

五、规范及要求

（1）现场操作时，由老师给出开关柜运行方式，学员按现场要求进行高压开关柜的停送电操作，模拟演示。
（2）按照作业任务要求正确填写操作票。

（3）按照作业任务要求正确选择安全用具，做好个人防护工作。
（4）遵循安全操作规程，按照操作票的步骤正确操作。
（5）操作结束后，对操作质量进行检查。

六、作业流程及标准（表 5-1）

表 5-1　高压开关柜停送电操作流程及评分标准

班级		姓名		学号		考评员		成绩	
序号	作业名称	质量标准			分值/分	扣分标准		扣分	得分
1	正确填写操作票	按照作业任务要求正确填写操作票			30	操作票填写不规范，视情况扣2~30分			
2	安全意识	正确选取操作所需安全工器具； 正确做好个人防护； 正确进行模拟预演； 正确核对设备的位置、名称、编号和运行方式			20	未正确选择所需的安全用具，扣1~5分； 未做好个人防护，扣1~5分； 未执行模拟倒闸操作，扣1~5分； 未核对设备的位置、名称、编号和运行方式，扣1~5分			
3	操作技能	遵循安全操作规程，操作步骤正确； 停送电时，拉合隔离开关前应检查断路器是否处于分闸位置，停电时，要把接地开关合上； 停送电时，操作隔离开关和断路器的顺序应正确，每操作完一步，要检查隔离开关和断路器的操作状态； 全部操作顺序要正确流畅； 储能装置储能完毕后电源开关要复位； 送电时，要通过电压转换开关来检查电压； 在规定时间内操作完成，并检查一次设备			50	停送电时，拉合隔离开关前应检查断路器是否处于分闸位置，停电时，要把接地开关合上，操作正确，少检查、少操作一个扣5分； 停送电时，操作隔离开关和断路器的顺序应正确，若错误扣50分； 每操作完一步，要检查隔离开关和断路器的操作状态，少检查一个扣5分； 全部操作顺序要正确流畅，不流畅扣6分，出现大的操作错误扣50分； 储能装置储能完毕后电源开关要复位，未复位，扣5分； 送电时，要通过电压转换开关来检查电压，未检查扣5分； 在规定时间内操作完成，并检查一次设备，超时扣5~10分，未检查扣5分			

续表

序号	作业名称	质量标准	分值/分	扣分标准	扣分	得分
4	操作票填写	针对操作任务正确填写操作票,若操作票填写不正确,不准操作,考生该题得分为零分,终止该项目的考试	否定项			
合计			100			

Task Ⅲ Training in power interruption and supply operations of HV switch cabinets

Ⅰ. Operating tasks

A group of 3 persons, one playing the part of the person in charge of work and the others playing the part of team members, are required to work in a "110 kV intelligent substation", and perform power interruption and supply operations of the HV switch cabinet as required by the *Electric Power Safety Working Regulations (Power Transformation) of State Grid Corporation of China*.

Ⅱ. Referenced standards and documents

(1) *Electric Power Safety Working Regulations (Power Transformation) of State Grid Corporation of China*;

(2) *18 Major Anti-accident Measures for Power Grid of State Grid Corporation of China*.

Ⅲ. Operating conditions

The operations shall be carried out in good weather, and the operators shall be in a good mental state and familiar with safety measures, technical measures, and dangerous points of field work.

Ⅳ. Preparation before operation

1. Basic requirements and conditions for on-site operations

Conduct a survey of equipment on site and refer to relevant technical data, including historical data and related procedures.

2. Selection of tools and instruments and materials

A station, a set of standard job cards, a gel pen, 3 safety helmets, a pair of insulating shoes, a pair of insulating gloves, a voice recorder, an "Executed" stamp, a set of operation tools, a set of keys for patrol inspections, and a red vest.

3. Dangerous points and preventive measures

(1) High voltage electric shock.

Prevention and control measures: Properly have a simulation rehearsal and take measures to ensure personal safety.

(2) The use of unqualified safety tools and instruments.

Prevention and control measures: Before operating equipment, the operator shall select the correct safety tools and instruments and check whether they are qualified.

4. Division of labor among operators

Person in charge of on-site work (supervisor): ×××

On-site operator: ×××

V. Specifications and requirements

(1) During the on-site operation, the instructor will show the operation mode of the switch cabinet, and the trainees are required to perform the supply interruption/power operation of the HV switch cabinet according to the on-site requirements and simulate the demonstration.

(2) Fill in the operation ticket in accordance with the requirements of the operation task.

(3) Choose proper safety appliances in accordance with the requirements of the operation task to protect personal safety.

(4) Abide by the safety operating procedures and follow the steps of the operation ticket.

(5) After the operation, check if the operation is well performed.

VI. Operation processes and standards (see Tab. 5-1)

Tab. 5-1 Power interruption and supply operation processes and scoring standards of switch cabinets

Class		Name		Student ID		Examiner		Score	
S/N	Operation name	Quality standard		Points	Deduction criteria			Deduction	Score
1	Properly completing the operation ticket	Fill in the operation ticket in accordance with the requirements of the operation task		30	Deduct 2-30 points if failing to fill in the operation ticket				
2	Safety consciousness	Select proper safety tools and instruments required for operation; Take proper measures to ensure personal safety; Correctly carry out a simulation rehearsal; Check the location, name, number, and operation mode of the equipment in a proper way		20	Deduct 1-5 points if failing to choose proper safety appliances; Deduct 1-5 points if failing to take proper measures to ensure personal safety; Deduct 1-5 points if failing to simulate switching operations; Deduct 1-5 points if failing to check the location, name, number, and operation mode of the equipment				
3	Operating skill	Abide by the safety operating procedures and follow the correct operation steps; When it comes to power interruption and supply, check if the circuit breaker is in the opening position before closing the isolating switch. To interrupt the power supply, close the grounding switch; When it comes to power interruption and supply, operate the isolating switch and circuit breaker in the correct sequence.		50	When it comes to power interruption and supply, check if the circuit breaker is in the opening position before closing the isolating switch. To interrupt the power supply, close the grounding switch. Deduct 5 points if omitting to do such a check or operation; Deduct 50 points if failing to operate the isolating switch and circuit breaker in the correct sequence when it comes to power interruption and supply;				

Continued

S/N	Operation name	Quality standard	Points	Deduction criteria	Deduction	Score
3	Operating skill	After each step, check the operating state of the isolating switch and circuit breaker; All operations shall be conducted smoothly in the correct sequence; After storing energy in the energy storage device, reset the power switch. When supplying power, check the voltage using the voltage change-over switch within the specified time, and check the primary equipment		After each step, the operating state of the isolating switch and circuit breaker shall be checked. Deduct 5 points if failing to the isolating switch or the circuit breaker each time; Deduct 6 points if failing to conduct all operations smoothly and 50 points if making a major misoperation; Deduct 5 points if failing to reset the power switch after storing energy in the energy storage device. Deduct 5 points if failing to check the voltage using the voltage change-over switch when supplying power; Deduct 5-10 points if failing to check the voltage within the specified time; Deduct 5 points if failing to check the primary equipment		
4	Completion of operation ticket	Fill in the operation ticket correctly according to the operation task. If the operation ticket is not filled in correctly, the examinee will not be allowed to operate and score zero for this question, and the exam of this item will be terminated.	Negation item			
Total			100			

任务四　变电站设备巡视检查实训

一、作业任务

3人一组：1人扮演工作负责人，2人扮演班组成员，以"110 kV智能变电站"为工作环境，按照《国家电网公司电力安全工作规程（变电部分）》中的要求，完成变电站设备巡视检查，并填写设备巡视记录卡。

二、引用标准及文件

（1）《国家电网公司电力安全工作规程（变电部分）》。
（2）《国家电网公司十八项电网重大反事故措施》。

三、作业条件

应在良好的天气进行设备巡视；作业人员精神状态良好，熟悉工作中安全措施、技术措施以及现场工作危险点。

四、作业前准备

1. 现场设备巡视的基本要求及条件

勘察现场设备情况，查阅相关技术资料，包括历史数据及相关规程。

2. 工器具及材料选择

工位1个、标准化巡视卡1套、中性笔1支、值班移动电话1部、安全帽、电筒、听音器、测温枪。

3. 危险点及预防措施

（1）高压触电。
危险点：巡视设备均应视为带电设备。
预防措施：巡视时应走规定巡视路线，并与带电设备保持足够的安全距离。
（2）使用不合格的安全工器具。
预防措施：巡视设备前，应正确选取安全工器具，并检查是否合格。

4. 作业人员分工

现场负责人（监护人）：×××
现场作业人员：×××

五、作业规范及要求

（1）现场操作时，由老师给出运行方式，学员按现场规程判断处理，模拟演示。
（2）在主变压器、电流互感器、电压互感器、接地刀闸、隔离开关、断路器、避雷器中抽选 2 种设备进行巡视检查。

（3）必须按巡视路线图巡视，如有异常应及时汇报。

六、作业流程及标准（表 5-2）

表 5-2 变电站巡视检查流程及评分标准

班级		姓名		学号		考评员		成绩	
序号	作业名称	质量标准			分值/分	扣分标准		扣分	得分
1	工作准备								
1.1	着装穿戴	穿工作服、工作鞋，戴安全帽、线手套			10	未穿工作服、工作鞋，戴安全帽、线手套，每缺项扣 1 分；着装穿戴不规范，每处扣 1 分			
2	工具检查								
2.1	取安全帽、常用工器具并检查	安全帽：检查外观、标签、合格证、下颌带是否完好； 电筒外观、标签、照明度足够； 值班移动电话外观、通信效果完好； 听音器外观、使用完好； 测温枪外观、性能完好； 携带控制柜钥匙			10	未取安全帽扣 2 分，未检查安全帽外观、标签、合格证、下颌带是否完好扣 1 分，未佩戴扣 5 分； 未检查电筒外观、标签、照明度扣 1 分，未拿手电筒扣 5 分； 未检查值班移动电话外观和通信效果扣 1 分，未拿值班移动电话扣 5 分； 未检查听音器外观和使用完好情况扣 1 分，未拿听音器扣 5 分； 未检查测温枪外观和性能是否完好扣 1 分，未拿测温枪扣 5 分； 未携带控制柜钥匙，扣 5 分			
3	工作过程（抽选其中两种设备）								
3.1	电流互感器	电流互感器标示清晰； 高压套管油位、油色正常，无渗、漏油现象，整体无渗、漏油现象； 高压套管及绝缘瓷瓶完好、清洁，套管无裂纹、无破损和放电现象；			40	未检查出名称、标志不完备，每处扣 3 分； 未检查出套管油位、油色异常，扣 5 分； 未检查出套管脏污、破损、放电，每处扣 5 分； 未检查出接地线松脱，每处扣 5 分			

续表

序号	作业名称	质量标准	分值/分	扣分标准	扣分	得分
3.1	电流互感器	接地线连接完好、牢固、无松动脱落和断线脱漆现象； 进出线连接牢固，接点无发热变色、无断线现象，无松动放电以及变色现象，用红外线测温仪测量其温度正常； 内部声音正常，无放电声及剧烈振动声音，电流互感器无异常焦味； 二次连接线各元器件接线完好，无焦臭味、松动、脱落现象； 端子排无变形、变位现象		未检查出进出线发热变色、松动，扣3分； 未检查出有异常声音、气味，扣5分； 未检查出二次端子松动、发热、异味，每处扣5分		
3.2	电压互感器	编号及名称标示清晰，与设备相符，设备卫生清洁； 电压互感器外壳清洁、完整、无破损及裂纹、无放电痕迹； 听内部声音正常，无放电声及剧烈振动声音； 高压引线连接完好，牢固无松动、无断股断线、无放电现象，用红外线测温仪测量其温度正常； 接地线连接完好，牢固无松动、无脱落及脱漆现象，无锈蚀； 二次连接线各元器件接线完好，无焦臭味、松动、脱落现象； 端子排无变形、变位现象	40	未检查出名称、标志不完备，每处扣3分； 未检查出外壳脏污、破损、放电，每处扣5分； 未检查出有异常声音，扣5分； 未检查出引线发热变色、松动，扣5分； 未检查出接地线松脱，每处扣5分； 未检查出二次端子松动、发热、异味，每处扣5分		
3.3	接地刀闸	接地刀闸编号及名称标示清晰，与设备相符； 地刀触头合闸时接触良好、严密； 刀片和刀嘴无脏污、无烧伤痕迹、无变形、无锈蚀、无倾斜； 闭锁装置良好，机械闭锁销子牢固； 操作机构各元器件完好无损，无变形、无变位及松动脱落现象，传动连接销子无脱落	40	未检查出名称、标志不完备每处，扣3分； 未检查出地刀接触不良，扣5分； 未检查出刀片或刀嘴脏污锈蚀、烧伤变形，每处扣5分； 未检查出闭锁装置损坏失效，扣5分； 未检查出操作机构变形、动作不灵活，扣5分		

续表

序号	作业名称	质量标准	分值/分	扣分标准	扣分	得分
3.4	各侧避雷器	瓷瓶无裂纹，构架正常； 避雷器计数器密封良好、指示正确； 接地引线接头无锈蚀、焊接良好； 避雷器内部无异常声音； 检查泄漏电流值是否正常	40	未检查出瓷瓶有裂纹，构架不正常，每处扣5分； 未检查出避雷器计数器密封不良、指示不正确，每处扣5分； 未检查出接地引线接头有锈蚀、焊接不良好，每处扣5分； 未检查出避雷器内部有异常声音，每处5分； 未正确检查，每处扣5分		
3.5	隔离开关	编号及名称标示清晰，与设备相符，设备卫生清洁； 绝缘子外观完好，无裂纹、破损、电晕和放电现象，无灰尘积淀； 隔离开关触头接触良好、严密，无发热变形现象； 刀片和刀嘴无脏污、无烧伤痕迹、无变形、无锈蚀、无倾斜； 闭锁装置良好，机械闭锁销子牢固； 操作机构各元器件完好无损、无变形、无变位及松动脱落现象； 电机无过热、无异常振动、无异音，电机电源接线完好，接地线正确无误、完好无损； 端子箱二次连接线各元器件接线完好，无焦臭味、松动脱落； 端子排无变形、变位现象，手感无过热	40	未检查名称、标志，每处扣3分； 未检查绝缘子，每处扣5分； 未检查触头，每处扣5分； 未检查刀片刀嘴，每处扣5分； 未检查闭锁装置，每处扣5分； 未检查操作机构，每处扣5分； 未检查电机，每处扣5分； 未检查端子箱格元器件，每处扣5分		
3.6	断路器	编号及名称标示清晰，与设备相符，设备卫生清洁； 操作机构箱完好无损，无脱漆、生锈现象，手感箱门动作灵活，箱门关好并上锁，无锈蚀，柜门玻璃罩完好，操作机构密封完好，通风良好，外壳接地牢固可靠，电缆孔封堵严密；	40	未检查出名称、标志不完备，每处扣3分； 未检查操作机构箱，扣5分； 未检查灭弧室内设备，每处扣5分； 未检查出引线松动、发热变形，每处扣5分；		

续表

序号	作业名称	质量标准	分值/分	扣分标准	扣分	得分
3.6	断路器	灭弧室完好无损； 上引线连接牢固，无断股，无过热、变色现象； 支柱瓷瓶完好无损，无放电痕迹和裂缝； 下引线连接牢固，无断股，无过热、变色现象； 密封装置密封完好，耳听无漏气声； 引出线无断股，无过热、振动、变形； 分合闸指示器完好无损，位置指示与实际运行工况相一致； 合闸弹簧完好无损，无变形变位现象； 分合闸线圈完好无损，手感无过热，嗅感无焦臭味； 空压机电机完好无损，无异常振动，电机电源接线正确、完好，接地线良好，空压机油位在中间位置，无甩油、漏油、渗油现象（此项可根据现场规定确定，采用液压操作机构和弹簧操作机构者检查相应部位）； 二次控制各元器件接线完好，无变形、变位、松动脱落现象，无过热、变色、断线现象，远方就地转换开关处于远方位置，各电源空气开关在合闸位置，无焦臭味； 操作压力在额定压力内变化，操作空气压力在额定值范围内变化，各压力表表阀位置正确无误		未检查出指示器损坏或与实际不一致，扣5分； 未检查出合闸弹簧变形变位，扣5分； 未检查出分合闸线圈损坏或有异味，扣5分； 未检查密封，每处扣5分； 未检查出二次回路接线松脱、发热变色、开关位置有误，每处扣5分； 未检查操作空气、储气罐，每处扣5分		

Task Ⅳ Training in routine inspections of substation equipment

Ⅰ. Operating tasks

A group of 3 persons, one playing the part of the person in charge of work and the others playing the part of team members, are required to work in a "110 kV intelligent substation", and perform a routine inspection of the substation equipment as required by the *Electric Power Safety Working Regulations (Power Transformation) of State Grid Corporation of China*.

Ⅱ. Referenced standards and documents

(1) *Electric Power Safety Working Regulations (Power Transformation) of State Grid Corporation of China*;

(2) *18 Major Anti-accident Measures for Power Grid of State Grid Corporation of China.*

Ⅲ. Operating conditions

The routine inspection shall be carried out in good weather, and the operators shall be in a good mental state and familiar with safety measures, technical measures, and dangerous points of field work.

Ⅳ. Preparation before operation

1. Basic requirements and conditions for on-site routine inspection of equipment

Conduct a survey of equipment on site and refer to relevant technical data, including historical data and related procedures.

2. Selection of tools and instruments and materials

A workstation, a set of standard routine inspection cards, a gel pen, a duty mobile phone, safety helmet, flashlight, listening gear, temperature gun.

3. Dangerous points and preventive measures

(1) High voltage electric shock.

Dangerous points: All the equipment under routine inspection shall be regarded as live equipment.

Prevention and control measures: The specified routine inspection route shall be followed and a sufficient safe distance shall be maintained.

(2) The use of unqualified safety tools and instruments.

Prevention and control measures: Before routine inspection, the operator shall select correct safety tools and instruments and check whether they are qualified.

4. Division of labor among operators

Person in charge of on-site work (supervisor): ×××

On-site operator: ×××

Ⅴ. Operating specifications and requirements

(1) When it comes to on-site operation, the instructor will show the operation mode, and the trainees are required to make a judgment and do accordingly according to the on-site procedures and simulate the demonstration.

(2) Select 2 from the main transformer, current transformer, voltage transformer, grounding knife-switch, isolating switch, circuit breaker, and lightning arrester for the routine inspection.

(3) Carry out the routine inspection according to the route map for routine inspection, and timely report anomalies if any.

Ⅵ. Operation processes and standards (see Tab. 5-2)

Tab. 5-2 Routine inspection processes and scoring standards of substations

Class		Name		Student ID		Examiner		Score	
S/N	Operation name	Quality standard		Points	Deduction criteria		Deduction		Score
1	Work preparation									
1.1	Wearing	Wear work clothes, work shoes, safety helmets, and cotton gloves		10	Deduct 1 point if failing to wear work clothes, work shoes, safety helmet, or cotton gloves; Deduct 1 point if wearing does not meet the requirements				
2	Tool inspection									
2.1	Getting and checking safety helmets and common tools and instruments	Safety helmet: Check the appearance, label, and certificate of conformity and confirm that the chin strap is intact; Flashlight: Check the appearance and label, and check if the flashlight has sufficient illumination; Duty mobile phone: It shall be intact and have good communication performance; Listening gear: It shall be intact and function well; Temperature gun: It shall be intact and have good performance; Bring the key to the control cabinet		10	Deduct 2 points if failing to take the safety helmet; deduct 1 point if failing to check the appearance, label and certificate of conformity of the helmet and the integrity of chin strap; deduct 5 points if failing to wear the safety helmet; Deduct 1 point if failing to check the appearance, label and illuminance of the flashlight; Deduct 5 points if failing to take the flashlight; Deduct 1 point if failing to check the appearance and communication performance of the duty mobile phone; Deduct 5 points if failing to bring the duty mobile phone;				

Continued

S/N	Operation name	Quality standard	Points	Deduction criteria	Deduction	Score
2.1	Getting and checking safety helmets and common tools and instruments			Deduct 1 point if failing to check the appearance and performance of the listening gear; Deduct 5 points if failing to bring the listening gear; Deduct 1 point if failing to check the appearance and performance of the temperature gun; Deduct 5 points if failing to bring the temperature gun; Deduct 5 points if failing to bring the key to the control cabinet		
3		Working process (select two of the equipment)				
3.1	Current transformer	The current transformer shall be clearly marked; For the HV bushing, the oil level and oil color shall be normal, and there shall be no oil seepage and oil leakage. Overall, there shall be free of oil seepage and oil leakage; The HV bushing and porcelain insulator shall be intact and clean, and the bushing shall be free of any cracks, damage, and discharge phenomena; The grounding wire connection shall be firm, intact, and free of loosening, coming off, snapping, or paint-shedding; The incoming and outgoing lines shall be firmly connected, and their connections shall be free of heating, becoming red, discoloring, loosening, and discharging. Test them with an infrared thermodetector and their temperature shall be normal; The sound inside the current transformer shall be normal, and there shall be no discharge sound and violent vibration sound, and no abnormal burnt smell from the current transformer;	40	Deduct 3 points if failing to notice that a name or a marking is incomplete; Deduct 5 points if failing to notice that the oil level or oil color of the bushing is abnormal; Deduct 5 points if failing to notice that a bushing is dirty, damaged, or discharging; Deduct 5 points if failing to notice that a grounding wire is loosened; Deduct 3 points if failing to notice that the incoming and outgoing lines are heated, discolored, and loosened; Deduct 5 points if failing to notice any abnormal noises and smell; Deduct 5 points if failing to notice that the secondary terminal is loosened, and heated and that it emits an unpleasant odor		

Continued

S/N	Operation name	Quality standard	Points	Deduction criteria	Deduction	Score
3.1	Current transformer	The connections of parts and components of the secondary connecting line shall be intact and free of burnt smell, loosening, and coming off. The terminal board shall not be deformed and displaced				
3.2	Voltage transformer	The number and name shall be clearly marked and consistent with the equipment, and the equipment shall be hygienic and clean; The voltage transformer housing shall be shell clean, complete, and free of any damage cracks, and discharge marks; The sound inside shall be normal, and there shall be no discharge sound and violent vibration sound; The HV lead connection shall be firm, intact, and free of loosening, coming off, strand breakage, snapping, or discharging; Test it with an infrared thermodetector and their temperature shall be normal; The grounding wire connection shall be firm, intact, and free of loosening, coming off, paint-shedding, or rusting; The connections of parts and components of the secondary connecting line shall be intact and free of burnt smell, loosening, and coming off. The terminal board shall not be deformed and displaced	40	Deduct 3 points if failing to notice that a name or a marking is incomplete; Deduct 5 points if failing to notice that a housing is dirty, damaged, or discharging; Deduct 5 points if failing to notice any abnormal noises; Deduct 5 points if failing to notice that the incoming and outgoing lines are heated, discolored, and loosened; Deduct 5 points if failing to notice that a grounding wire is loosened; Deduct 5 points if failing to notice that the secondary terminal is loosened, and heated and that it emits an unpleasant odor		
3.3	Grounding knife-switch	The number and name of the grounding knife-switch shall be clearly marked and consistent with the equipment; The contact terminals shall come into a nice and tight contact when the grounding knife-switch closes; The blade and knife point shall be free of any dirty, burn marks, deformation, rusting, and tilting;	40	Deduct 3 points if failing to notice that a name or a marking is incomplete; Deduct 5 points if failing to notice that the contact terminals of the grounding knife-switch are in poor contact; Deduct 5 points if failing to notice a dirty, rusty, burnt, or deformed spot of the blade or knife point;		

Continued

S/N	Operation name	Quality standard	Points	Deduction criteria	Deduction	Score
3.3	Grounding knife-switch	The locking device shall be in good condition and the mechanical locking pin shall be firmly secured; The parts and components of the operating mechanism shall be intact and shall not be deformed, displaced, or loosened, and the transmission connection pin shall not come off		Deduct 5 points if failing to notice that the locking device is damaged and can not work; Deduct 5 points if failing to notice that the operating mechanism is deformed and can not operate flexibly		
3.4	Lightning arresters on each side	The porcelain insulator shall be free of any cracks, and the frame shall be normal; The lightning arrester counter shall be well-sealed and give correct indications; The grounding lead connector shall be free of rusting and well-welded. There shall be no abnormal sound inside the lightning arrester; Check if the leakage current value is normal	40	Deduct 5 points if failing to notice that a porcelain insulator is cracked or a frame is abnormal; Deduct 5 points if failing to notice that a lightning arrester counter is poorly sealed and gives incorrect indications; Deduct 5 points if failing to notice that a grounding lead connector is rusty or poorly welded; Deduct 5 points if failing to notice any abnormal noises inside a lightning arrester; Deduct 5 points for each incorrect inspection		
3.5	Isolating switch	The number and name shall be clearly marked and consistent with the equipment, and the equipment shall be hygienic and clean; The insulator shall be intact and free of any cracks, damage, corona, discharge phenomena and dust; The contact terminals of the isolating switch shall be in nice and tight contact and free of heating and deformation; The blade and knife point shall be free of any dirty, burn marks, deformation, rusting, and tilting; The locking device shall be in good condition and the mechanical locking pin shall be firmly secured;	40	Deduct 3 points if failing to check a name or a marking; Deduct 5 points if failing to check an insulator; Deduct 5 points if failing to check a contact terminal; Deduct 5 points if failing to check a blade or knife point; Deduct 5 points if failing to check a locking device; Deduct 5 points if failing to check an operating mechanism; Deduct 5 points if failing to check a motor; Deduct 5 points if failing to check a part or a component of the terminal box		

Continued

S/N	Operation name	Quality standard	Points	Deduction criteria	Deduction	Score
3.5	Isolating switch	The parts and components of the operating mechanism shall be intact and free of deformation, displacement, or loosening; The motor shall be free of overheating, abnormal vibration and noises; The power cable of the motor shall be intact, and the grounding wire shall be correct and intact; The wiring of parts and components of the secondary connecting line of the terminal box shall be intact and free of burnt smell, loosening, and coming off. The terminal board shall not be deformed, displaced, and overheated				
3.6	Circuit breaker	The number and name shall be clearly marked and consistent with the equipment, and the equipment shall be hygienic and clean; The operating mechanism case shall be intact and free of paint-shedding and rusting, and the door shall be opened and closed flexibly. The door shall be closed and locked and shall not be rusty. The glass cover of the cabinet door shall be intact, the operating mechanism shall have an intact seal and shall be well ventilated, and the housing shall be grounded firmly and reliably; The cable holes shall be tightly sealed; The arc extinguishing chamber shall be intact, and the upper lead shall be firmly connected, and free of strand breakage, overheating, and discoloring; The porcelain insulator of the post shall be intact, and free of any discharge marks and cracks, and the lower lead shall be firmly connected; There shall be no stand breakage, overheating, and discoloring; The sealing device shall have an intact seal; There shall be no leak sound;	40	Deduct 3 points if failing to notice that a name or a marking is incomplete; Deduct 5 points if failing to check an operating mechanism; Deduct 5 points if failing to check a piece of equipment in the arc extinguishing chamber; Deduct 5 points if failing to notice that a lead is loosened, and heated and that it emits an unpleasant odor; Deduct 5 points if failing to notice that the indicator is damaged or the indication is inconsistent with the actual condition; Deduct 5 points if failing to notice the deformation or displacement of the closing spring; Deduct 5 points if failing to notice the damage or unpleasant odor of the closing coil; Deduct 5 points if failing to check a seal; Deduct 5 points if failing to notice that a secondary circuit connection is loosened, heated, or discolored and that a switch position is wrong;		

Continued

S/N	Operation name	Quality standard	Points	Deduction criteria	Deduction	Score
3.6	Circuit breaker	The lead wire shall be free of stand breakage, overheating, vibration, and deformation; The opening and closing indicators shall be intact, and the position indication shall be consistent with the actual working conditions; The closing spring shall be intact without deformation or displacement; The closing coil shall be intact and free of overheating and burnt smell; The air compressor motor shall be intact and free of abnormal vibration, and the motor's power cable shall be correctly connected and shall be intact; The grounding wire shall be in good condition. The oil level of the air compressor shall be in the middle, there shall be no oil shedding, oil leakage, or oil seepage (which can be determined according to the on-site regulations; and use the hydraulic operating mechanism and spring operating mechanism to check the corresponding parts); The connections of parts and components of the secondary control shall be intact, and free of deformation, displacement, loosening, coming off, overheating, discoloring, and snapping; The remote-local change-over switch shall be put in the remote position; The air switches of each power supply shall be in the closing position and free of burnt smell; The operating pressure shall vary within the rated pressure, and the operating air pressure shall vary within the rated range; The pressure gauge valve shall be in the correct position;		Deduct 5 points if failing to check operating air or an air storage tank		

Continued

S/N	Operation name	Quality standard	Points	Deduction criteria	Deduction	Score
3.6	Circuit breaker	The air storage tank shall be free of any cracks, paint-shedding, rusting, or leaking sound; The blowdown valve shall be free of any air leakage				
3.7	Main transformer	The name and markings shall be complete and intact; The upper oil temperature of the proper shall not exceed the specified value; No noise shall be heard from inside the proper when listening with the help of an object, and each component of the proper shall not leak; The explosion-proof pipe (or pressure relief valve) shall not leak; The grounding body of the proper shall be intact and free of rusting; The gas relay shall be intact; The porcelain insulator of the bushing on each side shall be free of any cracks, damage and discharge marks; The cooler of the cooling device shall be put into operation correctly and keep operating properly; The supporting porcelain insulator of the lead or cable on each side shall be free of any cracks, the frame shall be normal, and the joints shall be free of rusting and well-welded. The cable heads, phase-indicating colors, and joint temperature shall be normal; The oil level and color of the oil conservator shall be normal, and the oil conservator and the oil line connected to it shall not leak	40	Deduct 5 points if failing to notice double names; Deduct 3 points if a marking is incomplete; Deduct 5 points if failing to check the temperature of the proper; Deduct 5 points if failing to use an object to or properly hear the noise from inside the proper; Deduct 5 points if failing to notice that a component of the proper is leaking; Deduct 5 points if failing to notice that the explosion-proof pipe is leaking; Deduct 5 points if failing to notice that a grounding body of the proper is incomplete or rusty; Deduct 5 points if failing to check the gas relay; Deduct 3 points if failing to notice that a porcelain insulator is abnormal; Deduct 3 points if failing to notice that a cooler is not put into operation correctly or operating properly; Deduct 5 points if failing to notice that a lead or cable on each side is abnormal; Deduct 5 points if failing to notice an anomaly of the conservator		
Total			100			

任务五　10 kV 柱上变压器的停送电操作实训

一、作业任务

2 人一小组，1 人扮演监护人，1 人扮演操作人，以"10 kV 柱上变压器"为工作对象，按照《国家电网公司电力安全工作规程（变电部分）》中的要求，完成 10 kV 柱上变压器的停送电操作。

二、引用标准及文件

（1）《国家电网公司电力安全工作规程（变电部分）》。
（2）《国家电网公司十八项电网重大反事故措施》。

三、作业条件

应在良好的天气操作；作业人员精神状态良好，熟悉工作中的安全措施、技术措施以及现场工作危险点。

四、作业前准备

1. 现场操作的基本要求及条件

勘察现场设备情况，查阅相关技术资料，包括历史数据及相关规程。

2. 工器具及材料选择

工位 1 个、中性笔 1 支、安全帽 2 顶、绝缘手套 1 双、线手套 2 双、操作杆 1 个、"禁止合闸，有人工作"标识牌 1 个、已执行印章 1 枚。

3. 危险点及预防措施

（1）高压触电。
预控措施：正确进行模拟预演，做好个人防护。
（2）使用不合格的安全工器具。
预控措施：操作设备前，应正确选取安全工器具并检查合格。

4. 作业人员分工

现场工作负责人（监护人）：×××
现场作业人员：×××

五、作业规范及要求

（1）现场操作时，由老师给出 10 kV 柱上变压器运行方式，学员按现场要求进行 10 kV 柱上变压器的停送电操作，模拟演示。
（2）按照作业任务要求正确填写操作票。

配电倒闸操作票（运行转检修）

操作开始时间：	年 月 日 时 分	操作结束时间：	年 月 日 时 分

操作任务：10 kV 智青线 07 号杆由运行转检修

顺序	操 作 项 目	√
1	拉开 10 kV 智青线 07 号杆低压侧开关	
2	检查 10 kV 智青线 07 号杆低压侧开关在分闸位置	
3		
4		
5	拉开 10 kV 智青线 07 号杆 C 相跌落式熔断器	
6	检查 10 kV 智青线 07 号杆 C 相跌落式熔断器在分闸位置	
7	拉开 10 kV 智青线 07 号杆 A 相跌落式熔断器	
8	检查 10 kV 智青线 07 号杆 A 相跌落式熔断器在分闸位置	
9		
10		
备注		

操作人：	监护人：

正确填写操作票，把选项填在上边的操作票中	
A	拉开 10 kV 智青线 07 号杆 B 相跌落式熔断器
B	在 10 kV 智青线 07 号杆下端处悬挂"禁止合闸，有人工作"标示牌
C	在 10 kV 智青线 07 号杆低压侧开关处悬挂"禁止合闸，有人工作"标示牌
D	检查 10 kV 智青线 07 号杆 B 相跌落式熔断器在分闸位置

配电倒闸操作票（检修转运行）

操作开始时间：	年 月 日 时 分	操作结束时间：	年 月 日 时 分

操作任务：10 kV 智青线 07 号杆由检修转运行

顺序	操 作 项 目	√
1	检查 10 kV 智青线 07 号杆低压侧开关在分闸位置	
2		
3	合上 10 kV 智青线 07 号杆 A 相跌落式熔断器	
4	检查 10 kV 智青线 07 号杆 A 相跌落式熔断器在合闸位置	
5		
6		
7	合上 10 kV 智青线 07 号杆 B 相跌落式熔断器	
8	检查 10 kV 智青线 07 号杆 B 相跌落式熔断器在合闸位置	
9	摘下 10 kV 智青线 07 号杆低压侧开关处"禁止合闸，有人工作"标示牌	

续表

顺序	操 作 项 目	√
10		
11	检查10kV智青线07号杆低压侧开关在合闸位置	
备注		
操作人：	监护人：	

	正确填写操作票，把选项填在上边的操作票中	
A	摘下10 kV智青线07号杆下端处"禁止合闸，有人工作"标示牌	
B	检查10 kV智青线07号杆C相跌落式熔断器在合闸位置	
C	合上10 kV智青线07号杆低压侧开关	
D	合上10 kV智青线07号杆C相跌落式熔断器	

（3）按照作业任务要求正确选择安全用具，做好个人防护工作。
（4）遵循安全操作规程，按照操作票的步骤正确操作。
（5）操作结束后，对操作质量进行检查。

六、作业流程及标准（表5-3）

表5-3　10 kV柱上变压器的停送电操作流程及评分标准

班级		姓名		学号		考评员		成绩	
序号	作业名称	质量标准			分值/分	扣分标准		扣分	得分
1	正确填写操作票	按照作业任务要求正确填写操作票			30	按照作业任务要求正确填写操作票，操作票填写不规范的，视情况扣2~30分			
2	安全意识	准备好该项操作所需的安全用具并进行检验；做好个人防护，戴上安全帽、护目镜、绝缘手套和穿上绝缘靴；正确持操作票在模拟系统模拟操作一次，核对设备的位置、名称、编号和运行方式			20	未能准备好该项操作所需的安全用具并进行检验，扣2~5分；未能做好个人防护，未戴上安全帽、护目镜、绝缘手套和未穿绝缘靴，扣2~5分；要持操作票在模拟系统模拟操作一次，未核对设备的位置、名称、编号和运行方式，扣5~10分			

续表

序号	作业名称	质量标准	分值/分	扣分标准	扣分	得分
3	操作技能	拉、合跌落式熔断器的操作顺序正确；操作跌落式熔断器合闸时检查合闸牢固；在拉、合跌落式熔断器操作时，不允许跌落式熔断器落地；正确安装跌落式熔断器的熔丝；全部操作顺序正确流畅；不出现大的操作错误；在规定时间内操作完成，并检查一次设备	50	遵循安全操作规程，按照操作步骤正确拉、合跌落式熔断器的操作顺序不正确，扣5~50分；操作跌落式熔断器合闸时要检查合闸牢固，不检查扣10分；在拉、合跌落式熔断器操作时，不允许跌落式熔断器落地，落地一个扣10分；跌落式熔断器的熔丝安装工艺要正确，不正确扣5~10分；全部操作顺序要正确流畅，不流畅扣5分；出现大的操作错误扣50分；在规定时间内操作完成，并检查一次设备，不检查确认扣5分，超时扣5~10分		
4	操作票填写	针对操作任务正确填写操作票，若操作票填写不正确，不准操作，考生该题得分为零分，终止该项目的考试	否定项			
合计			100			

Task Ⅴ Training on Power-on/off Operation of 10 kV Pole-mounted Transformers

Ⅰ. Tasks

Let every two persons form a group, with one acting as a supervisor and the other an operator. Each group is required to complete the power-on/off operation of a 10 kV pole-mounted transformer in accordance with the requirements in the Electric Power Safety Working Regulations (Power Transformation) of State Grid Corporation of China.

Ⅱ. Reference standards and documents

(1) Electric Power Safety Working Regulations (Power Transformation) of State Grid Corporation of China;

(2) Eighteen Major Anti-accident Measures for Power Grids of State Grid Corporation of China.

Ⅲ. Operating conditions

Operations should be carried out in good weather. Operators should be in a good state of mind, and be familiar with the safety measures, technical measures, and dangerous points of the work on the site.

Ⅳ. Preparation for work

1. Basic requirements and conditions for on-site operation

Operators should check the condition of on-site equipment, and understand relevant technical data, including historical data and relevant procedures.

2. Selection of tools and instruments and materials

1 workstation, 1 gel pen, 2 safety helmets, 1 pair of insulating gloves, 2 pairs of cotton gloves, 1 operating lever, 1 "No Closing, Work in Progress" sign board, and 1 "Executed" seal.

3. Dangerous points and preventive and control measures

(1) High voltage electric shock.

Prevention and control measures: Carry out simulation and take proper personal protection measures.

(2) Use of non-conforming safety tools and instruments.

Prevention and control measures: Before operating the equipment, operators should select and check the safety tools and instruments properly.

4. Division of labor among operators

Person in charge of on-site work (supervisor): ×××

On-site operator: ×××

V. Operating procedures and requirements

(1) When operating on the site, the teacher introduces the operation mode of 10 kV pole-mounted transformers, and the students perform the power-off/on operation of a 10 kV pole-mounted transformer according to the requirements proposed on the site for simulation.

(2) Fill out the operation ticket correctly according to the requirements of the operation task.

Power distribution switching operation ticket (operation-to-overhaul)

Start time: year/month/day, hour/minute		End time: year/month/day, hour/minute	
Operation task: Operation-to-overhaul of Pole No. 07 of 10 kV Zhiqing Line			
Sequence	Operation item		√
1	Open the switch on the low voltage side of pole No. 07 of 10 kV Zhiqing Line		
2	Ensure that the switch on the low voltage side of pole No. 07 of the 10 kV Zhiqing Line is in the opening position		
3			
4			
5	Open the C-phase dropout fuse on pole No. 07 of the 10 kV Zhiqing Line		
6	Ensure that the C-phase dropout fuse on pole No. 07 of the 10 kV Zhiqing Line is in the opening position		
7	Open the A-phase dropout fuse on pole No. 07 of the 10 kV Zhiqing Line		
8	Ensure that the A-phase dropout fuse on pole No. 07 of the 10 kV Zhiqing Line is in the opening position		
9			
10			
Remarks			
Operator:		Supervisor:	

Correctly fill out the operation ticket. Select the options to fill in the suitable fields of the operation ticket above	
A	Open the B-phase dropout fuse on pole No. 07 of the 10 kV Zhiqing Line
B	Hang a "No Closing, Work in Progress" sign board at the lower end of pole No. 07 of the 10 kV Zhiqing Line
C	Hang a "No Closing, Work in Progress" sign board at the switch on the low voltage side of pole No. 07 of the 10 kV Zhiqing Line
D	Ensure that the B-phase dropout fuse on pole No. 07 of the 10 kV Zhiqing Line is in the opening position

Power distribution switching operation ticket (overhaul-to-operation)

Start time: year/month/day, hour/minute		End time: year/month/day, hour/minute	
Operation task: Overhaul-to-operation of Pole No. 07 of the 10 kV Zhiqing Line			
Sequence	Operation item		√
1	Ensure that the switch on the low voltage side of pole No. 07 of the 10 kV Zhiqing Line is in the opening position		
2			
3	Close the A-phase dropout fuse on pole No. 07 of the 10 kV Zhiqing Line		
4	Ensure that the A-phase dropout fuse on pole No. 07 of the 10 kV Zhiqing Line is in the closing position		
5			
6			
7	Close the B-phase dropout fuse on pole No. 07 of the 10 kV Zhiqing Line		
8	Ensure that the B-phase dropout fuse on pole No. 07 of the 10 kV Zhiqing Line is in the closing position		
9	Remove the "No Closing, Work in Progress" sign board at the switch on the low voltage side of pole No. 07 of the 10 kV Zhiqing Line		
10			
11	Ensure that the switch on the low voltage side of pole No. 07 of the 10 kV Zhiqing Line is in the closing position		
Remarks			
Operator:		Supervisor:	

Correctly fill out the operation ticket. Select the options to fill in the suitable fields of the operation ticket above

A	Remove the "No Closing, Work in Progress" sign board at the lower end of pole No. 07 of the 10 kV Zhiqing Line
B	Ensure that the C-phase dropout fuse on pole No. 07 of the 10 kV Zhiqing Line is in the closing position
C	Close the switch on the low voltage side of pole No. 07 of 10 kV Zhiqing Line
D	Close the C-phase dropout fuse on pole No. 07 of the 10 kV Zhiqing Line

(3) Select safety appliances correctly according to the requirements of the operation task, and take proper personal protection measures.

(4) Follow the safety operating procedures and operate correctly according to the steps in the operation ticket.

(5) Check the quality of operation after the operation.

Ⅵ. Operating procedures and standards (Tab. 5-1)

Tab. 5-1 Operation process and scoring criteria for power-on/off operation of 10 kV pole-mounted transformer

Class			Name		Student ID		Examiner		Score	
S/N	Description		Quality standard		Score (points)	Standard for deduction (points)		Deduction		Score
1	Fill out the operation ticket correctly		Fill out the operation ticket correctly according to the requirements of the operation task		30	Fill out the operation ticket correctly according to the requirements of the operation task; otherwise, 2–30 points will be deducted as appropriate				
2	Safety consciousness		Prepare and check the safety appliances required for the operation; Take proper personal protection measures and wear a safety helmet, safety glasses, insulating gloves and insulating boots; Correctly hold the operation ticket to simulate the operation in the simulation system, and check the location, name, number and operation mode of the equipment		20	Failure to prepare and check the safety appliances required for the operation will result in a deduction of 2–5 points; Failure to take proper personal protection measures or to wear a safety helmet, safety glasses, insulating gloves and insulating boots will result in a deduction of 2–5 points; Correctly hold the operation ticket to simulate the operation in the simulation system. Failure to check the location, name, number and operation mode of the equipment will result in a deduction of 5–10 points				
3	Operational skills		The operational sequence of opening and closing of the dropout fuse is correct; When the dropout fuse is closed, check the fuse for security; When the dropout fuse is opened and closed, it is not allowed to fall to the ground; Correctly install the dropout fuse; All operations are performed correctly and smoothly; No major operational errors occur; Complete the operations within the specified time and check the primary equipment		50	Follow the safety operating procedures and operate correctly according to the steps in the operation ticket, if the operational sequence of opening and closing of the dropout fuse is not correct, 5-50 points will be deducted; When the dropout fuse is closed, check the fuse for security; otherwise, 10 points will be deducted. When the dropout fuse is opened and closed, it is not allowed to fall to the ground, and 10 points will be deducted for each fall;				

Continued

S/N	Description	Quality standard	Score (points)	Standard for deduction (points)	Deduction	Score
3	Operational skills			The dropout fuse must be installed correctly; otherwise, 5–10 points will be deducted; All operations must be performed correctly and smoothly. Five points will be deducted if there is unsmooth operation, and 50 points will be deducted for major operating errors; Complete the operations within the specified time and check the primary equipment; 5 points will be deducted for failure to check. Overtime operation will result in a deduction of 5–10 points		
4	Operation ticket filling	Fill out the operation ticket properly for the operation task; otherwise, no operation is allowed, the examinee will score zero for the question, and the examination will be terminated	Negative			
Total			100			

模块五　电气设备安全运行（Module Ⅴ　Security operation of electrical equipment）

模块 六　电气设备安全检修

电气设备的健康状态不仅影响供电安全，还有可能危及电网的安全运行以及人身安全，因此，对电气设备定期进行检修，或根据设备健康状态安排检修，可有效地提高电气设备的安全性。但是由于发电厂、变电站电气设备种类繁多，结构性能各异，因此检修作业安全技术也不尽相同，同时由于变电站内设备多，运行方式复杂；因此，检修人员在工作过程中应严格执行《国家电网公司电力安全工作规程》，同时与现场运行值班人员密切配合，自觉遵守现场作业安全措施，才能有效保证设备安全和人身安全。本模块将介绍变电站主要电气设备检修安全技术及电气试验安全技术等相关内容。

学习目标：

（1）能说出主要电气设备检修安全技术。

（2）能说出电气试验安全技术要求。

（3）能安全正确进行变压器绝缘电阻试验。

Module VI Maintenance of electrical equipment

The health of electrical equipment may not only affect the safety of power supply but also endanger the safe operation of the power grid and personal safety. So, periodic maintenance or maintenance dictated by health can make electrical equipment safer. However, there is a wide range of electrical equipment in a power plant or a substation, and they have different structures and performances, so the maintenance safety technology is not the same. In addition, there is a lot of equipment in the substation and they have a complex operation mode, so the maintainer shall strictly implement the *Electric Power Safety Working Regulations of State Grid Corporation of China* at work, closely cooperate with the on-duty operator on site, and proactively take the on-site operation safety measures to ensure equipment safety and personal safety. This chapter introduces the safety technology for the maintenance of main electrical equipment and electrical tests of the substation.

Learning objectives:

(1) Be able to tell the safety technology for maintenance of main electrical equipment.

(2) Be able to tell the safety technology requirements for the overhauls of electrical tests.

(3) Be able to safely and correctly carry out insulation resistance tests of transformers.

任务一 认识主要电气设备检修安全技术

一、变压器检修安全措施

长期运行和新安装的变压器,由于受到电磁力、热应力、电腐蚀、运输振动、机械损伤、受潮和化学腐蚀等影响,会发生各种故障和隐患。为了保证变压器安全运行,对不符合规定和要求的部件和零件应及时更换和修复;对检测和检查发现隐患的部件要定期进行检修。变压器通过检修,可以消除隐患和故障,保证其安全运行。

变压器检修可分为大修和小修两类,是以吊芯与否为分界线的。变压器大修是指变压器吊芯或吊开钟罩的检查和检修;变压器小修是指变压器不吊芯或不吊开钟罩的检查和修理。当变压器临时发生故障时,有可能随时决定吊芯或吊开钟罩进行检修。正常运行的主变压器在投运后的第 5 年内和以后每 5~10 年内应吊芯大修一次;一般变压器及线路配电变压器如果未曾过载运行,一般是 10 年大修一次。对于新安装的变压器或运输后投入变压器运行满 1 年时,均应吊芯检修一次,以后每隔 5~10 年大修一次。

(一)变压器大修安全措施

1. 变压器大修项目

(1)打开变压器油箱盖,吊开钟罩(吊芯)检查器身。
(2)检查铁芯、线圈、分接开关和引出线。
(3)检修箱盖、油枕、防爆管、散热管、油阀门和高低压套管。
(4)检修冷却装置和滤油装置,如冷却器、油泵、水泵、风扇等。
(5)清扫外壳,必要时再补喷漆。
(6)检查控制测量仪表、信号和保护装置,进行瓦斯继电器的校验。
(7)进行变压器油的处理或更换。
(8)必要时干燥变压器铁心。
(9)装配变压器。
(10)进行规定的测量和试验。
(11)进行全部密封胶垫的更换和组件试漏。

2. 变压器大修注意事项

(1)起吊之前做好起吊准备工作,起吊设备吨位足够;所用钢丝绳应经检查合格,否则不能使用;吊绳与铅垂线之间的夹角不大于 30°;先试吊合格后才能正式吊芯。
(2)吊芯时要选择无风晴天,相对湿度不大于 75%,器身在空气中停留时间尽可能短,以防绕组绝缘受潮。环境温度应高于-15 ℃,器身低于环境温度时,应使器身加热温度高于大气温度 10 ℃ 以上。当相对湿度大于 85%时,要采取防护措施,如周围加温、绕组加温或缩短检修时间。器身在空气中暴露的时间,是从开始放油时器身与外界空气相接触时算起,注油时间不包含在内。当空气相对湿度大于 95%时不允许吊芯检查。
(3)检修前预先将变压器油放掉一部分,盛油容器必须清洁干燥,容器盛油后必须加盖以防尘防潮,油要进行化验,并符合标准。若油不够,需添补同型号的化验合格的新油。

(4)吊芯时,要有专人负责,油箱四角要有人监视,防止器身与油箱相撞。钟罩吊起时不可在空中摆动,以防撞坏器身。钟罩吊起100 mm时暂停,检查吊绳有无偏斜,如有偏斜时放下找正后再起吊。

(5)器身吊到油箱箱沿处时暂停,开始用油冲洗线圈,使冲洗的油落入油箱内。

(6)冲干净后,要将器身底部吊起比油箱沿较高时,在箱沿上放上方木,将器身先放在箱沿的方木上,使器身上的残油慢慢流下(吊车不能摘钩)。

(7)当器身的油基本空净时(20~40 min),可将油箱推开,将器身放在大油盆上,放稳后,才能将吊车钢丝绳拆掉。

(8)变压器大修应安排在检修间内进行。当施工现场无检修间时,需做好防雨、防潮、防尘和消防措施,清理现场及进行其他准备工作。

(9)大修前进行电气试验,按预防性试验进行,包括测量绕组的绝缘电阻和吸收比或极化指数,测量绕组连同套管一起的泄漏电流,测量绕组连同套管一起的介质损失角正切值$\tan\delta$,进行本体及套管中绝缘油的试验;测量绕组连同套管一起的直流电阻(所有分接头位置);进行套管试验;测量铁心对地绝缘电阻。必要时可增加其他试验项目供大修后进行比较。

(10)使用的工具要有专人保管,事先登记件数。

(二)变压器小修安全措施

变压器小修包括的项目及其安全措施如下:

(1)清扫套管,检查瓷套有无放电痕迹及破损现象。

(2)检查套管引线的接触螺栓是否松动,接头是否过热。

(3)清扫变压器油箱及储油柜、安全气道、净油器、调压装置等附件。

(4)检查安全气道防爆膜是否完好,清除压力释放阀阀盖内沉积的灰尘等杂物。

(5)检查储油柜油位是否正常,油位计是否完好、明净,并排出集污盆内的油污,必要时对套管、变压器本体和有载分接开关补油,进行各部油阀和油堵的检修。

(6)检查呼吸器,更换失效变色的干燥剂。

(7)补充变压器本体及充油套管的绝缘油;检查各部密封胶垫及密封情况,处理渗油。

(8)检查散热器有无渗油现象,冷却风扇是否正常。

(9)检查测量上层油温的温度计。

(10)检查气体继电器有无渗油现象,阀门开闭是否灵活、可靠,控制电缆绝缘是否良好;检查气体继电器集气盒有无气体及有无渗油现象。

(11)检查处理变压器外壳接地线及中性点接地装置。

(12)检查有载分接开关操作控制回路、传动部分及其接点动作情况,并清扫操作箱内部。

(13)从变压器本体、充油套管及净油器内取油样做简化分析,自变压器本体及电容式套管内取油样进行色谱分析。

(14)处理渗、漏油等能就地消除的缺陷。

(15)按规程要求进行测量和试验。

二、开关设备检修安全措施

开关设备是指应用于0.4~1 000 kV电力系统中,对电路进行接通和分断的电气机械装置。开关设备的质量好坏对电力系统是否能正常安全运行起着至关重要的作用。常用的开关设备有断路器、隔离开关和负荷开关。

(一) 断路器

断路器(俗称开关),不仅可以切断或闭合电路中的空载电流和负荷电流,而且当系统发生故障时可通过继电器保护装置的作用切断过负荷电流和短路电流。它具有相当完善的灭弧结构和足够的断流能力。断路器按电压等级分为高压断路器和低压断路器。低压断路器可以分为装置式断路器(塑料外壳式断路器)和框架式断路器(万能式断路器)两大类;高压断路器按灭弧介质可分为油断路器(多油断路器、少油断路器)、六氟化硫断路器(SF_6断路器)、真空断路器、压缩空气断路器等。随着新技术的发展和应用,目前油断路器和压缩空气断路器已基本被淘汰。

1. 断路器作用

(1)控制作用:根据电网运行要求,通过断路器操作使一部分电气设备及线路处于投入或退出运行状态,转为备用或检修状态。

(2)保护作用:在电气设备或线路发生故障时,通过继电保护装置及自动装置使断路器动作,将故障部分从电网中迅速切除,防止事故扩大,保证电网的无故障部分得以正常运行。

2. 断路器功能

(1)导电性能:不仅对正常的电流,而且对规定的短路电流也应能承受其发热和电动力的作用,保持可靠的接通状态。

(2)绝缘性能:相与相之间、相对地之间及断口之间具有良好的绝缘性能。

(3)开断功能:在闭合状态的任何时刻,应能在不发生危险过电压的条件下,在尽可能短的时间内安全地开断规定的短路电流。

(4)关合功能:在开断状态条件允许的时刻,应能在断路器触头不发生熔焊的条件下,在短时间内安全地闭合规定的短路电流。

3. 断路器结构

断路器结构如图6-1所示,其结构通常由通断元件、操动机构、传动机构、绝缘支撑元件、基座组成。

(1)通断元件:执行接通或断开电路的任务。其核心部分是触头和灭弧装置。

(2)操动机构:向通断元件提供分、合闸操作的能量,实现各种规定的顺序操作,并维持开关的合闸状态。

(3)传动机构:把操动机构提供操作能量及发出的操作命令传递给通断元件。

(4)绝缘支撑元件:支撑固定通断元件,实现与各结构部分之间的绝缘。

(5)基座:支撑、固定和安装开关电器的各结构部分,使之成为一个整体。

图 6-1 断路器结构

4. 真空断路器

真空断路器是在真空容器中进行电流开断与关合的断路器。断路器在开断电流时，随着触头的分离，触头接触面积迅速减小，其电流密度非常大，温度急剧升高，使接触点的金属熔化并蒸发出大量的金属蒸气。由于金属蒸气温度很高，同时又存在很强的电场，导致强电场发射和金属蒸气的电离，从而发展成真空电弧。真空断路器就是利用在真空电弧中生成的带电粒子和金属蒸气具有很高扩散速度的特性，在电弧电流过零、电弧暂时熄灭时，使触头间隙的介质强度能很快恢复而实现灭弧。真空灭弧室的结构如图 6-2 所示。

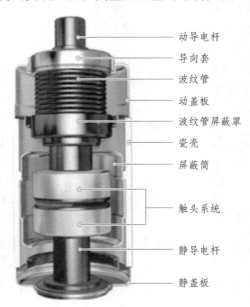

图 6-2 真空灭弧室的结构

（1）真空断路器的优点：

① 真空灭弧室内触头间隙小，整机体积特别小，对操动机构功率要求较小，质量轻。
② 操作时噪声小，不会污染周围环境。
③ 开断能力强，开断电流大，熄弧时间短，开断次数多，使用寿命可达20年。
④ 电弧开断后，介质强度恢复迅速，适合于频繁操作，并且具有多次重合闸功能。
⑤ 介质不会老化，也不需要更换，维护工作量小。
⑥ 使用安全，基本无火灾和爆炸的危险，能适用于各种不同的场合，特别是危险场所。
⑦ 触头部分为完全密封结构，不受外界的影响，工作可靠，通断性能稳定。
⑧ 灭弧室作为独立的元件，安装调试简单、方便。

（2）真空断路器的缺点：

① 开断感性负载或容性负载时，由于截流、振荡、重燃等，容易引起过电压。
② 真空断路器的触头采用对接式结构，操动机构使用弹簧，容易产生合闸弹跳与分闸反弹。合闸弹跳会产生较高的过电压，影响电网的稳定运行，还会使触头烧损，甚至熔焊，特别是在投入电容器组产生涌流时及短路关合的情况下更加严重。分闸反弹会减小弧后触头间距，导致灭弧后的重击穿，后果十分严重。
③ 对密封工艺、制造工艺要求很高，价格相对较高。

5. SF_6 断路器

SF_6 断路器的优良特性得益于 SF_6 气体的物理、化学特性。在标准条件下，SF_6 为无色、无味、无毒的气体，难溶于水和油；相对分子质量比空气重，有向低处积聚的倾向；导热性能强，比空气强 2~5 倍；声音传递速度较慢，但是它带来的温室效应是 CO_2 气体的 22 800 倍。SF_6 气体在常温下是一种化学性能非常稳定的惰性气体，在通常条件下对电气设备中常用的金属和绝缘材料是不起化学作用的，不侵蚀与它接触的物质。SF_6 具有优异的绝缘性能，这是由于 SF_6 气体及其分解物具有极强的电负性。在均匀电场中，SF_6 气体的绝缘强度为空气的 2.5~3 倍；具有优异的灭弧性能，SF_6 气体的灭弧能力为空气的 100 倍，开断能力为空气的 2~3 倍。这不仅是因为它具有优良的绝缘特性，还因为它具有独特的热特性和电特性，电流过零前的截流小，能避免产生较高的过电压。但是，SF_6 气体在断路器操作中和出现内部故障时，会产生不同量的、有腐蚀性、高毒性的分解物（如 HF 及 SO_2），会刺激皮肤、眼睛、黏膜，如果大量吸入，还会引起头晕和肺水肿。纯 SF_6 气体无腐蚀，但其分解物遇水后会变成腐蚀性强的电解质，会对设备内部某些材料（玻璃、瓷、绝缘纸及类似材料）造成损害，或制造成运行故障。为此，可以采用吸附剂（如氧化铝、碱石灰、分子筛或它们的混合物）清除设备内的潮气和 SF_6 气体的分解物。

SF_6 断路器按其结构特点分为瓷柱式（图6-3）和落地罐式（图6-4）两种。

（1）瓷柱式 SF_6 断路器。

其灭弧室安装在高强度瓷套中，用空心瓷柱支承和实现对地绝缘。灭弧室和绝缘瓷柱内腔相通，充有相同压力的 SF_6 气体，通过控制柜中的密度继电器和压力表进行控制和监视。穿过瓷柱的绝缘拉杆，把灭弧室的动触头和操动机构的驱动杆连接起来，通过绝缘拉杆带动触头完成断路器的分合操作。

按其整体布置形式，瓷柱式 SF_6 断路器可分为"I"形布置、"Y"形布置及"T"形布置三种。

"I"形布置：一般用于 220 kV 及以下的单柱单断口断路器，三级安装在一个或三个支架上，如 LW25 等系列的 110 kV 及以下电压级的断路器和 LW31A 等系列的 220 kV 断路器。

"Y"形布置：一般用于 220 kV 及以上的单柱双断口断路器，如 LW6 等系列的 220 kV 断路器、ABB 公司的 ELFSP4-2 型 220 kV 断路器。

"T"形布置：一般用于 220 kV 及以上，特别是 500 kV 的单柱双断口断路器，如 LW7 系列的 220 kV 断路器，日本三菱的 SFM 型 500 kV 断路器，西门子的 3AQ2 型 245 kV、3AT3 型 252 kV 断路器和 3AT2EI 型 550 kV 断路器，ABB 公司的 ELFSP7-21 型 500 kV 断路器，等。

图 6-3　瓷柱式 SF_6 断路器

图 6-4　罐式 SF_6 断路器

优点：瓷柱式 SF_6 断路器系列性强（可以用不同数量的标准灭弧单元及支柱瓷套组成不同电压级的产品），结构简单，用气量少，单断口电压高、开断电流大、运动部件少，价格相对便宜,运行可靠性高,检修维护工作量小,是目前生产和使用较多的一种。

缺点:重心高,抗震能力较差,且不能加装电流互感器,使用场所受到一定限制。

(2)罐式SF_6断路器。

罐式SF_6断路器外观结构如图6-5所示。

结构特点:灭弧室安装在接地的金属罐中,高压带电部分用绝缘子支持,对箱体的绝缘主要靠SF_6气体。绝缘操作杆穿过支承绝缘子,把动触头与机构驱动轴连接起来,在两个出线套管的下部都可安装电流互感器。

图6-5 罐式SF_6断路器外观结构

优点:结构重心低,抗震性能好,灭弧断口间电场较好,断流容量大,可以加装电流互感器,还能与隔离开关、接地开关、避雷器等融为一体,组合成复合式开关设备。借助于套管引线,基本上不用改装就可用于全封闭组合电器之中。

缺点:罐体耗用材料多,用气量大,系列性差,制造难度较高,造价比较昂贵。

110~500 kV罐式SF_6断路器外形基本相似,大多是引进日本三菱公司SFMT型或日立公司OFPTB技术的产品,如OFPTB-500-50LA型、国产LW12系列的220 kV和500 kV断路器。

6. 高压断路器的操动机构

操动机构是独立于断路器本体以外的对断路器进行操作的机械操动装置。其主要任务是将其他形式的能量转换成机械能,使断路器准确、迅速地进行分、合闸操作。

高压断路器的操动系统包括操动机构、传动机构、提升机构、缓冲装置和二次控制回路等几个部分。

7. 高压断路器检修注意事项

(1)在室内检修断路器时,若与附近的带电设备距离较近,或相邻运行中的断路器故障可能伤及检修人员时，应增设临时遮栏。临时遮栏可用干燥木材、橡胶或其他坚韧绝缘材料制成，装设牢固。

（2）在室外需搭脚手架进行检修作业时，脚手架要搭牢，板要铺好绑扎牢固，正确使用安全带，防止工作人员高空跌落受伤。上下脚手架要使用专用爬梯，架牢并做好防滑措施。工作人员要戴安全帽，防止高空落物伤人。

（3）真空断路器检修要注意触头磨损情况的检测，磨损严重的要及时更换真空泡，保证真空断路器有足够的开断能力。

（4）SF_6 断路器检修要防止发生工作人员中毒事故。需要进入 SF_6 配电装置低位区和电缆沟进行工作，首先要检测含氧量（不低于 18%），含氧量符合要求方能进入。断路器解体检修前要对 SF_6 气体进行检验，检修人员应穿着防护服，根据需要佩戴防毒面具，并对 SF_6 气体进行回收。打开设备封盖后，工作人员应暂时撤离现场 30 min，取出吸附剂和清扫粉尘时，要戴防毒面具和防护手套。设备解体后用氮气对断路器进行清洗。在室内检修时要注意通风换气（工作人员应站在上风口），工作现场要布置 SF_6 气体泄漏警报仪。重新安装时要注意对吸附剂的清洗或更换，SF_6 气体充气压力是否正常、有无泄漏，SF_6 气体的水分检测是否合格，以及是否进行干燥处理。

（二）隔离开关

高压隔离开关是目前我国电力系统中用量最大、使用范围最广的高压开关设备。隔离开关在分闸状态有明显可见的间隙，并具有可靠的绝缘；在合闸状态能可靠地通过正常工作电流和短路电流。但是隔离开关没有专门的灭弧装置，不能用来开断负荷电流和短路电流，通常与断路器配合使用。隔离开关的外观结构如图 6-6 所示。

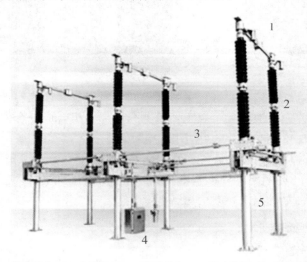

1—导电部分；2—绝缘部分；3—传动机构；4—操动机构；5—支持底座

图 6-6　隔离开关外观结构

1. 隔离开关的种类

（1）按极数可分为单极隔离开关和三极隔离开关。

（2）按绝缘支柱数目可分为单柱式隔离开关、双柱式隔离开关和三柱式隔离开关。

（3）按动作方式可分为闸刀式、水平旋转式、插入式、水平（垂直）伸缩式隔离开关。

（4）按有无接地刀闸可分为带接地刀闸隔离开关和不带接地刀闸隔离开关。

（5）按操动机构可分为手动式、电动式、气动式、液压式。

（6）按用途可分为一般用、快分用和变压器中性点接地用隔离开关。

2. 户内高压隔离开关

（1）GN19-10系列户内高压隔离开关。

GN19-10系列户内高压隔离开关采用三相共底架结构，如图6-7所示，主要由静触头、基座、支柱绝缘子、拉杆绝缘子、动触头组成。

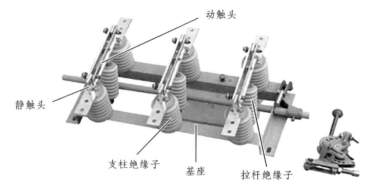

图6-7　GN19-10系列户内高压隔离开关结构

隔离开关导电部分由动、静触头组成，每相导电部分通过两个支柱绝缘子固定在基座上，三相平行安装。

（2）GN30-10型旋转式户内高压隔离开关。

开关主体通过两组绝缘子固定在开关底架上下两个面上，上下两个面之间由固定在开关架上的隔板完全分开，通过旋转触刀实现开关的合闸与分闸。其外观如图6-8所示。

该型高压隔离开关静触头分别安装在开关柜的上下两个面上，使其带电部分与不带电部分在开关柜内完全隔开，从而保证了维修人员的安全，特别适用于安装在高压开关柜内。

图6-8　GN30-10系列户内高压隔离开关外观

3. 户外高压隔离开关

户外高压隔离开关类型非常多，例如：采用双柱单断口水平旋转式结构的 GW4-110 型隔离开关，广泛用于 10～110 kV 配电装置中，其外观如图 6-9 所示；广泛用于 35～110 kV 电压等级中的 GW5-110D 隔离开关；采用三柱双断口水平旋转开启式结构的 GW7-220 型三柱式隔离开关；GW11-252 型隔离开关；GW6-220GD 型单柱式隔离开关；GW10-252 型单柱式隔离开关等。

图 6-9　GW4-110 型隔离开关外观

4. 隔离开关的操动机构

隔离开关的操动机构主要有手动操动机构和电动操动机构两种。

（1）手动操动机构。

杠杆式：一般适用于额定电流小于 3000 A 的隔离开关。

蜗轮式：一般适用于额定电流大于 3000 A 的隔离开关，如 CS6 型手动杠杆式操动机构、CS9 型手动蜗轮式操动机构。

（2）电动操动机构。

CJ2-XG 型电动操动机构属于户外用动力式机构，用于 GW4、GW7 等高压隔离开关或接地开关分、合闸操作，可进行远方控制，也可就地电动控制或利用手柄进行人力操作。

5. 隔离开关检修安全注意事项

（1）交接验收时检查：操动机构、传动装置、辅助切换开关及闭锁装置应安装牢固，动作灵活可靠，位置指示正确；三相不同期值应符合产品的技术规定；相间距离及分闸时触头打开角度和距离应符合产品的技术规定；触头应接触紧密良好；油漆应完整，相色标志正确，接地良好。

（2）运行中检查：绝缘子完整，无裂纹，无放电现象；操作连杆及机械各部分无损伤，无锈蚀，各机件紧固，位置正确，无歪斜、松动、脱落等不正常现象；闭锁装置良好，隔离开关的电磁闭锁或机械闭锁的销子、辅助触点的位置应正确；刀片和刀嘴的消弧角应无烧伤、过热、变形、锈蚀、倾斜现象，触头接触应良好，接头和触头不应有过热现象，其温度不应超过 70 ℃；刀片和刀嘴应无脏污、烧伤痕迹，弹簧片、弹簧及铜辫子应无断股、折断现象；接地开关接地应良好，特别是易损坏的可挠部分应无异常。

（3）操作前检查断路器的分、合位置，严防带负荷操作隔离开关。

（4）手动合闸迅速果断，不能用力过猛。

（5）分闸时迅速拉开隔离开关，以便尽快灭弧。

（6）若出现带负荷误合隔离开关，不准将隔离开关再拉开；若发生错拉隔离开关，刀片刚离开固定触头时，应立即合上。如隔离开关刀片已离开固定触头，则不得将误拉的隔离开关再合上。

（7）合闸操作后，应检查接触是否紧密；拉闸操作后，应检查每相是否均已在断开位置。

（8）操作完毕后，应立即锁住操作把手。

（三）负荷开关

负荷开关是一种带有简单灭弧装置，能开断和关合额定负荷电流的开关。负荷开关有一定的灭弧能力，可用来开断和关合负荷电流和小于一定倍数（通常为 3~4 倍）的过载电流；也可以用来开断和关合比隔离开关允许容量更大的空载变压器，更长的空载线路，有时也用来开断和关合大容量的电容器组。负荷开关还可以与限流熔断器串联组合（负荷开关-熔断器组合电器）代替断路器使用，即由负荷开关承担开断和关合小于一定倍数的过载电流，而由限流熔断器承担开断较大的过载电流和短路电流。熔断器可以装在负荷开关的电源侧，也可以装在负荷开关的受电侧。

负荷开关按其灭弧方式目前常用的有压气式负荷开关、产气式负荷开关、六氟化硫负荷开关和真空负荷开关。

（1）产气式负荷开关，利用固体产气材料在电弧作用下产生气体来进行灭弧的负荷开关，属于自能灭弧方式。

在产气式灭弧室中，灭弧材料气化形成局部高压力，电弧受到强烈吹弧和冷却作用，产生去游离使电弧熄灭。当电流较小时，主要靠产气壁冷却效应或电动力驱使电弧运动，拉长并熄灭电弧。

（2）压气式负荷开关，利用活塞和气缸在开断过程中相对运动将空气压缩，再利用被压缩的空气而熄弧的负荷开关。

（3）真空负荷开关，利用真空灭弧室作为灭弧装置的负荷开关，开断电流大，适宜于频繁操作。

（4）SF_6 负荷开关，利用 SF_6 气体作为绝缘和灭弧介质的负荷开关，在城网和农网中已大量使用。

① 合闸：先辅助刀闸，后主刀闸，主固定触头与主刀片可靠接触；分闸：先主刀闸，后辅助刀闸，三相灭弧刀片同时跳离固定灭弧触头。

② 灭弧筒内产生气体的有机绝缘物应完整无裂纹，灭弧触头与灭弧筒的间隙应符合要求。

③ 三相触头接触同期性和分闸状态时触头间净距及拉开角度应符合产品的技术规定。刀闸打开的角度，可通过改变操作杆的长度和操作杆在扇形板上的位置来达到。

④ 合闸时，在主刀闸上的小塞子应正好插入灭弧装置的喷嘴内，不应对喷嘴有剧烈碰撞的现象。

（四）熔断器

熔断器是电力系统中使用最早的、结构最简单的一种保护电器，主要用于线路及电力变压器等电气设备的短路及过载保护。当电力系统由于过载引起电流超过某一数值、电气设备或线路发生短路事故时，熔断器应能在规定的时间内迅速动作，切断电源，以起到保护设备的作用。它具有结构简单、体积小、质量轻、价格低廉、维护方便、使用灵活等特点。

1. 高压熔断器的运行与维护

（1）按规程要求选择合格产品及配件，运行中经常检查接触是否良好，加强接触点的温升检查。

（2）不可将熔断后的熔体联结起来再继续使用。

（3）更换熔断器的熔管（体），一般应在不带电情况下进行；若需带电更换，则应使用合格的绝缘工具。

（4）操作仔细，拉、合熔断器时不要用力过猛。

（5）定期巡视，每月不少于一次夜间巡视，查看有无放电火花和接触不良现象。

2. 高压熔断器的操作

（1）操作时由两人进行，戴试验合格的绝缘手套，穿绝缘靴，戴护目眼镜，使用与电压等级相匹配的合格绝缘棒操作，在雷电或者大雨的气候下禁止操作。

（2）拉闸时，先拉中相，再拉背风边相，最后拉迎风边相。

之所以这样操作是因为：拉闸时三相变两相运行，中间线产生电弧最小，因此先拉中相；再拉背风侧，因为一旦拉弧，电弧随风飘动不会飘到另外两相；最后分开的一相最不容易拉弧，所以拉最危险的一相。

（3）合闸时操作顺序与拉闸时相反，先合迎风边相，再合背风边相，最后合中间相。

（五）互感器

互感器分为电压互感器（TV）和电流互感器（TA），是电力系统中一次系统和二次系统之间的联络元件，用以变换电压或电流，分别为测量仪表、保护装置和控制装置提供电压或电流信号，从而反映电气设备的正常运行和故障情况。

1. 电流互感器

电流互感器结构与双绕组变压器相似，由铁心和一次、二次绕组两个主要部分构成，0.5 kV 电流互感器的一、二次绕组都套在同一铁心上，结构最简单。10 kV 及以上的电流互感器，为了使用方便和节约材料，常用多个没有磁联系的独立铁心和二次绕组组成一台有多个二次绕组的电流互感器，这样一台互感器可同时供测量和保护用。通常 10~35 kV 有 2 个二次绕组，63~110 kV 有 3~5 个二次绕组，220 kV 及以上有 4~7 个二次绕组。

为适应线路电流的变化，63 kV 及以上的电流互感器，常将一次绕组分成几段，通过串联或并联以获得两种或三种电流比。各种电流互感器外型如图 6-10~图 6-13 所示。

图 6-10　LMZD1-10 型环氧树脂浇注绝缘单匝母线式电流互感器

图 6-11　LDZJ1-10 型环氧树脂浇注绝缘单匝式电流互感器

图 6-12　LFZB-12 型环氧树脂浇注绝缘有保护级复匝式电流互感器

图 6-13　LB7-220 型户外电流互感器

（1）电流互感器的一次绕组串联于被测量电路内，二次绕组与测量仪表和继电器的电流线圈串联。

（2）一次绕组的电流完全取决于被测电路的负荷电流，而不是由二次电流的大小决定的。
（3）在正常运行中，电流互感器是在接近于短路的状态下工作的。
（4）运行中的电流互感器二次侧不得开路。

2. 电压互感器

电压互感器和变压器类似，是用来变换线路上电压的仪器。但是变压器变换电压的目的是输送电能，因此容量很大，一般都是以千伏安或兆伏安为计算单位；而电压互感器变换电压的目的，主要是用来给测量仪表和继电保护装置供电，用来测量线路的电压、功率和电能，或者用来在线路发生故障时保护线路中的贵重设备、电机和变压器，因此电压互感器的容量很小，一般都只有几伏安、几十伏安，最大也不超过 $1 kV·A$。电压互感器外观如图 6-14 所示。

图 6-14 JDZ-10 型浇注式单相电压互感器

电压互感器检修应注意：

（1）电压互感器在投入运行前要按照规程规定的项目进行试验检查，例如测极性、连接组别、摇绝缘、核相序等。

（2）电压互感器的接线应保证其正确性，一次绕组和被测电路并联，二次绕组应和所接的测量仪表、继电保护装置或自动装置的电压线圈并联，同时要注意极性的正确性。

（3）接在电压互感器二次侧负荷的容量应合适，一般不应超过其额定容量；否则，会使互感器的误差增大，难以达到测量的正确性。

（4）电压互感器二次侧不允许短路。由于电压互感器内阻抗很小，若二次回路短路时，会出现很大的电流，将损坏二次设备甚至危及人身安全。电压互感器可以在二次侧装设熔断器以保护其自身不因二次侧短路而损坏。在可能的情况下，一次侧也应装设熔断器以保护高压电网不因互感器高压绕组或引线故障危及一次系统的安全。

（5）为了确保人在接触测量仪表和继电器时的安全，电压互感器二次绕组必须有一点接地。因为接地后，当一次和二次绕组间的绝缘损坏时，可以防止仪表和继电器出现高电压危及人身安全。

Task I Know the safety technology for maintaining the main electrical equipment

I. Safety measures for transformer maintenance

Transformers that have been operating for a long time and that are newly installed are affected by electromagnetic force, thermal stress, electrical corrosion, transportation vibration, mechanical damage, moisture and chemical corrosion, so they will have a variety of faults and hidden dangers. To ensure the safe operation of the transformers, the parts and components that do not meet the regulations and requirements shall be replaced and repaired in time. If testing and inspections suggest that a component has a hidden danger, it shall be maintained periodically. Maintenance can eliminate hidden dangers and faults of transformers and ensure their safe operation.

Transformer maintenance can be divided into overhauls and minor repairs, according to whether the iron core is lifting or not. The overhauls of transformers refer to the inspections and maintenance in which the transformer's iron core or bell jar is lifted. The minor repairs of transformers refer to the inspections and repairs in which the transformer's iron core or bell jar is not lifted. When a transformer temporarily fails, it may be decided at any time to lift the iron core or bell jar for maintenance. A main transformer that operates properly shall be overhauled in the 5th year into operation and thereafter once every 5 to 10 years by lifting its iron core. Generally, transformers and line distribution transformers, if not overloaded, shall be overhauled once every 10 years. New transformers or transformers that have been put into operation after transportation shall be maintained once at the time of one year into operation by lifting their iron cores. Thereafter, they shall be overhauled once every 5 to 10 years.

(I) Safety measures for transformer overhauls

1. Transformer overhauls include the following items

(1) Open the transformer oil tank cover and lift the bell jar (iron core) to inspect the transformer body.

(2) Check the iron core, coil, tap changer, and lead wire.

(3) Maintain the tank cover, oil conservator, explosion-proof pipe, radiating pipe, oil valve, and high- and low-pressure bushings.

(4) Maintain the cooling devices and oil filters, such as coolers, oil pumps, water pumps, and fans.

(5) Clean the housing and repaint it if necessary.

(6) Check and control the calibration of measuring instruments, signal devices, protection devices, and gas relays.

(7) Dispose or replace the transformer oil.

(8) Dry the transformer's iron core if necessary.

(9) Assemble the transformer.

(10) Perform specified measurements and tests.

(11) Replace all sealing gaskets and test whether the assemblies leak.

2. Precautions for overhauls of transformers

(1) Prepare for lifting. The lifting equipment shall have a tonnage that is enough to lift the iron core, and the steel wire rope shall be checked to see if it is qualified, or it can not be used. And the angle between the lifting rope and the plumb line shall not be greater than 30°. A trial lifting shall be done before formally lifting the iron core.

(2) It is necessary to choose a windless sunny day to lift the iron core, and the relative humidity shall not exceed 75%. Make sure that the transformer body stays in the air as short as possible to prevent the winding insulation from moisture. The ambient temperature shall be higher than −15 °C, and when the transformer body is lower than the ambient temperature, the transformer body shall be heated at a temperature that is at least 10 °C higher than the atmospheric temperature. When the relative humidity is greater than 85%, protective measures shall be taken, such as heating on the periphery, heating windings, or shortening the maintenance time. The time for which the transformer body is exposed to air is calculated from the time when the body is exposed to outside air at the beginning of the oil discharge, and the oil filling time is not included. When the relative humidity of the air is greater than 95%, iron core lifting inspections are not allowed.

(3) Part of the transformer oil must be discharged before the maintenance, the oil container must be clean, dry, and covered to protect against dust and moisture, and the oil shall be tested and meet the standards. If the oil is not enough, new oil of the same type shall be refilled.

(4) When lifting the iron core, there shall be a specially assigned person to take responsibility, and the four corners of the oil tank shall be monitored to prevent the transformer body from colliding with the oil tank. When the bell jar is lifted, it shall not swing in the air to prevent damaging the transformer body. The bell jar shall be stopped when it is lifted at a height of 100 mm, and the lifting rope shall be checked to see if it is slant. If it is slant, it shall be corrected before proceeding with lifting.

(5) When the transformer body is lifted to the oil tank rim, it shall be stopped, and at this moment, the coils shall be washed with oil, which shall then flow right into the oil tank.

(6) After the coils are washed clean, when the bottom of the transformer body is lifted higher than the oil tank rim, square timbers shall be put on the oil tank rim, and the transformer body shall be put on the square timbers, so that the residual oil on the transformer body can slowly flow down (the crane hook can not be removed).

(7) When the oil of the transformer body is basically emptied (20 to 40 min), the oil tank can be pushed away, and the transformer body can be placed on the large oil basin. After the transformer body is stable, the steel wire rope can be removed from the crane.

(8) Transformers shall be overhauled in the maintenance room. If there is no maintenance

room at the construction site, measures shall be taken to protect against rain, moisture, dust, and fire, the site shall be cleaned, and other preparations shall be made.

(9) Before overhauls, electrical tests shall be conducted as if they were preventive tests, including measurement of insulation resistance and absorption ratio or polarization index of the windings, measurement of the leakage current of the windings and bushings, measurement of the tangent value tanδ of dielectric loss angle of the windings and bushings, and the test of insulating oil in the proper and bushings; measurement of the DC resistance of the windings and bushings (at all tapping positions); the bushing test; measurement of insulation resistance of the iron core to the ground. If necessary, additional test items can be added for comparison after overhaul.

(10) The tools used shall be kept by a specially designated person and the number of such tools shall be registered in advance.

(Ⅱ) Safety measures for minor repairs of transformers

Minor repairs of transformers include the following items:

(1) Clean the bushing and check if the porcelain bushing has any discharge marks and damage.

(2) Check if the contact bolts of the bushing lead are loose and if the connectors are overheated.

(3) Clean the transformer oil tank, oil storage tank, safety airway, oil purifier, pressure regulator, and other accessories.

(4) Check if the explosion-proof membrane of the safety airway is intact, and remove dust and other debris deposited in the pressure relief valve cover.

(5) Check if the oil level of the oil storage tank is normal and if the oil level gauge is intact and clear. Drain the oil out of the catch basin. If necessary, replenish the oil of the bushing, transformer proper, and on load tap changer; Maintain the oil valves and oil plugs of all parts.

(6) Check the breather and replace the inefficacious and discolored desiccant.

(7) Replenish the insulating oil of the transformer proper and the oil filled bushing; Check the sealing gaskets of all parts and their sealing conditions, and deal with oil seepage.

(8) Check if the radiator leaks oil and if the cooling fan is working properly.

(9) Check the thermometer that is used to measure the upper oil temperature.

(10) Check if oil seeps from the gas relay, if the valve is opened and closed flexibly and reliably, and if the insulation of the control cable is in good condition; Check the collector box of the gas relay for any gas leaks and oil seepage.

(11) Check the grounding wire of the transformer housing and the grounding device of its neutral point.

(12) Check the operations of the operation control circuit, transmission part, and contact of the on load tap changer, and clean the inside of the operation box.

(13) Take oil samples from the transformer proper, oil filled bushing, and oil purifier for simplified analysis, and take oil samples from the transformer proper and condenser type bushing for chromatographic analysis.

(14) Deal with oil seepage, oil leakage, and other defects that can be eliminated on the spot.

(15) Measure and test according to the specifications.

II. Safety measures for switchgear maintenance

A switchgear is an electromechanical device that is used in a 0.4–1,000 kV electric power system to make and break a circuit. The quality of the switchgear plays a crucial role in the normal and safe operation of the electric power system. Common switchgears are circuit breakers, isolating switches, and load switches.

(I) Circuit breakers

Circuit breakers (commonly known as switches) can not only break or make the no-load current and load current in a circuit but also cut off the overload current and short circuit current via the relay protection device when the system fails. They have a fairly perfect arc extinguishing structure and adequate breaking capacity. By voltage class, circuit breakers can be divided into HV circuit breakers and low-voltage circuit breakers. Low-voltage circuit breakers can be divided into molded case breakers (moulded case circuit breakers) and frame-type circuit breakers (universal circuit breakers). By arc extinguishing medium, HV circuit breakers can be divided into oil circuit breakers (bulk-oil circuit breakers, and low oil content circuit breakers), sulphur hexafluoride circuit breakers (SF_6 circuit breakers), vacuum circuit breakers, compressed air circuit breakers, etc. With the development and application of new technologies, oil circuit breakers and compressed air circuit breakers have been basically phased out.

1. Roles of circuit breakers

(1) Control: According to the operation requirements of the power grid, circuit breakers are used to put some electrical equipment and lines into operation, or to put them out of operation to change to the standby or maintenance state.

(2) Protection: When the electrical equipment or line fails, the circuit breaker is actuated by the relay protection device and the automatic device to remove the faulty part quickly from the power grid. In this way, it prevents the accident from escalating and ensures the fault-free part of the power grid can work properly.

2. Functions of circuit breakers

(1) Conductivity: It can not only reliably make the normal current but also can withstand the heating and electrodynamic force of the specified short circuit current to make it.

(2) Insulating property: It provides good insulating property between phases, between phase to ground and between breaks.

(3) Breaking function: At any moment in the closed state, it can safely break the specified short circuit current in the shortest possible time without causing dangerous overvoltage.

(4) Making function: When the conditions permit in the breaking state, it can safely make the specified short circuit current in a short time without causing fusion welding of the circuit breaker's

contact terminals.

3. Circuit breaker structure

As shown in Fig. 6-1, the circuit breaker structure is usually composed of a break-make element, operating mechanism, transmission mechanism, insulating support element, and base.

(1) Break-make element: It makes or breaks a circuit. Its core part is the contact terminals and the arc extinguishing device.

(2) Operating mechanism: It provides the break-make element with energy for opening and closing, enables specified sequential operations, and keeps the switch in the closing state.

(3) Transmission mechanism: It transmits the operating energy provided by the operating mechanism and the operation command issued by the operating mechanism to the break-make element.

(4) Insulating support element: It supports the fixed break-make element and achieves insulation between structural parts.

(5) Base: It supports, fixes, and is fitted with the various structural parts of the switching device to make it a whole.

Fig. 6-1　Circuit breaker structure

4. Vacuum circuit breakers

A vacuum circuit breaker is a circuit breaker for current breaking and making in a vacuum container. When the circuit breaker breaks the current, as the contact terminals are separated, their contact area decreases rapidly, the current density is very large, and the temperature rises sharply, so that the metal at the contact point melts and evaporates a large amount of metal vapor. Due to the high temperature of metal vapor and the existence of a strong electric field, the strong electric field emits and the metal vapor ionizes, causing a vacuum arc. The vacuum circuit breaker, taking advantage of the high diffusion speed of charged particles and metal vapor generated in the vacuum arc, quickly restores the dielectric strength of the contact terminal clearance and extinguishes the arc at the time of arc current zero crossing and temporarily arc extinguishing. The appearance and structure of a vacuum arc extinguishing chamber are as shown in Fig. 6-2.

Fig. 6-2　Structure of vacuum arc extinguishing chamber

(1) Advantages of a vacuum circuit breaker:

① The contact terminals in the vacuum arc extinguishing chamber have a small clearance. The whole vacuum circuit breaker is particularly small, so the power required for the operating mechanism is small. Besides, it is lightweight.

② It makes a low noise when operating, causing no noise pollution to the ambient environment.

③ It is characterized by strong breaking capacity, large breaking current, short arc distinguishing time, many breaking times, and long service life (up to 20 years).

④ After the arc is broken, it restores the dielectric strength quickly. Therefore, it is suitable for frequent operations and is able to reclose a circuit multiple times.

⑤ The medium does not age, nor does it need to be replaced so the maintenance workload is small.

⑥ It is safe to use, basically free from any fire and explosions, and can be used on different occasions, especially in dangerous places.

⑦ The contact terminals are a completely sealed structure immune to outside influence, which is reliable and has a stable break-make performance.

⑧ As an independent element, the arc extinguishing chamber is easy to install and commission.

(2) Disadvantages of a vacuum circuit breaker:

① When the vacuum circuit breaker is breaking the inductive load or capacitive load, overvoltage is likely to be caused by cut-off current, oscillation, reignition and other reasons.

② The contact terminals of the vacuum circuit breaker have a butting structure, and the operating mechanism uses a spring, which is prone to closing bouncing and opening rebounding. Closing bouncing will produce a high overvoltage, affecting the stable operation of the power grid, and causing burns or even fusion welding of the contact terminals, especially in the case of an inrush current generated after the capacitor bank is put into operation and in the case of short circuit closing. Opening rebounding will reduce the spacing between back-arc contact terminals, resulting in restrike after arc extinguishing and other very serious consequences.

③ It has high requirements for the sealing process and manufacturing process and is more expensive.

5. SF_6 circuit breakers

SF_6 gas gives SF_6 circuit breakers excellent traits. Under standard conditions, SF_6 is a colorless, odorless, non-toxic gas that is insoluble in water and oil. It is heavier than air and tends to accumulate in low places; It has a high thermal conductivity, which is 3–6 times higher than air. Sound travels more slowly through it, but the gas's greenhouse effect is 22,800 times that of CO_2. SF_6 gas is a chemically stable noble gas at room temperature, which does not react chemically to metals and insulation commonly used in electrical equipment under normal conditions, and does not erode any substances it contacts. It has excellent insulation properties, which is due to the strong electronegativity of SF_6 gas and its decomposed substances. In a uniform electric field, the dielectric strength of SF_6 gas is 2.5 to 3 times higher than air. SF_6 gas excels in arc extinguishing, its arc extinguishing capacity is 100 times as large as air, and its breaking capacity is about 2 to 3 times as large as air. This is not only because it has excellent insulation properties, but also because it has unique thermal and electrical properties. The cut-off current before current zero crossing is low and can avoid high overvoltage. During circuit breaker operation and in the event of an internal fault, SF_6 gas will produce different amounts of corrosive, highly toxic decomposed substances (such as HF and SO_2), which are irritative to the skin, eyes, and mucous membranes, and cause dizziness and pulmonary edema if a large amount of inhaling a lot. Pure SF_6 gas is non-corrosive, but its decomposed substances will become corrosive electrolytes when they encounter water, which will damage some materials inside the equipment (glass, porcelain, insulating paper, and similar materials), or cause an operational fault. Adsorbents (such as aluminium oxide, soda lime, molecular sieve, or a mixture thereof) can be used to discharge moisture and decomposed substances of SF_6 gas out of the equipment.

By structure characteristic, SF_6 circuit breakers can be divided into live tank SF_6 circuit breakers (as shown in Fig. 6-3) and dead tank SF_6 circuit breakers (as shown in Fig.6-4).

Fig. 6-3　Live tank SF_6 circuit breaker

Fig. 6-4　Dead tank SF_6 circuit breaker

(1) Live tank SF_6 circuit breakers.

The arc extinguishing chamber of this type of circuit breaker is installed in a high-strength porcelain bushing, supported by a hollow porcelain insulator and insulated against ground. The arc extinguishing chamber is connected to the inner cavity of the porcelain insulator, and both are filled with SF_6 gas of the same pressure and are controlled and monitored by the density relay and pressure gauge in the control cabinet. The insulating tie rod passing through the porcelain insulator connects the moving contact terminal of the arc extinguishing chamber to the drive rod of the operating mechanism, and it drives the contact terminals to open and close the circuit breaker.

By the overall layout, Live tank SF_6 circuit breakers can be divided into the "I-shaped" layout, "Y-shaped" layout, and "T-shaped" layout.

"I-shaped" layout: This is a general layout of single-post single-break circuit breakers of 220 kV and below, in which three poles are installed on one or three supports. LW25- series circuit breakers of 110 kV and below and LW31A- series 220 kV circuit breakers are examples.

"Y-shaped" layout: This is generally used for single-post double-break circuit breakers of 220 kV and above, such as LW6-series 220 kV circuit breakers, and ABB ELFSP4-2 220 kV circuit breakers.

"T-shaped" layout: This is generally used for single-post double-break circuit breakers of 220 kV and above, especially 500 kV, such as LW7-series 220 kV circuit breakers, Mitsubishi SFM 500 kV circuit breakers, Siemens 3AQ2 245 kV, 3AT3 252 kV, and 3AT2EI 550 kV circuit breakers, and ABB ELFSP7-21 500 kV circuit breakers.

Advantages: The live tank SF_6 circuit breaker series covers a wide range (different quantities of standard arc extinguishing units and post porcelain bushings can be used to make products of different voltage classes). Such circuit breakers are more commonly produced and used at present by virtue of simple structure, less gas consumption, high single-break voltage, high breaking current, fewer moving components, relatively low price, high operational reliability, and small maintenance workload.

Disadvantages: They have a high center of gravity, poor seismic capacity, and can not be retrofitted with a current transformer, so where they can be used is limited.

(2) Dead tank SF_6 circuit breakers. The appearance and structure of a dead tank SF_6 circuit breaker are as shown in Fig. 6-5.

Structure characteristic: The arc extinguishing chamber of this type of circuit breaker is installed in a grounded metal tank, the HV live part is supported by an insulator, and the insulation of the tank mainly depends on SF_6 gas. The insulating bar passes through the mounting insulator to connect the moving contact terminals to the drive shaft of the mechanism, and the current transformer can be installed in the lower part of the two lead-out bushings.

Fig. 6-5 Dead tank SF_6 circuit breaker

Advantages: A dead tank SF_6 circuit breaker has a low center of gravity, good seismic capacity, good electric field between arc extinguishing breaks, large breaking capacity, can be retrofitted with a current transformer, and can also be integrated with an isolating switch, grounding switch, lightning arrester, etc., to form a combined type switchgear. By using the bushing lead, it can be used in a fully enclosed composite apparatus without modification by and large.

Disadvantages: A tank of circuit breakers of this type takes more materials to make, consumes a lot of gas, and belongs to a small series. Such a circuit breaker is difficult to make and is more expensive.

110-500 kV dead tank SF_6 circuit breakers are basically similar in appearance, most of them are Mitsubishi SFMT circuit breakers or products using Hitachi OFPTB technology, such as OFPTB-500-50LA circuit breakers, and domestic LW12-series 220 kV and 500 kV circuit breakers.

6. Operating mechanism of HV circuit breaker

The operating mechanism is a mechanical operating device that operates the circuit breaker, which is separate from the circuit breaker proper. It's mainly used to convert other forms of energy into mechanical energy so that the circuit breaker can be accurately and quickly opened and closed.

The operating system of the HV circuit breaker includes several parts, such as the operating mechanism, transmission mechanism, lifting mechanism, buffer device, and secondary control circuit.

7. Precautions for maintenance of HV circuit breakers

(1) To maintain a circuit breaker indoors, if it is close to the live equipment nearby, or if an adjacent circuit breaker in operation is faulty and may injure the maintainer, temporary barriers shall be provided. Temporary barriers can be made of dry wood, rubber, or other tough insulating materials and shall be installed firmly.

(2) When scaffolding is needed for outdoor maintenance, the scaffold shall be set up firmly, the boards shall be laid and fastened firmly, and the safety belt shall be used correctly to prevent the workers from falling from height. A special ladder shall be used to ascend and descend a scaffold. The scaffolds shall be set up firmly and anti-slip measures shall be taken. Workers shall wear safety helmets to prevent injuries caused by falling objects.

(3) For maintenance of a vacuum circuit breaker, wear of contact terminals shall be tested. If the wear is severe, the vacuum arc extinguishing chamber shall be replaced in time to ensure that the vacuum circuit breaker has sufficient breaking capacity.

(4) Measures shall be taken to prevent workers from getting poisoned during SF_6 circuit breaker maintenance. When it's required to work in the low area of an SF_6 power distribution unit or a cable trench, first, the oxygen content shall be detected (it shall not be less than 18%). Do not enter such places unless the oxygen content meets the requirement. Prior to disassembling and maintaining a circuit breaker, SF_6 gas shall be detected, and the maintainers shall wear protective suits, wear gas masks as needed, and recover SF_6 gas. After the cover of the equipment is opened, workers shall temporarily leave the site for 30 min. When removing the adsorbent and cleaning the dust, they shall wear gas masks and protective gloves. After the equipment is disassembled, the circuit breaker shall be purged with nitrogen. Ventilation is required during indoor maintenance (workers shall stand on the windward side), and an SF_6 gas leakage alarm shall be provided at the

work site. When re-installing the circuit breaker, be sure to clean or replace the adsorbent, and check if the SF_6 gas inflation pressure is normal, if there are any leaks, and if the moisture of SF_6 gas is acceptable through detection. Dry the SF_6 gas if the moisture is not acceptable.

(Ⅱ) Isolating switches

HV isolating switches are the number one, both in terms of quantity and application range, HV switchgear in China's electric power systems. An isolating switch has an obvious clearance in the opening state and provides reliable insulation; in the closing state, it is an electrical switchgear where normal operating current and short circuit current reliably passes through. However, the isolating switch does not have a special arc extinguishing device so it can not be used to break the load current and short circuit current. Usually, it is used in conjunction with a circuit breaker. The appearance and structure of an isolating switch are as shown in Fig. 6-6.

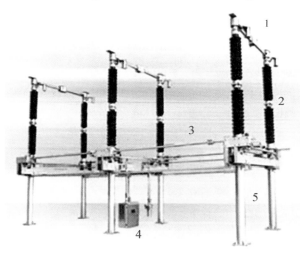

1—Conductive part; 2—Insulating part; 3—Transmission mechanism; 4—Operating mechanism; 5—Supporting base.

Fig. 6-6　Appearance and structure of isolating switch

1. Types of isolating switches

(1) By the number of poles, isolating switches can be divided into single-pole isolating switches and three-pole isolating switches.

(2) By the number of insulating posts, isolating switches can be divided into single-post isolating switches, double-post isolating switches, and three-pole isolating switches.

(3) By the operation mode, isolating switches can be divided into knife type, center rotating, plug-in, and horizontally (vertically) retractable isolating switches.

(4) According to whether there is a grounding knife-switch, isolating switches can be divided into isolating switches with grounding knife-switch and isolating switches without grounding knife-switch.

(5) By the operating mechanism, isolating switches can be divided into manual, electric, pneumatic, and hydraulic isolating switches.

(6) By use, isolating switches can be divided into general isolating switches, quick-breaking isolating switches and isolation switches for grounding the transformer neutral point.

2. Indoor HV isolating switches

(1) GN19-10 indoor HV isolating switches

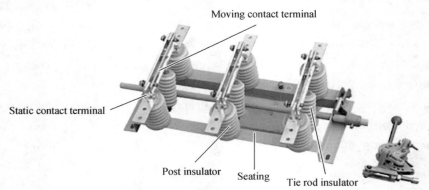

Fig. 6-7   Structure of GN19-10 indoor HV isolating switch

A GN19-10 indoor HV isolating switch adopts a three-phase underframe structure, as shown in Fig. 6-7, which mainly consists of static contact terminals, seating, post insulators, tie rod insulators, and moving contact terminals.

The conductive part of the isolating switch is composed of moving and static contact terminals, and the conductive part of each phase is fixed on the seating by two post insulators, and the three phases are installed in parallel.

(2) GN30-10 rotary indoor HV isolating switches

The main part of the switch is fixed on the upper and lower surfaces of the switch's underframe by two groups of insulators, and the upper and lower surfaces are completely separated by the partition fixed on the switch frame. The closing and opening of the switch are enabled by rotating the contact knife. Its appearance is as shown in Fig. 6-8.

Fig. 6-8   GN30-10 indoor HV isolating switch

The static contact terminals are installed on the upper and lower surfaces of the switch cabinet, so the live part and the non-live part are completely separated in the switch cabinet, thus ensuring the safety of the maintainers. Such an isolating switch is especially suitable for installation in an HV switch cabinet.

3. Outdoor HV isolating switches

There is a wide variety of outdoor HV isolating switches. Examples are GW4-110 isolating switches of double-post, single-break, horizontal, and rotary structure, which are widely used in the 10–110 kV power distribution units (whose appearance are as shown in Fig. 6-9); GW5-110D isolating switches, which are widely used for a voltage class of 35–110 kV; GW7-220 three-post isolating switches, which have a three-post, double-break, horizontal, rotary, and open structure; GW11-252 isolating switches; GW6-220GD single-post isolating switches; GW10-252 single-post isolating switches, etc.

Fig. 6-9　Appearance of GW4 -110 isolating switches

4. Operating mechanisms of isolating switches

The operating mechanisms of the isolating switches mainly are manual operating mechanisms and electric operating mechanisms.

(1) Manual operating mechanisms

Lever operating mechanisms: generally suitable for isolating switches with a rated current of less than 3,000 A;

Worm gear operating mechanisms: generally suitable for isolating switches with a rated current of greater 3,000 A, such as the CS6 manual lever type operating mechanism, and the CS9 manual worm gear operating mechanism.

(2) Electric operating mechanisms

The CJ2 -XG operating mechanism is an outdoor power mechanism, which is used for opening and closing the GW4, GW7, and other HV isolating switches or grounding switches. It can be used for remote control, local electric control, or manual control using the handle.

5. Safety precautions for maintenance of isolating switches

(1) Inspection at the time of handover and acceptance: The operating mechanism, transmission device, auxiliary change-over switch, and locking device shall be firmly installed, operate flexibly and reliably, and give correct position indications. The three-phase asynchroneity value shall comply with the technical specifications of the product. The interphase distance, and the angle at which the contact terminals open and the distance of the contact terminals when opening shall comply with the technical specifications of the product. Contact terminals shall be able to come into a nice and tight contact. The paintcoat shall be intact, the phase-indicating color marks shall be correct, and the isolating switch shall be well grounded.

(2) Inspection during operation: The insulators shall be intact and free of any cracks and discharging. The operating connecting rod and the mechanical parts shall be free of damage and rusting. Each part shall be fastened in correct positions, and free of skewing, loosening, coming off, and other abnormal phenomena. The locking device shall be in good condition, and the pins and auxiliary contacts of the electromagnetic or mechanical locking device of the isolating switch shall be in the correct positions. The arc extinction angle of the blade and the knife point shall be free of burns, overheating, deformation, rusting, and skewing, the contact terminals shall be in good contact, and the joints and the contact terminals shall not be overheated to a temperature above 70°C. The blade and knife point shall be free of dirt and burn marks, and the spring leaf, spring, and copper braid shall be free of broken strands or fractures. The grounding switch shall be well grounded, and the flexible part that is prone to damage in particular shall be free of any anomalies.

(3) Check the opening and closing positions of the circuit breaker before operating it, and prevent operating an isolating switch with load.

(4) Manual closing is quick and decisive, but do not exert too much force.

(5) Open the isolating switch quickly to extinguish the arc as soon as possible.

(6) In the event of closing the isolating switch with load by mistake, do not open it again. If an isolating switch is opened by mistake, it shall be immediately closed when the blade is just disconnected from the fixed contact terminal. If the isolating switch blade has disconnected from the fixed contact terminal, the isolating switch opened by mistaken shall not be reclosed.

(7) After closing, the contact shall be checked for tightness. After opening, make sure that all phases are at open position.

(8) After the operations, the operating handle shall be locked without delay.

(Ⅲ) Load switch

Load switch is a switch with simple arc extinguishing device for rated load current opening and closing. A load switch has certain arc extinguishing capability and can be used to break and close load current and overload current less than a certain multiple (usually 3-4 folds); It can also be used to open and close no-load transformers with higher allowable capacity than an isolating switch, and longer no-load lines are sometimes used for opening and closing high-capacity capacitor banks. A load switch can also be combined in series with current-limiting fuse (load switch-fuse combination) to serve as a circuit breaker. That is to say, a load switch is for opening and closing overload current less than a certain multiple, and a current-limiting fuse is for opening higher overload current and short circuit current. Fuses can be installed on either power supply side or power receiving side of a load switch.

Commonly used load switches are classified into pneumatic load switch, air evolving load switch, sulfur hexafluoride load switch and vacuum load switch based on types of arc extinguishing.

(1) An air evolving load switch is a load switch that generates air by the act of arc using solid air producing materials for arc extinguishing, and it is of self-energized arc extinguishing type.

In an air evolving arc extinguishing chamber, vaporization of arc extinguishing materials results in local high pressure, and the arc gives rise to deionization to allow arc extinguishing by the effect of intense blow-out and cooling. When current is low, the arc is driven by cooling effect of the air producing wall or electric power, and arc is elongated and extinguished.

(2) Pneumatic load switch is a load switch that compresses air utilizing relative motion during piston and cylinder opening and then uses the compressed air for arc extinguishing.

(3) Vacuum load switch is a load switch that uses vacuum arc extinguishing chamber as a load switch of arc extinguishing device. It has high breaking current and is suitable for frequent operation.

(4) SF_6 load switch is a load switch that uses SF_6 as insulation and arc extinguishing medium. It has been widely used in urban and rural power grids.

① Closing: auxiliary switch is closed and then main knife switch is closed, and the main fixed contact terminal is in reliable contact with the main blade; Opening: main knife switch is opened and then auxiliary knife switch is opened, and three-phase arc extinguishing blades simultaneously trip off the fixed arc extinguishing contact terminal.

② The organic insulation material that generates gas inside the arc extinguishing cylinder shall be intact and free of crack, and the clearance between arc extinguishing contact terminal and arc extinguishing cylinder shall be as required.

③ Net distance and opening angle between three-phase contact terminals during contact simultaneity and at opening status shall comply with technical specifications for the product. Knife switch opening angle can be controlled by changing operating rod length and position of the operating rod on the fan-shaped plate.

④ At closing state, the small plug on the main knife switch shall be inserted right into nozzle of the arc extinguishing device, and violent collision with the nozzle is not allowed.

(Ⅳ) Fuses

As the earliest and simplest protective devices used in electric power systems, fuses are used for short circuit and overload protection for electrical equipment such as lines and power transformers. When current in an electric power system is higher than a certain value due to overload or when short circuit occurs in electrical equipment or lines, the fuses shall be capable of quick action within specified time to cut off power supply and protect the equipment. It has simple structure, small size, light weight and low price and allows easy maintenance and flexible use.

1. Operation and maintenance of high voltage fuses

(1) Qualified products and fittings shall be selected according to the Regulations, frequent check is required during operation to make sure that they are in close contact, and contact points shall be checked for temperature rise.

(2) Connecting fuse elements that are blown for reuse is not allowed.

(3) Replacing fusion tube (fuse element) of a fuse at charged state is not advisable. When replacement at charged state is required, qualified insulation tools shall be used.

(4) Operation with care is required, and use of excessive force for opening or closing a fuse is not allowed.

(5) Regular routine inspection is required, and routine inspection at night is required at least once a month to check for presence of discharge sparks and poor contact.

2. Operation of high voltage fuses

(1) The operation is to be completed by two persons which shall wear qualified insulating gloves, insulating boots and safety goggles and use qualified insulating rods matching the voltage class, and operation in thunderstorm or heavy rain weather is forbidden.

(2) For opening, mid phase is opened, and then leeward side phase is opened, and finally windward phase is opened.

The reason for the operation is that three-phase power is converted into two-phase power during opening, and the middle line generates the minimum arc, hence mid phase is opened at first; and then the phase at leeward side is opened because upon occurrence of arcing, the arc fluttering with the wind will not drift to the other two phases; the phase disconnected at last is at the lowest risk of arcing, so the most dangerous phase is disconnected at first.

(3) Operation sequence in closing is opposite to that in opening. The phase at windward side is closed at first, and then the phase at leeward side phase is closed and finally the mid phase is closed.

(Ⅴ) Mutual inductor

Mutual inductor includes voltage transformer (TV) and current transformer (TA). It is a connecting element between primary system and secondary system in an electric power system and is used to transform voltage or current, provide voltage or current signals for measuring instruments, protection devices and control devices to allow reflection of normal operation and fault conditions of electrical equipment.

1. Current transformer

A current transformer has structure similar with that of two-winding transformer and consists of iron core and primary and secondary windings. Primary and secondary windings of a 0.5 kV current transformer are on a same iron core and the current transformer has the simplest structure. To allow easy use and save materials, multiple independent iron cores without magnetic connection and secondary windings are usually used for 10 kV and higher current transformer with multiple secondary windings. Such a current transformer can be used for both measurement and protection. A 10-35 kV current transformer usually has 2 secondary windings. A 63-110 kV current transformer has 3-5 secondary windings. A 220 kV and higher current transformer has 4-7 secondary windings.

To adapt to changes in line current, 63 kV and higher current transformers usually have primary winding composed of a few sections which are in series connection or parallel connection to provided two or three current ratios. Current transformers have appearances as shown in Fig. 6-10, Fig. 6-11, Fig. 6-12 and Fig. 6-13.

Fig. 6-10 LMZD1-10 epoxy resin insulated single-turn busbar current transformer

Fig. 6-11 LDZJ1-10 epoxy resin insulated single-turn current transformer

Fig.6-12 LFZB-12 epoxy resin insulated double-turn current transformer with degree of protection

Fig. 6-13 LB7 - 220 outdoor current transformer

(1) Primary windings of a current transformer are connected in series inside the measured circuit; Secondary winding is connected in series with current coil of measuring instrument and relay.

(2) Current in primary winding entirely depends on load current in the circuit tested rather than magnitude of secondary current.

(3) In normal operation, current transformers operate at a state close to short circuit.

(4) Secondary side of a current transformer in operation shall not have open circuit.

2. Voltage transformer

Just like a transformer, voltage transformer is used to convert voltage on the line. But as voltage transformation by a transformer is for electric energy transmission, a transformer has high transmission capacity which is usually measured in kVA or MVA; and voltage transformation by a voltage transformer is for supplying power for measuring instruments and relay protection devices and measuring line voltage, power and electric energy or for protecting valuable equipment, motors and transformers in the line in the case of line fault. Therefore, a voltage transformer has low capacity which is usually a few volt amperes, dozens of volt amperes, and the maximum capacity is no more than 1,000 VA. Appearance of a voltage transformer is as shown in Fig. 6-14.

Fig. 6-14 JDZ-10 cast-resin single-phase voltage transformer

In voltage transformer inspection, attention shall be paid to:

(1) A voltage transformer must undergo test and inspection before put into operation as specified in the specifications. Inspection items include polarity test, connection group, insulation resistance measurement with tramegger, phase sequence check, etc.

(2) Voltage transformers shall be correctly wired. Primary windings shall be connected in parallel with the circuit to be tested, and secondary windings shall be connected in parallel with voltage coil of measuring instrument, relay protection device or automatic device, and attention shall be paid to correctness of polarity.

(3) Capacity of load connected to secondary side of a voltage transformer shall be appropriate and usually no more than its nameplate capacity, otherwise, transformer error will be higher and accurate measurement is impossible.

(4) Short circuit at secondary side of a voltage transformer is not allowed. As a voltage transformer has very low internal impedance, when there is short circuit in a secondary circuit, high current will occur which will cause damage to secondary equipment and even endanger personal safety. A fuse can be provided at secondary side of a voltage transformer to protect it from damage due to short circuit at secondary side. If possible, a fuse shall also be installed on primary side to protect high-voltage power grid so that fault in high-voltage windings or leads of a voltage transformer will not endanger primary system.

(5) To protect safety of personnel in contact with measuring instruments and relay, one point of secondary winding of a voltage transformer must be earthed. The earthing can prevent high voltage from the instrument and relay which may endanger personnel when the insulation between primary and secondary windings are damaged.

任务二　认识电气试验安全技术

一、高压试验

（1）高压试验应填用变电站（发电厂）第一种工作票。在高压试验室（包括户外高压试验场）进行试验时，按《电力安全工作规程　高压试验室部分》（GB 26861—2011）的规定执行。

在同一电气连接部分，许可高压试验工作票前，应先将已许可的检修工作票收回，禁止再许可第二张工作票。如果试验过程中，需要检修配合，应将检修人员的名字填写在高压试验工作票中。

在一个电气连接部分同时有检修和试验时，可填用一张工作票，但在试验前应得到检修工作负责人的许可。

如加压部分与检修部分之间的断开点，按试验电压有足够的安全距离，并在另一侧有接地短路线时，可在断开点的一侧进行试验，另一侧可继续工作。但此时在断开点应挂有"止步，高压危险！"的标示牌，并设专人监护。

（2）高压试验工作不得少于两人。

试验负责人应由有经验的人员担任，开始试验前，试验负责人应向全体试验人员详细布置试验中的安全注意事项，交待邻近间隔的带电部位，以及其他安全注意事项。

（3）因试验需要断开设备接头时，拆前应做好标记，接后应进行检查。

（4）试验装置的金属外壳应可靠接地；高压引线应尽量缩短，并采用专用的高压试验线，必要时用绝缘物固定牢固。

试验装置的电源开关，应使用明显断开的双极刀闸。为了防止误合刀闸，可在刀刃上加绝缘罩。

试验装置的低压回路中应有两个串联电源开关，并加装过载自动跳闸装置。

（5）试验现场应装设遮栏或围栏，遮栏或围栏与试验设备高压部分应有足够的安全距离，向外悬挂"止步，高压危险！"的标示牌，并派人看守。被试设备两端不在同一地点时，另一端还应派人看守。

（6）加压前应认真检查试验接线，使用规范的短路线，表计倍率、量程、调压器零位及仪表的开始状态均正确无误，经确认后，通知所有人员离开被试设备，并取得试验负责人许可，方可加压。加压过程中应有人监护并呼唱。

高压试验工作人员在全部加压过程中，应精力集中，随时警戒异常现象发生，操作人应站在绝缘垫上。

（7）变更接线或试验结束时，应首先断开试验电源、放电，并将升压设备的高压部分放电、短路接地。

（8）未装接地线的大电容被试设备，应先行放电再做试验。高压直流试验时，每告一段落或试验结束时，应将设备对地放电数次并短路接地。

（9）试验结束时，试验人员应拆除自装的接地短路线，并对被试设备进行检查，恢复试验前的状态，经试验负责人复查后，进行现场清理。

（10）变电站、发电厂升压站发现有系统接地故障时，禁止进行接地网接地电阻的测量。

（11）特殊的重要电气试验，应有详细的安全措施，并经单位批准。

直流换流站单极运行，对停运的单极设备进行试验，若影响运行设备安全，应有措施，并经单位批准。

二、使用携带型仪器的测量工作

（1）使用携带型仪器在高压回路上工作，至少由两人进行。需要高压设备停电或做安全措施的，应填用变电站（发电厂）第一种工作票。

（2）除使用特殊仪器外，所有使用携带型仪器的测量工作，均应在电流互感器和电压互感器的二次侧进行。

（3）电流表、电流互感器及其他测量仪表的接线和拆卸，需要断开高压回路者，应将此回路所连接的设备和仪器全部停电后，方可进行。

（4）电压表、携带型电压互感器和其他高压测量仪器的接线和拆卸无须断开高压回路者，可以带电工作；但应使用耐高压的绝缘导线，导线长度应尽可能缩短，不准有接头，并应连接牢固，以防接地和短路，必要时用绝缘物加以固定。

使用电压互感器工作时，先应将低压侧所有接线接好，然后用绝缘工具将电压互感器接到高压侧。工作时应戴手套和护目眼镜，站在绝缘垫上，并应有专人监护。

（5）连接电流回路的导线截面积，应适合所测电流数值。连接电压回路的导线截面积不得小于 1.5 mm^2。

（6）非金属外壳的仪器，应与地绝缘，金属外壳的仪器和变压器外壳应接地。

（7）测量用装置必要时应设遮栏或围栏，并悬挂"止步，高压危险！"的标示牌。仪器的布置应使工作人员距带电部位不小于表 3-1 规定的安全距离。

三、使用钳形电流表的测量工作

（1）运行人员在高压回路上使用钳形电流表的测量工作，应由两人进行。非运行人员测量时，应填用变电站（发电厂）第二种工作票。

（2）在高压回路上测量时，严禁用导线从钳形电流表另接表计测量。

（3）测量时若需拆除遮栏，应在拆除遮栏后立即进行。工作结束，应立即将遮栏恢复原状。

（4）使用钳形电流表时，应注意钳形电流表的电压等级。测量时戴绝缘手套，站在绝缘垫上，不得触及其他设备，以防短路或接地。

观测表计时，要特别注意保持头部与带电部分的安全距离。

（5）测量低压熔断器和水平排列低压母线电流时，测量前应将各相熔断器和母线用绝缘材料加以包护隔离，以免引起相间短路，同时应注意不得触及其他带电部分。

（6）在测量高压电缆各相电流时，电缆头线间距离应在 300 mm 以上，且绝缘良好、测量方便。当有一相接地时，严禁测量。

（7）钳形电流表应保存在干燥的室内，使用前要擦拭干净。

四、使用兆欧表测量绝缘的工作

（1）使用兆欧表测量高压设备绝缘，应由两人进行。

（2）测量用的导线，应使用相应的绝缘导线，其端部应有绝缘套。

（3）测量绝缘时，应将被测设备从各方面断开，验明无电压，确实证明设备无人工作后，方可进行。在测量中禁止他人接近被测设备。

在测量绝缘前后，应将被测设备对地放电。测量线路绝缘时，应取得许可并通知对侧后方可进行。

（4）在有感应电压的线路上测量绝缘时，应将相关线路同时停电，方可进行。雷电时，严禁测量线路绝缘。

（5）在带电设备附近测量绝缘电阻时，测量人员和兆欧表安放位置，应选择适当，保持安全距离，以免兆欧表引线或引线支持物触碰带电部分。移动引线时，应注意监护，防止工作人员触电。

五、直流换流站阀厅内的试验

（1）进行晶闸管（可控硅）高压试验前，应停止该阀塔内其他工作，并撤离无关人员；试验时，作业人员应与试验带电体位保持 0.7 m 以上距离，试验人员禁止直接接触阀塔屏蔽罩，防止被可能产生的试验感应电伤害。

（2）地面加压人员与阀体层作业人员应通过对讲机保持联系，防止高处作业人员未撤离阀体时误加压。阀体工作层应设专责监护人（在与阀体工作层平行的升降车上监护、指挥），加压过程中应有人监护并呼唱。

（3）换流变压器高压试验前应通知阀厅内高压穿墙套管侧试验无关人员撤离，并派专人监护。

（4）阀厅内高压穿墙套管试验加压前应通知阀厅外侧换流变压器上试验无关人员撤离，确认其余绕组均已可靠接地，并派专人监护。

（5）高压直流系统带线路空载加压试验前，应确认对侧换流站相应的直流线路接地刀闸（地刀）、极母线出线隔离开关（刀闸）、金属回线隔离开关（刀闸）在拉开状态；单极金属回线运行时，禁止对停运极进行空载加压试验；背靠背高压直流系统一侧进行空载加压试验前，应检查另一侧换流变压器是否处于冷备用状态。

Task II Learn safety technologies for electric testing

I. HV test

(1) For HV test, the first type of work ticket for substation (power plant) shall be completed. During test in a HV test room (including outdoors HV test site), *Safety Code of Electric Power Industry - Part of High Voltage Laboratory* (GB 26861-2011) shall be followed.

For a same electric connection part, before issuing a HV test work ticket, current work ticket shall be withdrawn, and issuing a second work ticket is forbidden. If assistance from a maintainer is required during the test, name of the maintainer shall be filled in HV test work ticket.

When repair and testing are carried simultaneously on an electric connection part, a work ticket can be issued and used, but prior permission from the person in charge of repair is required.

If there is adequate safe distance between voltage applied part and repair part at test voltage and there is grounding short circuit line on the other side, test can be carried out on one side of the disconnection point, and operation of the other side can be continued. And a sign board bearing characters "Stop! High voltage, danger!" shall be provided, and persons shall be specially assigned for supervising.

(2) HV test requires no less than two persons.

The person in charge of test shall be experienced which shall provide the test staff detailed description of precautions in the test, tell adjacent live parts and other precautions for safety.

(3) When equipment is to be disconnected for test, they shall be marked before disassembly and checked after reconnection.

(4) Metal housings of the testing devices shall be reliably grounded; HV leads shall be as short as possible, special HV test wires shall be used, and insulation shall be used for fastening as required.

A test device shall have clearly disconnected bipolar knife switch as power switch. To prevent knife switch closing by mistake, an insulation cover can be provided on edge.

There shall be two power switches in LV circuit of the test device in series connection, and additional automatic overload tripping device shall be provided.

(5) The test site shall have barriers or fences between which there is adequate safe distance, and "Stop! High voltage, danger!" sign board shall be provided, and persons shall be specially assigned for guarding. When two ends of the equipment under test are not at a same location, the other end shall be guard by specially assigned person.

(6) Before applying voltage, test wires shall be carefully checked, standard short circuit lines shall be used, multiplying power and range of meters, zero position of voltage regulator and instrument starting status shall be checked, and all persons shall be asked to leave the equipment under test, and prior permit from the person in charge of the test is required. During voltage application, personnel shall be assigned for supervision and calling.

During voltage application, HV test staff shall concentrate on the test and stay alert to any abnormal condition, and operators shall stand on insulation cushion.

(7) During wiring change or at the end of the test, test power supply disconnection and discharging are required at first, and then high-voltage part of the step-up equipment shall undergo discharging and shorting to ground.

(8) High-capacitance equipment under test without grounding wires shall undergo discharging before test. During high-voltage DC test, the equipment shall undergo ground discharging for a few times and be shorted to ground.

(9) At the end of the test, the test staff shall remove the grounding short circuit line installed, check the equipment under test, restore the state before the test and carry out site cleaning after rechecking by the person in charge of the test.

(10) In substations and power plant step-up substations with system grounding fault, measurement of grounding resistance in a grounding grid is forbidden.

(11) For special important electric tests, detailed safety measures and approval from the organization are required.

A DC converter station operates with a single pole, and unipolar equipment at shutdown state is to be tested. If there is impact on safe operation of the equipment, measures shall be taken, and approval from the organization is required.

II. Measurement using portable instruments

(1) At least two persons are required for work on HV circuit using portable instruments. When HV equipment outage or safety precautions are required, the first type of work ticket for substations (power plants) shall be issued and used.

(2) All measurements using portable instruments other than special instruments shall be conducted at secondary side of a current transformer and voltage transformer.

(3) When HV circuit disconnection is required for wiring and disassembly of ammeters, current transformers and other measuring instruments, equipment and instruments connected to this circuit shall be powered off.

(4) When HV circuit disconnection is not required for wiring and disassembly of voltmeters, portable voltage transformers and other measuring instruments, live-wire work is allowed. But HV-resistant insulated conductors shall be used, and the conductors shall be as short as possible, have no joint and be firmly connected to prevent grounding and short circuit. Insulation shall be used for fastening as required.

Before start of work using voltage transformer, all wires at low-voltage side shall be connected, and then the voltage transformer shall be connected to HV side with insulation tools. During the work, operators shall wear gloves and goggles and stand on insulation cushions, and persons shall be specially assigned for supervision.

(5) Cross section of a conductor connecting current circuits shall be fit for measured current

values. Cross section of a conductor connecting voltage circuits shall be no less than 1.5 mm^2.

(6) An instrument with non-metal housing shall be ground insulated, and instruments with metal housings and transformer housings shall be grounded.

(7) Barriers or fences shall be provided for measuring devices as required, and sign board bearing characters "Stop! High voltage, danger!" shall be provided. Instruments shall be so provided that working personnel stay at no less than the safe distance specified in Table 3-1 from live parts.

III. Measurement using clip-on ammeter

(1) Measurement on HV circuit using clip-on ammeter requires two persons. In the case of measurement by persons other than operators, the second type of work ticket for substations (power plants) shall be issued and used.

(2) During measurement on HV circuit, use of a conductor for connecting another meter from clip-on ammeter for measurement is strictly forbidden.

(3) When barriers are to be removed for measurement, measurement shall be started upon removal of the barriers. After work is completed, the barriers shall be replaced.

(4) When clip-on ammeter is used, attention shall be paid to voltage class of clip-on ammeter. Measurement staff shall wear insulating gloves and stand on insulation cushion, and touching other equipment is not allowed, otherwise there will be short circuit or grounding.

When reading the meters, special care shall be taken to keep safe distance between head and the live parts.

(5) Before measuring current of LV fuses and horizontally arranged LV busbars, the fuses and busbars shall be isolated using insulating materials to prevent interphase short circuit, and care shall be taken to avoid touching other live parts.

(6) In measuring phase current of HV cable, cable heads shall have more than 300 mm wire spacing and be well-insulated to allow easy measurement. When a phase is grounded, measurement is forbidden.

(7) Clip-on ammeters shall be stored in a dry room and shall be wiped clean before use.

IV. Insulation measurement using megohmmeter

(1) HV equipment insulation measurement using megohmmeter shall be completed by two persons.

(2) Proper insulated conductor shall be used for measurement, and insulation sleeve shall be provided at the ends.

(3) Before start of insulation measurement, make sure that the equipment to be measured is disconnected, there is no voltage and there is no person working on the equipment. Efforts shall be made to prohibit others from approaching the equipment under test during measurement.

Before and after insulation measurement, the equipment measured shall undergo ground discharging. For measuring line insulation, prior permission is required and prior notice shall be issued.

(4) Before measuring insulation on lines carrying induced voltage, relevant lines shall be powered off simultaneously. Measuring line insulation in thunderstorm days is strictly forbidden.

(5) For insulation resistance measurement nearby electrified equipment, appropriate measuring staff and megohmmeter positions shall be selected and safe distance shall be kept to prevent megohmmeter lead or lead support from touching the live parts. For moving a lead, supervision is required to protect working personnel from electric shock.

V. Test in valve hall in DC convertor station

(1) Before HV test on thyristor (controllable silicon), other work in the valve tower shall be suspended and unrelated personnel shall be evacuated; During the test, the operators shall stay more than 0.7 m away from the charged body for test, and direct contact with valve tower shielding case by test personnel is forbidden, otherwise they may be injured by induced electricity in the test.

(2) Voltage application personnel on the ground shall keep in touch with operators at valve body floor through walkie-talkie to prevent voltage application before the operators leave valve body. Special supervisor shall be assigned for valve body operating floor (for supervision and command from lift truck parallel with valve body operating floor), and supervision and shouting are required during voltage application.

(3) Before HV test on a converter transformer, irrelevant personnel at HV wall-through bushing side in valve hall shall be asked to leave the site, and persons shall be specially assigned for supervision.

(4) Before start of HV wall-through bushing side test, irrelevant personnel on converter transformer outside valve hall shall be asked to leave the site, remaining windings shall be checked for reliable grounding, and persons shall be specially assigned for supervision.

(5) Before start of voltage application to HV DC system with no-load line, make sure that DC line grounding knife-switch, pole busbar outgoing line isolating switch (knife switch) and metal circuit isolating switch (knife switch) in opposite converter transformer station are at open state; When single-pole metal circuit is operating, no-load applied voltage test on shutdown poles is forbidden; Before no-load applied voltage test on one side of back-to-back HV DC system, make sure that the converter transformer on the other side is at cold standby state.

任务三　高压开关柜故障判断及处理实训

一、作业任务

3人一小组，1人扮演工作负责人，2人扮演班组成员，以"10 kV开关柜"为工作环境，按照《国家电网公司电力安全工作规程（变电部分）》中的要求，完成高压开关柜故障判断及处理。

二、引用标准及文件

（1）《国家电网公司电力安全工作规程（变电部分）》。
（2）《国家电网公司十八项电网重大反事故措施》。

三、作业条件

应在良好的天气作业；作业人员精神状态良好，熟悉工作中安全措施、技术措施以及现场工作危险点。

四、作业前准备

1. 现场操作的基本要求及条件

勘察现场设备情况，查阅相关技术资料，包括历史数据及相关规程。

2. 工器具及材料选择

工位、中性笔、护目镜、绝缘手套、绝缘靴、安全帽、电筒、扳手、万用表、常用工具1套。

3. 危险点及预防措施

（1）机械伤人。
预防措施：进行故障判断时严禁用手直接进行操作，以防机械伤人。
（2）使用不合格的安全工器具。
预防措施：操作前，应正确选取安全工器具并检查合格。

4. 作业人员分工

现场工作负责人（监护人）：×××
现场作业人员：×××

五、作业规范及要求

（1）现场操作时，由老师设置开关柜故障，学员按现场规程判断处理，模拟演示。
（2）学员判断出故障类型后必须向老师进行汇报，才能进行下一步操作。

六、作业流程及标准（6-1）

表 6-1 高压开关柜故障判断、处理流程及评分标准

班级		姓名		学号		考评员		成绩	
序号	作业名称	质量标准			分值/分	扣分标准		扣分	得分
1		安全意识之工作准备							
1.1	着装穿戴	穿工作服、工作鞋，戴安全帽、线手套			5	未穿工作服、工作鞋和戴安全帽、线手套，每缺少一项扣1分； 着装穿戴不规范，每处扣1分			
2		安全意识之工具检查							
2.1	取安全帽、常用工器具并检查	安全帽：检查外观、标签、合格证、下颌带完好； 电筒外观、标签、照明度足够； 绝缘手套完好； 绝缘靴完好； 扳手等其他常规工具外观、性能完好； 携带控制柜钥匙			10	未取安全帽扣2分，未检查安全帽外观、标签、合格证和下颌带完好扣1分，未佩戴安全帽扣5分； 未检查电筒外观、标签、照明度扣1分，未拿手电筒扣5分； 未检查绝缘手套完好扣1分，未检查绝缘手套气密性扣5分，未拿绝缘手套扣5分； 未检查绝缘靴完好扣1分，未拿绝缘靴扣5分； 未检查扳手外观、性能完好扣1分，未拿扳手扣5分； 未携带控制柜钥匙，扣5分			
3		工作过程							
3.1	模拟预演	持操作票在模拟系统上模拟操作一次，核对设备的位置、名称、编号和运行方式			5	要持操作票在模拟系统上模拟操作一次，未核对设备的位置、名称、编号和运行方式，扣1~5分			
3.2	故障分析	正确观察高压开关柜保护装置的报警信息； 依据信号指示正确判断故障类型			25	未能正确观察高压开关柜保护装置的报警信息，扣5分； 未能依据信号指示正确判断故障类型，扣15分			

续表

序号	作业名称	质量标准	分值/分	扣分标准	扣分	得分
3.3	故障处理	遵循安全操作规程，按照操作步骤正确操作，倒闸时，操作顺序应正确流畅；在转检修状态后应把合闸电源和控制电源的熔断器（控制开关）取下；在规定时间内完成操作	55	遵循安全操作规程，按照操作步骤正确操作，倒闸时，操作顺序应正确流畅，操作出现错误，视情况扣5~55分；在转检修状态后应把合闸电源和控制电源的熔断器（控制开关）取下，漏取一个扣10分；在规定时间内能正确操作完成，超时扣5~10分		
合计			100			

Task Ⅲ HV switch cabinet fault identification and practical training on fault handling

Ⅰ. Operating task

A group consists of 3 person, with one person playing the part of person in charge of work and the other two persons playing the part of team members. With "10 kV switch cabinet" as a work environment, they are required to complete HV switch cabinet fault identification and handling as required in *Electric Power Safety Working Regulations (Power Transformation) of State Grid Corporation of China*.

Ⅱ. Referenced standards and documents

(1) *Electric Power Safety Working Regulations (Power Transformation) of State Grid Corporation of China*;

(2) *18 Major Anti-accident Measures for Power Grid of State Grid Corporation of China*.

Ⅲ. Operating conditions

The activity shall be carried at in favorable weather, and the operators shall be at good mental state and familiar with safety measures, technical measures and dangerous point in on-site work.

Ⅳ. Preparation before operation

1. Basic requirements and conditions for on-site operations

Conduct a survey of equipment on site and refer to relevant technical data, including historical data and related procedures.

2. Selection of tools and instruments and materials

A set of work station, roller ball pens, safety glasses, insulating gloves, insulating boots, safety helmet, flashlight, wrench, multimeter and common tools.

3. Dangerous points and preventive measures

(1) Mechanical injury.

Prevention and control measures: operating with hands during fault diagnosis is strictly forbidden, otherwise mechanical injury may occur.

(2) The use of unqualified safety tools and instruments.

Prevention and control measures: before operation, correct safety tools and instruments shall be selected and checked for conformity.

4. Division of labor among operators

Person in charge of on-site work (supervisor): ×××

On-site operator: ×××

V. Operating specifications and requirements

(1) In on-site operation, switch cabinet faults are to be set by instructors, and the trainees will identify and handle the faults by following on-site procedures and complete simulation and demonstration.

(2) Trainees shall report to instructors after identifying faults, and then the next step can be started.

VI. Operation process and standards (see Tab. 6-1)

Tab. 6-1 HV switch cabinet fault identification, practical training on fault handling and evaluation standards

Class		Name		Student ID		Examiner		Score	
S/N	Operation name	Quality standard		Points	Deduction criteria		Deduction		Score
1		Safety consciousness - Preparations							
1.1	Wearing	Wear work clothes and shoes; wear safety helmet and cotton gloves		5	Deduct 1 point if failing to wear work clothes, work shoes, safety helmet or cotton gloves; Deduct 1 point if wearing does not meet the requirements				
2		Safety consciousness - Tools check							
2.1	Getting and checking safety helmets and common tools and instruments	Safety helmet: Check the appearance, label, and certificate of conformity and confirm that the chin strap is intact; Flashlight: Check the appearance and label, and check if the flashlight has sufficient illumination; Check insulating gloves for intactness; Check insulating boots for intactness; Check wrenches and other conventional tools for intact appearance and performance; Bring the key to the control cabinet		10	Deduct 2 points for failure to get safety helmet; deduct 1 point for failure to visually check safety helmet, label and certificate of conformity and intactness of jaw strip; deduct 5 points for failure to wear safety helmet; Deduct 1 point if failing to check the appearance, label and illuminance of the flashlight; Deduct 5 points if failing to take the flashlight; Deduct 1 point if insulating gloves are not checked for intactness; Deduct 5 points if insulating gloves are not checked for airtightness, deduct 5 points if no insulating gloves are taken; Deduct 1 point if insulating boots are not checked for intactness, deduct 5 points if no insulating boots are got; Deduct 1 point if wrenches are not checked for appearance and performance intactness; Deduct 5 points if no wrench is taken; Deduct 5 points if failing to bring the key to the control cabinet				

Continued

S/N	Operation name	Quality standard	Points	Deduction criteria	Deduction	Score
3		Work process				
3.1	Simulating rehearsal	Complete an operation on simulation system by operation ticket, check equipment position, name, number and operating mode	5	Deduct 1-5 points for failure to check equipment position, name, number and operating mode during an operation on simulation system by operation ticket		
3.2	Fault analysis	Watch alarm message from HV switch cabinet protection device; Identify fault types considering signal indication	25	Deduct 5 points for failure to properly check alarm message from HV switch cabinet protection device; Deduct 15 points if fault types are not identified based on signal indication		
3.3	Fault handling	Properly complete operation by following operating steps as required in safe operating procedures. When switching, operating sequence shall be correct and smooth; After going to inspection state, fuses of closing power supply and control power supply (control switch) shall be removed; Complete the operations in specified time	55	Properly complete operation by following operating steps as required in safe operating procedures. During switching, the operation sequence shall be correct and smooth. If there is an error in the operation, 5-55 points will be deducted depending on the situation; After going to inspection state, fuses of closing power supply and control power supply (control switch) shall be removed, and 10 points will be deducted for missing a fuse; Deduct 5-10 points for failure to properly complete operation within specified time		

任务四　变压器绝缘电阻测试实训

一、作业任务

本任务介绍变压器绝缘电阻的测试方法和技术要求。通过测试流程的介绍，学员应自设现场安全措施，正确选择仪器、仪表，测量 10 kV 双绕组变压器的绝缘电阻，清理并结束试验现场，对测试结果进行分析判断，引用正确标准及文件完成试验报告。

二、引用标准及文件

（1）《电气装置安装工程　电气设备交接试验标准》（GB 50150—2016）。
（2）《输变电设备状态检修试验规程》（Q/GDW 1168—2013）。
（3）《电力设备预防性试验规程》（DL/T 596—1996）。
（4）《国家电网公司五项通用制度变电检修管理规定》。
（5）《现场绝缘试验实施导则》（DL/T 474.1～474.5—2018）。
（6）《国家电网公司电力安全工作规程（变电部分）》。
（7）《国家电网公司十八项电网重大反事故措施》。
（8）《高电压试验技术》。

三、作业条件

进行绝缘试验时，被试品温度不应低于 5 ℃，户外试验应在良好的天气下进行，空气相对湿度一般不高于 80%，且应同时测量被试品的温度和周围空气的温度和湿度。作业人员精神状态良好，熟悉工作中安全措施、技术措施以及现场工作危险点。

四、作业前准备

1. 现场勘察的基本要求及条件

勘察现场设备情况，查阅相关技术资料，包括历史数据及相关规程。

2. 工器具及材料选择

2500 V/2500 MΩ 绝缘电阻表，温度计（酒精、水银），湿度计，接地线，放电棒，绝缘手套，安全围栏，"在此工作""止步，高压危险！""从此进出"标示牌若干，测试线（含高压屏蔽线）、裸铜线若干。

3. 危险点及预防措施

（1）高压触电。

危险点：被试设备及相应套管引线均视为带 10 kV 线电压。

预防措施：用围栏将被试设备与相邻带电设备（间隔）隔离，并且向作业现场外悬挂"止步，高压危险！"标示牌，在通道处设置唯一出入口，悬挂"从此进出"标示牌；工作时至少需要两人，一人监护，一人操作，听工作负责人指挥。拆、接试验接线前，应将被试设备对地充分放电，以防止剩余电荷、感应电压伤人以及影响测量结果。测试前与检修负责人协调，不允许有交叉作业，试验接线应正确、牢固，试验人员应精力集中。试验人员之间应分工明确，测量时应配合默契，测量过程中要大声呼唱。

（2）低压触电

危险点：试验电源为交流 220 V，搭接试验电源注意监护。

预防措施：绝缘电阻表若需要外接电源，搭接试验电源需要两人操作：一人监护，一人操作，听工作负责人指挥。

（3）设备损坏

危险点：未按照要求操作绝缘电阻表、野蛮操作，或者操作绝缘电阻表顺序错误，均可能会对设备和绝缘电阻表造成不同程度的损坏。

预防措施：禁止野蛮操作，操作过程中，必须按照正确操作顺序使用绝缘电阻表，若出现严重违反《安规》的现象，应当立即制止；工作时至少需要两人：一人监护，一人操作，听工作负责人指挥。

4. 作业人员分工

现场工作负责人（监护人）：×××

现场作业人员：×××

五、作业规范及要求

（1）非被测部位均应短路接地且接触良好。

（2）测试前后均应对被试品进行充分放电；摇表停止转动前或被试品未放电前，严禁用手触及；拆线时，也不要触及引线的金属部分。

（3）测量时摇动摇表手柄的速度均匀保持 120 r/min；读取数值后，应先断开兆欧表与被试品的连线，然后再将兆欧表停止转动，以免被试品的电容上所充的电荷经兆欧表放电而损坏兆欧表。

六、作业流程及标准（表6-2）

表6-2 变压器绝缘电阻测试流程及评分标准

班级		姓名		学号		考评员		成绩	
序号	作业名称	质量标准			分值/分	扣分标准		扣分	得分
1	着装穿戴	穿工作服、工作鞋，戴安全帽、线手套			2	未穿工作服、工作鞋，未戴安全帽、线手套，每缺少一项扣1分；着装穿戴不规范，每处扣1分			
2	现场安全措施	按现场标准化作业进行设置、检查安全措施			3	不设置、不检查、设置不正确、检查不充分扣 0.5~3 分			
3		准备工作							

续表

序号	作业名称	质量标准	分值/分	扣分标准	扣分	得分
3.1	仪器、仪表	检查使用仪器、仪表是否在使用有效期内； 检查兆欧表"0"位、"∞"、驱动部分和测试线是否正常	10	未检查仪器仪表有效期，扣2分； 未检查兆欧表，扣10分		
3.2	放电、接地	接测试线前必须对变压器充分放电（放电时间≤5 min）； 将被试设备外壳可靠接地	5	不放电扣5分，放电不够扣5分； 不接地扣5分		
4	测量变压器高压、低压及地，绝缘电阻及吸收比	防止人员高空坠落； 将变压器低压短路接地，高压侧短路； 使用高压屏蔽线，屏蔽层接地； 兆欧表"E"端接地，驱动兆欧表达120～140 r/min，指针达到"∞"后，将测试线搭上变压器高压侧，读取15 s、60 s绝缘电阻值； 读数完毕后，先断开加压端，再停止摇兆欧表； 测量后用带限流电阻放电杆充分放电； 加压前注意监护并大声呼唱	35	不系安全带扣5分，系安全带不规范扣2分； 低压侧未短路扣10分，短路、接地不良扣5分，高压侧未短路扣5分； 屏蔽层未接地扣2分； 兆欧表接线不正确扣5分，转速不正确扣5分，未读取15 s、60 s绝缘电阻值扣10分，测试线搭接不正确扣5分； 顺序错误扣5分； 不放电扣15分，不用带限流电阻放电杆放电扣2分； 未呼唱扣5分		
5	熟练程度	作业标准、流程规范、操作熟练	10	作业不标准、流程不规范、操作不熟悉扣1～10分		
6	试验结束	拆除试验接线，恢复被试品初始状态； 清理试验现场； 结束工作手续，试验人员、设备撤离现场	5	不符合要求扣0.5～5分		
7	试验报告	记录被试设备相关信息、试验日期、试验人员、试验地点、温度、湿度、变压器上层油温； 依据相关试验标准（判据）正确分析试验数据并得出试验结论	20	基本信息不齐全每处扣0.5分； 引用标准错误、分析错误或结论错误扣1～10分		
合计			100			

Task Ⅳ Practical training on measurement of transformer insulation resistance

Ⅰ. Operating tasks

In this module, methods and technical requirements for transformer insulation resistance measurement will be provided. Through presentation of testing process, the trainees shall know how to set their own on-site safety measures, correctly select instruments and meters, measure insulation resistance of 10 kV double-winding transformers, complete the test and clear the test site, analyze and evaluate test results and prepare test report by applying correct standards and documents.

Ⅱ. Referenced standards and documents

(1) *Electrical equipment installation engineering- standard for hand-over test of electric equipment* (GB 50150—2016);

(2) *Code of condition-based maintenance & test for electric equipment* (Q/GDW 1168—2013);

(3) *Preventive test code for electric power equipment* (DL/T 596—1996);

(4) *Regulations on management of substation repair under the five general regulations of State Grid Corporation of China*;

(5) *Guidelines for on-site insulation tests* (DL/T 474.1-474.5—2018);

(6) *Electric Power Safety Working Regulations* (*Power Transformation*) *of State Grid Corporation of China*;

(7) *The 18 major anti-accident measures of State Grid Corporation of China*;

(8) High voltage test techniques.

Ⅲ. Operating conditions

During insulation test, temperature of the test object shall be no less than 5 °C. Outdoor tests shall be conducted in good weather, and relative humidity of air shall be no more than 80% in general. Test object temperature and ambient air temperature and humidity shall be measured simultaneously. The operation personnel shall be in a good mental state and familiar with safety measures, technical measures, and dangerous points of field work.

Ⅳ. Preparation before operation

1. Basic requirements and conditions for on-site investigation

Conduct a survey of equipment on site and refer to relevant technical data, including historical data and related procedures.

2. Selection of tools and instruments and materials

2,500 V/2,500 MΩ insulation resistance meters, thermometers (alcohol, mercury), hygrometers, grounding wires, discharging rods, insulating gloves, safety fences, "Work here", "Stop! High voltage, danger!", "Enter and exit from here" sign boards, test wires (including HV shielding wire) and bare copper wires.

3. Dangerous points and preventive measures

(1) High voltage electric shock.

Dangerous point: The equipment under test and corresponding bushing leads are considered to have 10 kV line voltage.

Prevention and control measures: Use fences to isolate the equipment under test from adjacent electrified equipment (compartment), and provide "Stop! High voltage, danger!" sign board outside the work site, provide an only access at the passage, and provide "Enter and Exit from Here" sign board; Two persons are required for the work, one for supervision and the other for operation under commands of the person in charge of work. Before disconnecting or connecting test wires, the equipment under test shall be fully discharged to ground to prevent residual electric charge and induced voltage from injuring the people and affecting measurement results. Before test, coordinate the person in charge of the inspection, cross operations are not allowed, test wires shall be correct and firm, and testing personnel shall concentrate on their job. Division of labor among testing personnel shall be clear, and they shall be in tacit cooperation with one another during measurement, and shouting loudly during measurement is required.

(2) Low voltage electric shock.

Dangerous point: test power supply is AC 220 V, and supervision is required for connecting test power supply.

Prevention and control measures: If external power supply is required for insulation resistance meter, two persons are required for connecting test power supply, one is responsible for supervision and the other is responsible for operation by following commands of the person in charge of work.

(3) Equipment damage.

Dangerous point: failure to operate insulation resistance meter, rough handling or operating insulation resistance meter in a wrong sequence may cause damage to equipment and insulation resistance.

Prevention and control measures: rough handling is forbidden; insulation resistance meter must be used by following correct operating sequence; upon discovery of any behavior violating the Electric Power Safety Working Regulations, the behavior shall be stopped; two persons are required for the work, one is responsible for supervision and the other is responsible for operation by following commands of the person in charge of work.

4. Division of labor among operators

Person in charge of on-site work (supervisor): ×××

On-site operator: ×××

Ⅴ. Operating specifications and requirements

(1) Positions that are not to be measured shall be shorted to ground and have good contact.

(2) The test objects shall undergo full discharge before and after the test. Touching with hands is strictly forbidden before the megameter stops rotation or the test object is discharged; In disassembling wires, do not touch metal part of the lead.

(3) During measurement, speed of megameter handle shaking shall be kept at 120 r/min; After reading the values, the wire between megohmmeter and the test object shall be disconnected and then the megohmmeter shall be shut down to prevent electric charge on capacitor of the test object from discharging through megohmmeter and cause damage to megohmmeter.

Ⅵ. Operation process and standards (see Tab. 6-2)

Tab. 6-2   Transformer insulation resistance testing process and scoring standards

Class		Name		Student ID		Examiner		Score	
S/N	Operation name	Quality standard		Points	Deduction criteria			Deduction	Score
1	Wearing	Wear work clothes and shoes; wear safety helmet and cotton gloves		2	Deduct 1 point if failing to wear work clothes, work shoes, safety helmet or cotton gloves; Deduct 1 point for non-standard clothing				
2	Safety precautions on the site	Safety measures are to be provided and checked according to standardized on-site operation		3	Deduct 0.5-3 points for failure to provide, failure to check, failure to correctly provide and failure to adequately check the measures				
3	Preparations								
3.1	Instruments and apparatuses	Check that the instruments and apparatuses are within valid period; Check that "0" position, "∞", driving parts and test wires of megohmmeter are normal		10	Deduct 2 points for failure to check the above items; Deduct 10 points for failure to check the above items				
3.2	Discharge and grounding	Before connecting test wires, transformers must be electrically discharged (discharge time ≤ 5 min); Housing of the equipment under test shall be firmly grounded		5	Deduct 5 points for failure to discharge, deduct 5 points for inadequate discharging; Deduct 5 points if the equipment is not grounded				

Continued

S/N	Operation name	Quality standard	Points	Deduction criteria	Deduction	Score
4	Measure transformer high voltage, low voltage and voltage to ground, insulation resistance and absorption ratio	Protect personnel from falling from height; Short LV side of transformer to ground, short HV side; Use high-voltage shielding wires and shielding layer for grounding; Have "E" end of megohmmeter grounded, drive megohmmeter at up to 120–140 r/min; when the indicator is at "∞", connect test wire to HV side of transformer, read insulation resistance at 15 s and 60 s; After reading the values, disconnect the voltage applied end and then stop shaking the megohmmeter; After measurement, allow full discharge using discharging rod with current limiting resistor; Provide supervision and shout loudly before applying voltage	35	Deduct 5 points for failure to fasten safety belt, deduct 2 points for incorrect safety belt fastening; Deduct 10 points for failure to short LV side, deduct 5 points for poor grounding, deduct 5 points for failure to short HV side; Deduct 2 points if the equipment is not grounded; Deduct 5 points for incorrect megohmmeter wiring, deduct 5 points for incorrect speed, deduct 10 points for failure to read insulation resistance at 15 s and 60 s, deduct 5 points for incorrect test wire connection; Deduct 5 points if sequence is not correct; Deduct 15 points for failure to discharge, deduct 2 points for failure to discharge using discharging rod with current limiting resistor; Deduct 5 points for failure to shout loudly		
5	Proficiency level	Job standards, process specifications, and proficient operation	10	Deduct 1–10 points for non-standard operations, non-standard processes and unfamiliar operations		
6	End of test	Remove test wires, restore the test object to original state; Clean the test site; Complete work procedures and withdraw testing personnel and equipment from the site	5	Deduct 0.5–5 points if requirements are not met		
7	Test report	Information of the equipment under test, including test date, test personnel, test location, temperature, humidity and upper oil temperature of transformer; Correctly analyze test data and draw conclusions based on relevant test standards (criteria)	20	Deduct 0.5 point in each case when basic information is incomplete; Deduct 1–10 points for applying incorrect standard, incorrect analysis and wrong conclusions		
Total			100			

任务五　变压器分接开关的调整

一、作业任务

两人一小组：一人扮演监护人，一人扮演操作人，以"10 kV柱上变压器"为工作对象，按照《国家电网公司电力安全工作规程（变电部分）》中的要求，完成变压器分接开关的调整。

二、引用标准及文件

（1）《国家电网公司电力安全工作规程（变电部分）》。
（2）《国家电网公司十八项电网重大反事故措施》。

三、作业条件

应在良好的天气操作；作业人员精神状态良好，熟悉工作中的安全措施、技术措施以及现场工作危险点。

四、作业前准备

1. 现场操作的基本要求及条件

勘察现场设备情况，查阅相关技术资料，包括历史数据及相关规程。

2. 工器具及材料选择

工位1个、中性笔1支、安全帽2顶、绝缘手套1双、线手套2双、直流电阻测试仪1个、放电棒、验电器。

3. 危险点及预防措施

（1）高压触电。
预防措施：正确进行模拟预演，做好个人防护。
（2）使用不合格的安全工器具。
预防措施：操作设备前，应正确选取安全工器具并检查合格。

4. 作业人员分工

现场工作负责人（监护人）：×××
现场作业人员：×××

五、作业规范及要求

（1）现场操作时，老师指定柱上变压器当前低压侧电压为350 V或410 V，学员选择正确的挡位调节分接开关，并测量高压电阻偏差值，确定是否可以投运。
（2）按照作业任务要求正确选择安全用具，做好个人防护工作。
（3）按流程（表6-3）完成调节分接开关，测量高压电阻偏差值。

表 6-3　变压器分接开关的调整流程及说明

序号	工作流程	说明
1	放电	因为假设变压器为刚停运下来,所以其可能内部存储容性电荷,所以要放电
2	验电	通过验电确定变压器经放电后已无电荷
3	选择挡位,调节分接开关	低压侧 380 V 以上选 1 挡,380 V 选 2 挡,380 V 以下选 3 挡
4	测量 R_{AB}、R_{BC}、R_{AC}	使用直流电阻测量仪进行测量,每次测量后需要放电方可进行下次测量
5	计算 R_A、R_B、R_C	计算公式为: $R_A = (R_{AB} + R_{AC} - R_{BC})/2$ $R_B = (R_{BC} + R_{AB} - R_{AC})/2$ $R_C = (R_{AC} + R_{BC} - R_{AB})/2$
6	记录 R_{AB}、R_{BC}、R_{AC},计算 $R_{线}\%$	计算公式为 $R_{线}\% = (R_{线max} - R_{线min})/R_{线min}$
7	记录 R_A、R_B、R_C,计算 $R_{相}\%$	计算公式为 $R_{相}\% = (R_{相max} - R_{相min})/R_{相min}$
8	判断是否可以投运	$R_{线}\%$、$R_{相}\%$ 都小于 2% 则可以投运,否则不能投运
9	整理工器具,收拾现场	做到工完料尽场地清

(4)遵循安全操作规程,按照操作票的步骤正确操作。
(5)操作结束后,对操作质量进行检查。

六、作业流程及标准(表 6-4)

表 6-4　变压器分接开关的调整操作流程及评分标准

班级		姓名		学号		考评员		成绩	
序号	作业名称	质量标准			分值/分	扣分标准		扣分	得分
1	正确填写记录页	按照作业任务要求正确填写记录页			30	按照作业任务要求正确填写记录页,记录页填写不规范的,视情况扣 2~30 分。			
2	安全意识	准备好该项操作所需的安全用具并进行检验;做好个人防护,戴上安全帽、护目镜、绝缘手套,穿上绝缘靴;			20	未能准备好该项操作所需的安全用具并进行检验,扣 2~5 分;			

续表

序号	作业名称	质量标准	分值/分	扣分标准	扣分	得分
2	安全意识	正确持操作票在模拟系统上模拟操作一次，核对设备的位置、名称、编号和运行方式		未能做好个人防护，未戴上安全帽、护目镜、绝缘手套，未穿绝缘靴，扣2~5分；要持操作票在模拟系统上模拟操作一次，未核对设备的位置、名称、编号和运行方式，扣5~10分		
3	操作技能	直流电阻测量操作正确；计算 R_A、R_B、R_C 正确；计算 $R_线$% 正确；计算 $R_相$% 正确；判断是否能够投运正确	50	遵循安全操作规程，若直流电阻测量操作不正确，扣10分；计算 R_A、R_B、R_C 不正确，扣10分；计算 $R_线$% 不正确，扣10分；计算 $R_相$% 不正确，扣10分；判断是否能够投运不正确，扣10分		
4	选择挡位	选择错误挡位进行调节	否定项			
合计			100			

Task V Adjustment of Transformer Tap Changer

I. Tasks

Let every two persons form a group, with one acting as a supervisor and the other an operator. Each group is required to complete the adjustment of the tap changer of a 10 kV pole-mounted transformer in accordance with the requirements in the Electric Power Safety Working Regulations (Power Transformation) of State Grid Corporation of China.

II. Reference standards and documents

(1) *Electric Power Safety Working Regulations (Power Transformation) of State Grid Corporation of China*;

(2) *18 Major Anti-accident Measures for Power Grids of State Grid Corporation of China*.

III. Operating conditions

Operations should be carried out in good weather. Operators should be in a good state of mind, and be familiar with the safety measures, technical measures, and dangerous points of the work on the site.

IV. Preparation for work

1. Basic requirements and conditions for on-site operation

Operators should check the condition of on-site equipment, and understand relevant technical data, including historical data and relevant procedures.

2. Selection of tools and instruments and materials

1 workstation, 1 gel pen, 2 safety helmets, 1 pair of insulating gloves, 2 pairs of cotton gloves, 1 DC grounding resistance tester, 1 discharging rod, and 1 electroscope.

3. Dangerous points and preventive and control measures

(1) High voltage electric shock.

Prevention and control measures: Carry out simulation and take proper personal protection measures.

(2) Use of non-conforming safety tools and instruments.

Prevention and control measures: Before operating the equipment, operators should select and check the safety tools and instruments properly.

4. Division of labor among operators

Person in charge of on-site work (supervisor): ×××

On-site operator: ×××

V. Operating procedures and requirements

(1) During on-site operation, the teacher specifies that the voltage on the low voltage side of the pole-mounted transformer is 350 V or 410 V. The student is required to select the correct range to adjust the tap changer, and measure the deviation of the high voltage resistor to determine whether the transformer can be put into operation.

(2) Select safety appliances correctly according to the requirements of the operation task, and take proper personal protection measures.

(3) Complete the adjustment of the tap changer according to the procedures as shown in Tab. 6-3, and measure the deviation of the high voltage resistor.

Tab. 6-3 Adjustment procedures of Transformer tap charger

S/N	Work process	Description
1	Discharge	Because it is assumed that the transformer has just been shut down, it may store capacitive charges internally and needs to be discharged
2	Electricity test	Confirm that the transformer has been discharged and is free of charge by detecting the presence of an electric charge
3	Select the measuring range and adjust the tap changer	Select range 1 if the voltage on the low voltage side is above 380 V, range 2 if it is 380V, and range 3 if it is below 380 V
4	Measure R_{AB}, R_{BC}, and R_{AC}	Use a DC grounding resistance tester to measure, and discharge it after each measurement
5	Calculate R_A, R_B, and R_C	The calculation formula is $R_A = (R_{AB} + R_{AC} - R_{BC})/2$ $R_B = (R_{BC} + R_{AB} - R_{AC})/2$ $R_C = (R_{AC} + R_{BC} - R_{AB})/2$
6	Record R_{AB}, R_{BC}, and R_{AC}, and calculate R_{line}%	The calculation formula is $R_{line}\% = (R_{line\,max} - R_{line\,min})/R_{line\,min}$
7	Record R_A, R_B, and R_C, and calculate R_{phase}%	The calculation formula is $R_{phase}\% = (R_{phase\,max} - R_{phase\,min})/R_{phase\,min}$
8	Determine whether it can be put into operation	It can be put into operation only If R_{line}% and R_{phase}% are both less than 2%
9	Place the tools and instruments properly and clean up the site	Upon completion of the work, the materials shall run out and the site shall be clean

(4) Follow the safety operating procedures and operate correctly according to the steps in the operation ticket.

(5) Check the quality of operation after the operation.

VI. Operating procedures and standards (Tab. 6-4)

Tab. 6-4 Operation process and scoring criteria for adjustment of transformer tap changer

Class		Name		Student ID		Examiner		Score	
S/N	Description	Quality standard			Score (points)	Standard for deduction (points)		Deduction	Score
1	Fill out the record page correctly	Fill out the record page correctly according to the requirements of the operation task			30	Fill out the record page correctly according to the requirements of the operation task; otherwise, 2–30 points will be deducted as appropriate			
2	Safety consciousness	Prepare and check the safety appliances required for the operation; Take proper personal protection measures and wear a safety helmet, safety glasses, insulating gloves and insulating boots; Correctly hold the operation ticket to simulate the operation in the simulation system, and check the location, name, number and operation mode of the equipment			20	Failure to prepare and check the safety appliances required for the operation will result in a deduction of 2–5 points; Failure to take proper personal protection measures or to wear a safety helmet, safety glasses, insulating gloves and insulating boots will result in a deduction of 2–5 points; Correctly hold the operation ticket to simulate the operation in the simulation system. Failure to check the location, name, number and operation mode of the equipment will result in a deduction of 5–10 points			
3	Operational skills	DC resistance is measured correctly; The R_A, R_B, and R_C are calculated correctly; The $R_{line}\%$ is calculated correctly; The $R_{phase}\%$ is calculated correctly; Determine whether it can be put into operation properly			50	Follow the safety operating procedures. If the DC resistance is measured correctly, 10 points will be deducted; Calculate R_A, R_B, and R_C, If it is not correct, 10 points will be deducted; The $R_{line}\%$ is calculated correctly; If it is not correct, 10 points will be deducted;			

Continued

S/N	Description	Quality standard	Score (points)	Standard for deduction (points)	Deduction	Score
3	Operational skills			The $R_{phase}\%$ is calculated correctly; If it is not correct, 10 points will be deducted; Determine whether it can be put into operation properly. If it is not correct, 10 points will be deducted		
4	Range selection	Wrong range is selected for adjustment	Negative			
Total			100			

模块七　电力线路安全运检

现代生产和生活都离不开电力。电力线路，作为连接发电厂（火电厂、水电厂、核电厂、风力发电厂、潮汐发电厂等）、变电站（升压变电站、降压变电站、开关站）和电力用户之间的重要组成部分，其运行的稳定性影响着整个社会的发展。按照作用不同，电力线路可分为送电线路（输电线路）和配电线路。电力线路的作业，主要包括电力线路运行维护、线路检修和架设。为加强线路作业现场安全管理，规范各类工作人员的行为，保证人身、电网和设备安全，依据国家有关法律法规，结合电力线路作业实际，国家电网公司在电力安全规程中对电力线路作业有明确的安全要求。本模块主要介绍电力线路的安全要求和配电线路带电作业的安全措施。

学习目标：

（1）能阐述电力线路的作用和分类。
（2）能阐述对线路作业人员的一般要求。
（3）能说出电力线路巡视的分类和安全要求。
（4）能说出电力电缆的结构。
（5）能说出电力电缆作业的安全要求。
（6）能说出配电线路带电作业的安全措施。
（7）能对运行电力电缆进行停电摇测绝缘电阻的操作。

Module Ⅶ Safe operation and inspection of electric power lines

Electric power is essential for modern production and life. Power line is an important part connecting power plants (thermal power plants, hydropower plants, nuclear power plants, wind farms, tidal power plants, etc.), substations (step-up substations, step-down substations, switchyards) and power users. Stable operation of electric power lines has impact on development of the whole society. Electric power lines can be classified into transmission lines and distribution lines based different purposes. Operations on electric power lines include power lines operation and maintenance, line repair and erection. To strengthen management of safety in electric power line operation, behaviors of workers shall be managed and personnel, power grid, and equipment safety shall be protected in accordance with national laws and regulations, considering actual operations on power lines, State Grid Corporation of China provides clear safety requirements for operations on power line in electric power safety regulations. This chapter will provide requirements for electric power line safety and safety measures for live-wire work on distribution lines.

Learning objectives:

(1) Be able to explain roles and classification of electric power lines.

(2) Be able to explain general requirements for line operators.

(3) Be able to tell classification and safety requirements of electric power line routine inspection.

(4) Be able to tell power cable structure.

(5) Be able to tell safety requirements for operations on power cable.

(6) Be able to tell safety measures for live-wire work on distribution lines.

(7) Be able to interrupt power cable supply for measuring insulation resistance using tramegger.

任务一 电力线路运行与检修

一、电力线路分类

电力线路是指在发电厂、变电站和电力用户间用来传送电能的线路。它是供电系统的重要组成部分，担负着输送和分配电能的任务。

（一）按架设方式分

按架设方式来分，电力线路分为架空线路和电缆线路两种。

1. 架空线路

架空线路是指安装在室外的杆塔上，用来输送电能的线路。架空线路具有成本低、投资少、安装容易、维护和检修方便、易于发现和排除故障等优点，所以长期以来，架空线路一直是国内外电力线路的首选。

架空线路由导线、杆塔、绝缘子和线路金具等主要元件组成。为了防雷，有的架空线路上还装设有架空地线。为了加强杆塔的稳固性，有的杆塔还安装有拉线或扳桩。

导线是线路的主体，担负着输送电能的功能。导线架设在杆塔上面，要承受自身重量和各种外力的作用，并要承受大气中各种有害物质的侵蚀。因此，导线必须具有良好的导电性，同时要具有一定的机械强度和耐腐蚀性。导线按材质分有铜、铝、钢三种。常用导线有单股铝线（L）、多股铝绞线（LJ）、钢芯铝绞线（LGJ）（图7-1）、单股铜线（T）、多股铜绞线（TJ）、钢绞线（GJ）等，这几种导线均为裸导线，截面积为 10～300 mm^2。钢芯铝绞线的线芯是钢线，用以增强导线的抗拉强度，弥补铝线机械强度较差的缺点，而其外围用铝线，取其导电性较好的优点，因钢芯铝绞线在通过交流电时，由于交流电的集肤效应，电流实际从铝中流过，从而克服了钢线导电性差的缺点。超高压、高压架空线路基本上都采用裸导线，裸导线散热性能好、载流量大，又节省绝缘材料。而中压和低压架空线路主要是在市区内，近几年来已开始采用绝缘导线（图7-2）。绝缘导线比架空裸导线有明显的优点：能有效降低人身触电伤亡危险、线路沿线树木砍伐量、外物引起线路相间短路概率，减少导线腐蚀程度，可根据线路通道情况减小与物体之间距离，提高导线使用寿命和线路供电可靠率。

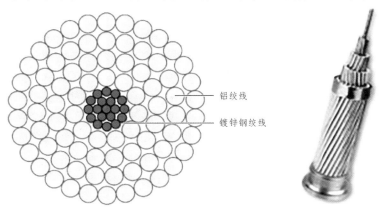

图7-1 钢芯铝绞线

图 7-2　绝缘导线

10 kV 绝缘导线的架设方式可以采用传统架空线路方式，亦可采用吊在钢索上成束架设的方式。当两根绝缘导线平行紧密接触架设时，一根绝缘导线加电压 80 kV，另一根绝缘导线接地，通过实验未发生击穿。实验结果证明，成束架设方式是可行的。

10 kV 架空绝缘导线，目前国内产品以单芯（铜芯或铝芯）、高密度聚乙烯或交联聚乙烯绝缘为主，标准截面积为 $10 \sim 300 \text{ mm}^2$。

1 kV 低压塑料绝缘线，目前国内产品有铜芯聚乙烯线、铝芯聚乙烯线、铜芯交联聚乙烯线、铝芯交联聚乙烯线等，标准截面积为 $16 \sim 300 \text{ mm}^2$。

2. 电缆线路

电力电缆是用于传输和分配电能的电缆。电力电缆常用于城市地下电网、发电站引出线路、工矿企业内部供电及过江海水下输电线。其主要优点有占地少、可靠性高、分布式电容大、维护工作量少、电击可能性小等；缺点是成本比较高，运行不灵活，发生故障不容易快速发现，接头工艺复杂。

随着城市发展，在电力线路中，电缆所占比重正逐渐增加。常用的电力电缆有下列几种：

（1）油纸绝缘电力电缆。它有黏性油浸纸绝缘电缆和不滴油浸渍纸绝缘电缆两种。其优点是允许运行温度较高、介质损耗低、耐电压强度高、使用寿命长。其缺点是绝缘材料弯曲性能差，不能在低温时敷设，否则易损伤绝缘。目前，已逐渐被其他绝缘材料电力电缆所替代。

（2）塑料绝缘电缆。它是绝缘层为挤压塑料的电力电缆，一般用于低压线路。绝缘材料有聚氯乙烯和聚乙烯两种。前者具有非燃性，化学稳定性高，安装维护方便，能满足高落差敷设的要求，但工作温度高时影响其机械性能；后者介电性能优良，便于加工，但受热易变形，易延燃，易发生应力龟裂。

（3）橡胶绝缘电缆。它的绝缘层为橡胶加上各种配合剂，经过充分混炼后挤包在导电线芯上，经过加温硫化而成。它柔软，富有弹性，适合于移动频繁、敷设弯曲半径小的场合。常用作绝缘的胶料有天然胶-丁苯胶混合物、乙丙胶、丁基胶等。此种电缆一般用于低压线路。其电气性能、机械性能和化学稳定性好，但耐电晕、耐热、耐油、耐臭氧性能较差。

（4）交联聚乙烯绝缘电缆。它是目前可应用于各级电压电网的一种电缆。其绝缘介质材料是通过利用高能辐照或化学方法对聚乙烯分子进行交联制成的，容许较高温度，具有载流量大、介电性能优良、质轻坚固等优点，适宜于环境恶劣、高落差敷设，但其抗电晕、游离放射性能较差。

（二）按用途分

电力线路按用途主要分为配电线路和输电线路（送电线路）两种。

（1）配电线路，指从降压变电站把电力送到配电变压器或将配电变电站的电力送到用电单位的线路，主要包括 35 kV 及以下配电网中的架空线路、电缆线路及其附属设备等。常见的配电线路电压等级为 35 kV、10 kV 和 380 V。

（2）输电线路，指输送电能，并联络各发电厂、变电站（所）使之并列运行，实现电力系统联网，并能实现电力系统间功率传递的线路。高压输电线路是电力工业的大动脉，是电力系统的重要组成部分。我国输电线路的电压等级有 110 kV、220 kV、330 kV、500 kV、750 kV、800 kV、1 000 kV。

二、电力线路的构成

（一）架空线路的构成

架空线路主要由导线和避雷线（架空地线）、杆塔、绝缘子、金具、杆塔基础、拉线和接地装置等构成。各部分作用如下：

（1）导线：用来传导电流、输送电能的元件。

（2）避雷线：也叫架空地线，保护设备，避免雷击，是消除雷击、保护建筑物或仪器的设施。

（3）电杆、铁塔：支撑输电线的支撑物。

（4）绝缘子：实现电气绝缘和机械固定。

（5）金具：用于支持、固定和接续导线及绝缘子连接成串，亦用于保护导线和绝缘子，如图 7-3 所示，包括耐张线夹、并沟线夹、悬垂线夹、防震锤等。

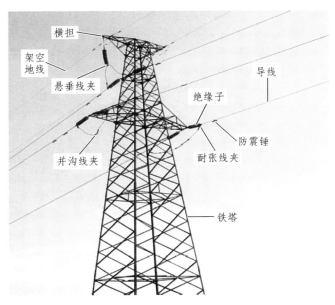

图 7-3　架空线路的构成

（6）杆塔基础：用于稳定杆塔，使杆塔不致因承受垂直荷载、水平荷载、事故断线张力和外力作用而上拔、下沉或倾倒。

（7）拉线：平衡作用于杆塔的横向荷载和导线张力，可减少杆塔材料的消耗量，降低线路造价。

（8）接地装置：降低杆塔顶电位，保护线路绝缘不致击穿闪络等。

（二）电力电缆的构成

电力电缆主要是由线芯、绝缘层、屏蔽层和保护层组成。

（1）线芯，电力电缆的导电部分，用来输送电能，是电力电缆的主要部分。

（2）绝缘层，是将线芯与大地以及不同相的线芯间在电气上彼此隔离的部分，用于保证电能输送，是电力电缆结构中不可缺少的组成部分。

（3）屏蔽层，为了均匀导电线芯和绝缘电场，6 kV 及以上的电力电缆一般都有导体屏蔽层和绝缘屏蔽层，部分低压电缆不设置屏蔽层。屏蔽层有半导电屏蔽和金属屏蔽两种。

（4）保护层，保护电力电缆免受外界杂质和水分的侵入，以及防止外力直接损坏电力电缆。

三、电力线路的安全要求

（一）作业人员基本条件

（1）经医师鉴定，无妨碍工作的病症（体格检查每两年至少一次）。

（2）具备必要的安全生产知识，学会紧急救护法，特别要学会触电急救。

（3）接受相应的安全生产知识教育和岗位技能培训，掌握配电作业必备的电气知识和业务技能，并按工作性质，熟悉本规程的相关部分，经考试合格后上岗。

（4）参与公司系统所承担电气工作的外单位或外来人员应熟悉相关规程；经考试合格，并经设备运维管理单位认可后，方可参加工作。

（5）作业人员应被告知其作业现场和工作岗位存在的危险因素、防范措施及事故紧急处理措施。作业前，设备运维管理单位应告知现场电气设备接线情况、危险点和安全注意事项。

（6）进入作业现场应正确佩戴安全帽，现场作业人员还应穿全棉长袖工作服、绝缘鞋。

（7）进出配电站、开闭所应随手关门。

（8）工作人员禁止擅自开启直接封闭带电部分的高压配电设备柜门、箱盖、封板等。

（9）作业人员对相关规程应每年考试一次。因故间断电气工作连续 3 个月及以上者，应重新学习相关规程，并经考试合格后，方可恢复工作。

（10）新参加电气工作的人员、实习人员和临时参加劳动的人员（管理人员、非全日制工等），应经过安全生产知识教育后，方可下现场参加指定的工作，并且不得单独工作。

（二）配电线路和设备的基本条件

（1）在多电源和有自备电源的用户线路的高压系统接入点，应有明显断开点。

（2）在绝缘导线所有电源侧及适当位置（如支接点、耐张杆处等）、柱上变压器高压引线，应装设验电接地环或其他验电、接地装置。

（3）高压配电站、开闭所、箱式变电站、环网柜等高压配电设备应有防误操作闭锁装置。

（4）柜式配电设备的母线侧封板应使用专用螺丝和工具，专用工具应妥善保存，柜内有电时禁止开启。

（5）封闭式高压配电设备进线电源侧和出线线路侧应装设带电显示装置。

（6）配电设备的操作机构上应有中文操作说明和状态指示。

（7）配电设备接地电阻应合格。

（8）环网柜、电缆分支箱等箱式配电设备宜装设验电、接地装置。

（9）柱上断路器应有分、合位置的机械指示。

（10）封闭式组合电器引出电缆备用孔或母线的终端备用孔应用专用器具封闭。

（11）待用间隔（已接上母线的备用间隔）应有名称、编号，并纳入调度控制中心管辖范围。其隔离开关（刀闸）操作手柄、网门应能加锁。

（12）高压手车开关拉出后，隔离挡板应可靠封闭。

（三）架空线路的安全要求

1. 绝缘强度

架空线路架设于户外，受环境影响很大。对于架空线路来说，首先应该保证足够的绝缘强度，满足相间绝缘和对地绝缘的要求；除此以外，还应该能满足各种操作过电压和接地过电压的要求，能经受大气过电压的考验。为了满足以上要求，架空线路应根据不同电压等级，使用相应的绝缘子串进行架设，同时保持足够的线间距离。

在海拔高度 1 000 m 及以下地区，不同电压等级下操作过电压及雷电过电压要求的悬垂绝缘子串的最少绝缘子片数分别是：10 kV 2 片，35 kV 3 片，110 kV 7 片，220 kV 13 片，330 kV 17 片，500 kV 25 片，750 kV 32 片。

2. 机械强度

架空线路应有很强的机械强度，一方面它要担负起自身的重力产生的拉力，另一方面还要经得起风吹、雨雪、覆冰等负荷。因此，架空导线的横截面积要足够大，一般导线的机械强度安全系数应不低于 2.5，即导线的截面电流要大于负荷电流的 2.5 倍。

3. 导电能力

导线最重要的作用就是导通、传送电能。按照导电能力的要求，导线的截面大小必须满足运行发热和运行电压损失的要求。运行发热主要受最大持续负荷电流的限制，如果负荷电流太大，导线过度发热，可能会引起导线熔断，造成停电或电力火灾事故。因此，线路运行时，应监视其运行温度，使其不超过规定值（一般裸导线、橡皮绝缘线不超过 70 ℃，塑料绝缘导线不超过 65 ℃）。运行电压损失主要是指线路运行时消耗在线路上的电压降，如果线路电压降过大，用电设备得不到合格的电压，无法正常运行，还可能造成事故。选择导线的横截面积应同时考虑经济电流密度、安全载流量、电压损失和机械强度。

（四）电缆线路的安全要求

埋设电缆应满足以下要求：

（1）电缆线相互交叉时，高压电缆应在低压电缆下方。如果其中一条电缆在交叉点前后 1 m

范围内穿管保护或用隔板隔开时，最小允许距离为 0.15 m。

（2）电缆与热力管道接近或交叉时，如有隔热措施，平行和交叉的最小距离分别为 0.5 m 和 0.15 m。

（3）电缆与铁路或道路交叉时应穿管保护，保护管应伸出轨道或路面 2 m 以外。

（4）电缆与建筑物基础的距离，应能保证电缆埋设在建筑物散水以外；电缆引入建筑物时应穿管保护，保护管亦应超出建筑物散水以外。

（5）直接埋在地下的电缆与一般接地装置的接地之间应相距 0.15 ~ 0.5 m；直接埋在地下的电缆埋设深度一般不小于 0.7 m，并应埋在冻土层下。

此外，为保证电缆安全运行，电缆通道如图 7-4 所示。其支架应牢固、无锈蚀；电缆应摆放整齐、固定牢固；电缆均应选择阻燃电缆或加装防火槽盒；电缆穿过孔洞处用防火堵料密实封堵；电缆沟每隔一定的距离应采取防火隔离措施；防火墙墙体应无破损，封堵良好。

图 7-4　电缆通道

四、电力线路的运行巡视

（一）线路巡视分类

（1）正常巡视，指线路巡视人员按一定的周期对线路所进行的巡视，包括对线路设备（指线路本体和附属设备）和线路保护区（线路通道）所进行的巡视。

（2）故障巡视，指运行单位为查明线路故障点、故障原因及故障情况等所组织的线路巡视。

（3）特殊巡视，在特殊情况下或根据需要，采用特殊巡视方法所进行的线路巡视。特殊巡视包括夜间巡视、交叉巡视、登杆检查、防外力破坏巡视以及无人机空中巡视等。

（4）监察巡视，一般由运检室领导或公司线路工程技术人员组织进行，了解本体和通道情况，发现问题提出对策，同时又可检查巡视人员质量，一般通过现场和 GPS 定位或巡检系统后台实施。

（5）夜间巡视，主要是为了弥补白天巡视过程中难以观察到的设备缺陷和异常情况，一般以线路设备的节点"热"缺陷及部件异常的火花放电、绝缘子放电、导线电晕等为重点的巡查。

（二）架空线路巡视要求

所谓线路巡视检查，就是沿着线路详细巡视线路设备的运行情况，及时发现设备存在的缺陷和故障点，并详细记录，以作为线路检修的依据。

为了按周期巡视、迅速发现问题，可根据线路的长度、线路所在的行政区、沿途地形和运行的方便性，将全线路划分几个巡线区段，分别由线路所在的供电部门承担巡视和维护运行。线路巡视区段要划分明确，不得形成无人管理的空白点。线路巡视应由有经验的人员进行定期巡视。

1. 巡线员应具备的要求

（1）必须熟悉所承担巡视线路的设备运行状况和特性。

（2）掌握线路设计、施工情况，以便于发现异常现象。

（3）掌握线路基础技术知识，熟悉有关规程规定。

（4）掌握线路运行中曾出现的故障和异常现象，以及采取的预防措施。

（5）熟悉沿线杆塔所处地点的地形、地貌、交通道路和村镇分布，以便检修时车辆能够顺利通行、尽快到达检修地点。

（6）了解沿线各种气象变化规律，以便在不同季节采取预防故障措施。

2. 《国家电网公司电力安全工作规程》对线路巡视的要求

（1）巡视工作应由有配电工作经验的人员担任。单独巡视人员应经工区批准并公布。

（2）电缆隧道、偏僻山区、夜间、事故或恶劣天气等巡视工作，应至少两人一组进行。

（3）正常巡视应穿绝缘鞋；雨雪、大风天气或事故巡线，巡视人员应穿绝缘靴或绝缘鞋；汛期、暑天、雪天等恶劣天气和山区巡线应配备必要的防护用具、自救器具和药品；夜间巡线应携带足够的照明用具。

（4）大风天气巡线，应沿线路上风侧前进，以免触及断落的导线。事故巡视应始终认为线路带电，保持安全距离。夜间巡线，应沿线路外侧进行。巡线时禁止泗渡。

（5）雷电时，禁止巡线。

（6）地震、台风、洪水、泥石流等灾害发生时，禁止巡视灾害现场。

（7）灾害发生后，若需对配电线路、设备进行巡视，应得到设备运维管理单位批准。巡视人员与派出部门之间应保持通信联络。

（8）单人巡视，禁止攀登杆塔和配电变压器台架。

（9）巡视中发现高压配电线路、设备接地或高压导线、电缆断落地面、悬挂空中时，室内人员应距离故障点 4 m 以外，室外人员应距离故障点 8 m 以外，并迅速报告调度控制中心和上级，等候处理。处理前应防止人员接近接地或断线地点，以免跨步电压伤人。进入上述范围人员应穿绝缘靴，接触设备的金属外壳时，应戴绝缘手套。

（10）无论高压配电线路、设备是否带电，巡视人员不得单独移开或越过遮栏；若有必要移开遮栏时，应有人监护，并保持表 7-1 规定的安全距离。

表 7-1　高压线路、设备不停电时的安全距离

电压等级/kV	安全距离/m	电压等级/kV	安全距离/m
10 及以下	0.7	330	4.0
20、35	1.0	500	5.0
66、110	1.5	750	8.0
220	3.0	1 000	9.5

注：表中未列电压应选用高一挡电压等级的安全距离。750 kV 数据按海拔 2 000 m 校正，其他电压等级数据按海拔 1 000 m 校正。

（11）进入 SF_6 配电装置室，应先通风。

（12）配电站、开闭所、箱式变电站等的钥匙至少应有 3 把：一把专供紧急时使用，一把专供运维人员使用，还有一把可以借给经批准的高压设备巡视人员和经批准的检修、施工队伍的工作负责人使用，但应登记签名，巡视或工作结束后应立即交还。

（13）低压配电网巡视时，禁止触碰裸露带电部位。

五、砍剪树木

人体的安全电压是 36 V，而在野外的高压线上的电压至少是 10 kV。电压等级在 35 kV 以上的高压线所用的裸铝线是没有绝缘层的，并且高电压在空气中会产生电离作用。因此不仅是高压线本身，高压线的周围都是危险区域。

高压线多为裸铝线，线下树木若触碰到高压线会对树产生放电现象。树木与高压线长时间摩擦，会对高压线强度产生影响，造成断线。如果遇到大风大雨天气，树木随风摆动，会造成高压线路接地或跳闸，引起大面积停电。

树木与高压线接触或距离高压线很近时，高压线会对树木打火放电，遇到大风大雨天气会导致跳闸，引发大面积短路停电。

我国《电力法》第五十三条规定：电力管理部门应当按照国务院有关电力设施保护的规定，对电力设施保护区设立标志。任何单位和个人不得在依法划定的电力设施保护区内修建可能危及电力设施的建筑物、构筑物，不得种植可能危及电力设施安全的植物，不得堆放可能危及电力设施安全的物品。在依法划定的电力设施保护区前已经种植的植物妨碍电力设施安全的，应当修剪或者砍伐。

我国《电力设施保护条例》第二十四条规定：新建、改建或扩建电力设施，需要损害农作物，砍伐树木、竹子或拆迁建筑物及其他设施的，电力建设企业应按照国家有关规定给予一次性补偿。在依法划定的电力设施保护区内种植的或自然生长的可能危及电力设施安全的树木、竹子，电力企业应依法予以修剪或砍伐。

电力线路作业人员需要定期开展线路巡视，掌握线路通道内树竹的生长情况，对于可能危及电力线路安全运行的树竹，要进行砍伐。在砍伐树竹的过程中，有如下安全要求：

（1）砍剪树木应有人监护。

（2）砍剪靠近带电线路的树木，工作负责人应在工作开始前，向全体作业人员说明电力线路有电；人员、树木、绳索应与导线保持表 7-2 规定的安全距离。

表 7-2　邻近或交叉其他高压电力线工作的安全距离

电压等级 /kV	安全距离/m	电压等级/kV	安全距离/m
10 及以下	1.0	330	5.0
20、35	2.5	500	6.0
66、110	3.0	750	9.0
220	4.0	1 000	10.5

（3）待砍剪的树木下方和倒树范围内不得有人逗留。

（4）为防止树木（树枝）倒落在线路上，应使用绝缘绳索将其拉向与线路相反的方向，绳索应有足够的长度和强度，以免拉绳的人员被倒落的树木砸伤。

（5）砍剪山坡树木应做好防止树木向下弹跳接近线路的措施。

（6）砍剪树木时，应防止马蜂等昆虫或动物伤人。

（7）上树时，应使用安全带，安全带不得系在待砍剪树枝的断口附近或以上。不得攀抓脆弱和枯死的树枝；不得攀登已经锯过或砍过的未断树木。

（8）风力超过 5 级时，禁止砍剪高出或接近带电线路的树木。

（9）使用油锯和电锯的作业，应由熟悉机械性能和操作方法的人员操作。使用时，应先检查所能锯到的范围内有无铁钉等金属物件，以防金属物体飞出伤人。

六、线路检修

（一）架空线路检修的安全措施

架空线路停电检修时，要保障人身、设备、电网的安全，需要做好以下措施：

（1）严格按照线路检修安全组织措施，履行好现场勘察制度、工作票制度、工作许可制度、工作监护制度和工作间断、转移、终结制度。现场勘察，须明确工作任务、停电范围起止杆号，记录需要保留的带电部位，记录作业地点地理情况等，编制组织措施、技术措施、安全措施。填用电力线路第一种工作票，办理工作许可手续。工作监护人应始终在工作现场，对工作人员的安全进行认真监护。工作间断后恢复工作，应先检查各项安全措施完整后方可开工。工作完毕时，应办理工作结束手续。工作许可人在接到工作负责人的完工报告后，办理工作结束手续，所有工作人员已撤离线路，所有接地线已拆除，并记录核对无误后，可下令拆除发电厂及变电站线路侧安全措施，恢复线路送电。

（2）严格按照安全技术措施的要求执行，包括停电、验电、装设接地线、悬挂标示牌和装设安全围栏。装设接地线需注意，凡是有可能送电到停电线路的各分支线上都要挂接地线，防止突然来电。应充分考虑用户反送电、交叉跨越断线或感应电、值班人员误操作等情况出现的可能性。

（3）《安规》中的相关要求。

杆塔作业应禁止以下行为：

① 攀登杆基未完全牢固或未做好临时拉线的新立杆塔。

② 携带器材登杆或在杆塔上移位。

③ 利用绳索、拉线上下杆塔或顺杆下滑。

杆塔上作业应注意以下安全事项：

① 作业人员攀登杆塔，在杆塔上移位及杆塔上作业时，手扶的构件应牢固，不得失去安全保护，并有防止安全带从杆顶脱出或被锋利物损坏的措施。

② 在杆塔上作业时，应使用有后备保护绳或速差自锁器的双控背带式安全带，安全带和保护绳应分挂在杆塔不同部位的牢固构件上。

③ 上横担前，应检查横担腐蚀情况、连接是否牢固，检查时安全带（绳）应系在主杆或牢固的构件上。

④ 在人员密集或有人员通过的地段进行杆塔上作业时，作业点下方应按坠落半径设围栏或其他保护措施。

⑤ 杆塔上下无法避免垂直交叉作业时，应做好防落物伤人的措施，作业时要相互照应，密切配合。

⑥ 杆塔上作业时不得从事与工作无关的活动。

在杆塔上使用梯子或临时工作平台，应将两端与固定物可靠连接，一般应由一人在其上作业。雷电时，禁止线路杆塔上作业。

（4）国家电网公司生产现场作业"十不干"。

① 无票的不干。

在电气设备上及相关场所的工作，正确填用工作票、操作票是保证安全的基本组织措施。无票作业易造成安全责任不明确、保证安全的技术措施不完善、组织措施不落实等问题，进而造成管理失控发生事故。倒闸操作应有调控值班人员、运维负责人正式发布的指令，并使用经事先审核合格的操作票；在电气设备上工作，应填用工作票或事故紧急抢修单，并严格履行签发许可等手续，不同的工作内容应填写对应的工作票；动火工作必须按要求办理动火工作票，并严格履行签发、许可等手续。

② 工作任务、危险点不清楚的不干。

在电气设备上的工作（操作），做到工作任务明确、作业危险点清楚，是保证作业安全的前提。工作任务、危险点不清楚，会造成不能正确履行安全职责、盲目作业、风险控制不足等问题。倒闸操作前，操作人员（包括监护人）应了解操作目的和操作顺序，对操作指令有疑问时应向发令人询问清楚，确认无误后再执行。持工作票工作前，工作负责人、专责监护人必须清楚工作内容、监护范围、人员分工、带电部位、安全措施和技术措施，清楚危险点及安全防范措施，并对工作班成员进行告知交底。工作班成员工作前要认真听取工作负责人、专责监护人交代，熟悉工作内容、工作流程，掌握安全措施，明确工作中的危险点，履行确认手续后方可开始工作。检修、抢修、试验等工作开始前，工作负责人应向全体作业人员详细交代安全注意事项，交代邻近带电部位，指明工作过程中的带电情况，做好安全措施。

③ 危险点控制措施未落实的不干。

采取全面有效的危险点控制措施，是现场作业安全的根本保障，分析出的危险点及预防措施也是"两票""三措"中的关键内容，在工作前向全体作业人员告知，能有效防范可预见性的安全风险。运维人员应根据工作任务、设备状况及电网运行方式，分析倒闸操作过程中的危险点，并制定防控措施，操作过程中应再次确认落实到位。工作负责人在工作许可手续完成后，组织作业人员统一进入作业现场，进行危险点及安全防范措施告知，结束后全体作业人员签字确认。全体人员在作业过程中，应熟知各方面存在的危险因素，随时检查危险点控制措施是否完备、是否符合现场实际，危险点控制措施未落实到位或完备性遭到破坏的，要立即停止作业，按规定补充完善后再恢复作业。

④ 超出作业范围未经审批的不干。

在作业范围内工作，是保障人员、设备安全的基本要求。擅自扩大工作范围、增加或变更工作任务，将使作业人员脱离原有安全措施保护范围，极易引发人身触电等安全事故。增加工作任务时，如不涉及停电范围及安全措施的变化，现有条件可以保证作业安全，经工作票签发人和工作许可人同意后，可以使用原工作票，但应在工作票上注明增加的工作项目，并告知作业人员。如果增加工作任务时涉及变更或增设安全措施，应先办理工作票终结手续，然后重新办理新的工作票，履行签发、许可手续后，方可继续工作。

⑤ 未在接地保护范围内的不干。

在电气设备上工作，接地能够有效防范检修设备或线路突然来电等情况。未在接地保护范围内作业，如果检修设备突然来电或邻近高压带电设备存在感应电，容易造成人身触电事故。检修设备停电后，作业人员必须在接地保护范围内工作。禁止作业人员擅自移动或拆除接地线。高压回路上的工作，必须要拆除全部或一部分接地线后才可进行工作，工作前应征得运维人员的许可（根据调控人员指令装设的接地线，应征得调控人员的许可），工作完毕后立即恢复。

⑥ 现场安全措施布置不到位、安全工器具不合格的不干。

悬挂标示牌和装设遮栏（围栏）是保证安全的技术措施之一。标示牌具有警示、提醒作用，不悬挂标示牌或悬挂错误存在误拉合设备、误登、误碰带电设备的风险。围栏具有阻隔、截断的作用，如未在工作地点四周装设至出入口的围栏、未在带电设备四周装设全封闭围栏或围栏装设错误，则存在误入带电间隔，将带电体视为停电设备的风险。安全工器具能够有效防止触电、灼伤、坠落、摔跌等，保障工作人员人身安全。合格的安全工器具是保障现场作业安全的必备条件，使用前应认真检查有无缺陷，确认试验合格并在试验期内，拒绝使用不合格的安全工器具。

⑦ 杆塔根部、基础和拉线不牢固的不干。

近年来，公司系统多次发生因倒塔导致的人身伤亡事故，教训极为深刻。确保杆塔稳定性，对于防范杆塔倾倒造成作业人员坠落伤亡事故十分关键。作业人员在攀登杆塔作业前，应检查杆根、基础和拉线是否牢固，铁塔塔材是否缺少、螺栓是否齐全、匹配和紧固。铁塔组立后，地脚螺栓应随即加垫板并拧紧螺母及打毛丝扣。新立的杆塔应注意检查杆塔基础，若杆基未完全牢固，回填土或混凝土强度未达标准或未做好临时拉线前，不能攀登。

⑧ 高处作业防坠落措施不完善的不干。

高坠是高处作业最大的安全风险，防高处坠落措施能有效保证高处作业人员人身安全。高处作业均应先搭设脚手架、使用高空作业车、升降平台或采取其他防止坠落措施方可进行。在没有脚手架或在没有栏杆的脚手架上工作，高度超过 1.5 m 时，应使用安全带，或采取其他可靠的安全措施。在高处作业过程中，要随时检查安全带是否拴牢。高处作业人员在转移作业地点过程中，不得失去安全保护。

⑨ 有限空间内气体含量未经检测或检测不合格的不干。

有限空间进出口狭小，自然通风不良，易造成有毒有害、易燃易爆物质聚集或含氧不足，在未进行气体检测或检测不合格的情况下贸然进入，可能造成作业人员中毒及有限空间燃爆事故。电缆井、电缆隧道、深度超过 2 m 的基坑、沟（槽）内等工作环境比较复杂，同时又是一个相对密闭的空间，容易聚集易燃易爆及有毒气体。在上述空间内作业，为避免中毒及氧气不足，应排除浊气，经气体检测合格后方可工作。

⑩ 工作负责人（专责监护人）不在现场的不干。

工作监护是安全组织措施的最基本要求，工作负责人是执行工作任务的组织指挥者和安全负责人，工作负责人、专责监护人应始终在现场认真监护，及时纠正不安全行为。作业过程中专责监护人临时离开时，应通知被监护人员停止工作或离开工作现场，专责监护人必须长时间离开工作现场时，应变更专责监护人。工作期间工作负责人若因故暂时离开工作现场时，应指定能胜任的人员临时代替，并告知工作班成员；工作负责人必须长时间离开工作现场时，应变更工作负责人，并告知全体作业人员及工作许可人。

（二）电缆线路检修的安全措施

（1）电力电缆停电检修作业应填写作业票。

（2）作业前，必须详细核对电缆名称，其标示牌是否与作业票所写的相符，特别是多根电缆一并敷设时，必须分清哪条是检修电缆。

（3）电缆的检修作业，不论是移动位置、拆除改装、更换接头或电缆终端头，必须在停电下进行。由于电缆的特殊性，除了采用一般的安全措施外，作业前，没有确切验明是否有电时，一律按有电对待。

（4）对于挖掘电缆或接头盒的作业，必须从始端按照标志桩逐步进行，不得用金属工具强行挖掘，以免破坏其钢甲或绝缘。

（5）对于已挖出的电缆或接头盒，当作业需要将其下面挖空以便作业时，必须将其悬挂保护；悬挂的电缆应每隔 1.0～1.5 m 悬挂一点，并保证其弯曲半径在允许范围以内，悬吊接头盒必须使其放平放牢，且不得使该接头受到拉力。移动电缆接头盒时，通常应停电进行，如必须带电移动时，应平正缓慢移动，以防绝缘损坏而发生爆炸或触电。移动中的电缆其弯曲部位必须保证其处于允许的最小弯曲半径。

（6）锯割电缆前，必须与电缆清册和电缆敷设平面图核实，并用仪器验证确切无电后才能进行。送电侧的开关必须挂警告牌且上锁，必要时应有专人看护，锯割电缆的锯刀应可靠接地，且有绝缘的手柄，操作者应戴绝缘手套并站在绝缘垫上，必要时戴护目镜。

（7）检修电缆头或电缆接头的作业应戴绝缘手套、穿绝缘靴。

（8）制作环氧树脂电缆头、调配环氧树脂时应有防毒、防火的措施。杆上制作时，应有防止坠落的措施。

（9）进入电缆隧道或入孔井进行检修作业时，应先排除井内的浊气，必要时使用通风设备。

（10）判断电缆是否有电及其规格型号时，记录不能作为绝对的依据，必须实测实量。

（11）运行的电缆有明显的外伤时，作业人员不应触及外伤部位，处理时应申请停电，没有正式停电前，不准锯割钢甲或用喷灯封补铅皮。

（12）检修电缆作业，有条件的应配备电缆漏电检测仪，以便准确判断电缆是否带电。

（13）电缆检修后必须经试验合格后，才能送电投入运行。试验前后，必须验电、放电。

（14）检修后的电缆沟、电缆隧道的保证防止水淹及防止动物进入的措施应齐全无损，否则应进行修复。雨季检修电缆挖开的电缆沟,作业完毕后应立即回填夯实;不能回填的应有防止洪水流进的措施,电缆隧道也应采用防洪措施。

(15)使用携带型火炉或喷灯作业的安全措施:

① 火焰与带电部分的安全距离:电压在 10 kV 及以下者,应大于 1.5 m;电压在 10 kV 以上者,应大于 3 m。

② 不得在带电导线、带电设备、变压器、油断路器(开关)附近以及在电缆夹层、隧道、沟洞内对火炉或喷灯加油、点火。

Task Ⅰ Electric power line operation and maintenance

Ⅰ. Electric power line classification

Electric power line refers to line for transmitting electric energy between power plants, substations and power users. As an important part of power supply system, it is for transmitting and distributing electric energy.

(Ⅰ) Based on types of erection

Electric power lines are classified into overhead line and cable line on erection type basis.

1. Overhead lines

Overhead lines are lines installed on outdoor towers to transmit electric energy. As overhead lines have advantages such as low cost, less investment, easy installation and maintenance and repair and easy fault identification and troubleshooting, they have been the first choice for electric power lines both domestically and internationally.

An overhead line is composed of main elements such as conductor, tower, insulator and line hardware. Some overhead lines have overhead grounding wires for lightning protection. To improve tower stability, stay wires or pulling piles are provided on some towers.

As the main body of a line, conductor is for transmitting electric energy. Conductors on towers must bear their own weight and various external forces and withstand erosion by harmful substances in the air. Therefore, conductors must be of high conductivity, certain mechanical strength and corrosion resistance. Conductors include three types based on materials: copper, aluminum and steel. Commonly used conductors include single-strand aluminum conductor (L), multi-strand aluminum wire (U), aluminum conductor steel reinforced (LGJ) (Fig. 7-1), single-strand copper wire (T), multi-strand copper wire (TJ) and steel strand (GJ). All of them are bare conductors with 10-300 mm^2 cross section. Aluminum conductor steel reinforced has steel core which provides higher tensile strength and compensates for low mechanical strength of aluminum wire. Aluminum wires of high conductivity are used for periphery; when AC current flows through aluminum conductor steel reinforced, it flows through aluminum due to skin effect of AC current, thus the disadvantage of low conductivity of steel wire is overcome. Bare conductors are usually used on UHV and HV overhead lines, which have high heat dissipation performance, high current carrying capacity and require less insulation materials. While MV and LV overhead lines are located in urban areas, and insulated conductors are used in recent years (as shown in Fig. 7-2). Insulated conductors have obvious advantages over overhead bare conductors: they can effectively reduce risk of electric shock, the number of trees to be felled along the line, probability of interphase short circuit caused by foreign objects; reduce wire corrosion; reduce the distance from objects considering condition of line channel and extend service life of wires and improve reliability of power supply on the lines.

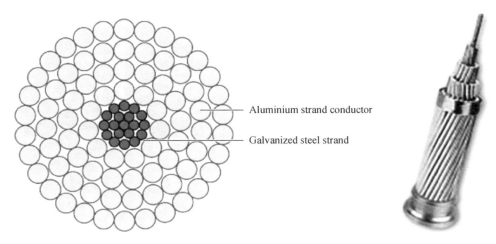

Fig. 7-1　Aluminum conductor steel reinforced

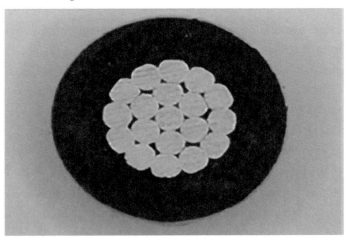

Fig. 7-2　Insulated conductor

　　10 kV insulated conductors can be installed by method of traditional overhead line installation or hung in bundles on cable wires. When two insulated conductors are installed parallelly and in close contact to each other, 80 kV voltage is applied to one conductor, and another conductor is grounded, and no breakdown occurs during the test. The experiment results show that installation in bundles is feasible.

　　At present, 10 kV overhead insulated conductors in our country are dominated by single-core (copper core or aluminum core), HDPE or XLPE insulated ones and have 10–300 mm^2 standard cross section.

　　At present, 1 kV low-voltage insulated wires in our country include copper-cored polyethylene insulated wire, aluminum-cored polyethylene insulated wire, copper-cored XLPE insulated wire, aluminum-cored XLPE insulated wire, etc. and have 16–300 mm^2 standard cross section.

2. Cable lines

　　Power cables are cables used for transmitting and distributing electric energy. Power cables are commonly used in urban underground power grids, power station lead wires, internal power supply for industrial and mining enterprises and underwater transmission lines crossing rivers and the sea. It has advantages such as small footprint, high reliability, high distributed capacitance, low maintenance workload and low possibility of electric shock. It has disadvantages such as high cost, inflexible operation, slow fault identification and complicated joint process.

With urban development, proportion of cables in electric power lines is gradually increasing. Commonly used power cables include:

(1) Oil paper insulated power cables. It includes adhesive oil immersed paper insulated cables and non-leaking oil paper insulated cables. It has advantages such as adaptability to high operating temperature, low dielectric loss, high withstand voltage and long service life. It has disadvantages such as poor bending performance and inadaptability to low temperatures which causes insulation damage. At present, it has been replaced by power cables with other insulating materials.

(2) Plastic insulated cable are Power cables with extruded plastic insulation layer. They are usually used for LV lines. Insulation materials include polyvinyl chloride and polyethylene. The former is of non-ignitibility, high chemical stability and allows easy installation and maintenance and meet requirements for high fall. But high operating temperature has impact on its mechanical properties. The latter has excellent dielectric properties and is easy to process, but it is easily deformable and combustible after heating and is prone to stress cracking.

(3) Rubber insulated cable. An insulation layer is made of rubber and additives which are thoroughly mixed and extruded onto conductive core and then heated and vulcanized. It is soft and elastic and is suitable for location where there is frequent movement and bending radius is small. Commonly used insulation materials include natural rubber-SBR mixture, ethylene propylene rubber, butyl rubber, etc. They are usually used for LV lines. The materials has high electrical and mechanical properties and chemical stability and poor resistance to corona, heat, oil and ozone.

(4) XLPE Insulated cables. They are cable that can be used for power grids of all voltage classes. Insulating medium material is made through cross linking polyethylene molecules by high-energy irradiation or chemical methods so that they allow high temperature, have high current-carrying capacity, excellent dielectric performance, light weight and are firm and can be used in harsh environments with high drop. But they have poor resistance to corona and low free radioactivity.

(Ⅱ) Classification on purpose basis

Electric power lines can be classified into transmission lines and distribution lines on purpose basis.

(1) Distribution lines are lines for transmitting electric power from step-down substations to distribution transformers or from distribution substations to consumers. They include overhead lines and cable lines and their auxiliary equipment in 35 kV and lower distribution networks. Common distribution line voltage class include 35 kV, 10 kV and 380 V.

(2) A transmission line is for transmitting electric energy and connecting power plants and

substations to allow parallel operation, electric power system interconnection and power transmission between power systems. HV transmission lines are the main arteries of electric power industry and important part of electric power system. Transmission lines in China have 110 kV, 220 kV, 330 kV, 500 kV, 750 kV, ±800 kV and 1,000 kV voltage classes.

II. Compositions of an electric power line

(I) Compositions of an overhead line

An overhead line consists of conductors and lightning wires (overhead grounding wires), towers, insulators, hardware, tower foundations, stay wires and grounding devices. Roles of the parts:

(1) Conductor: an element for conducting current and transmitting electric energy.

(2) Lightning wire: (also known as overhead grounding wire) a device for protection equipment, prevent lightning strike, protection buildings or instruments.

(3) Pole and tower: for supporting transmission lines.

(4) Insulator: for electric insulation and mechanical fixation.

(5) Hardware: for supporting, fixing and connecting conductors and insulators into strings; for protection conductors and insulators (Fig. 7-3), strain clamp, parallel groove clamp and suspension clamp and counterweight.

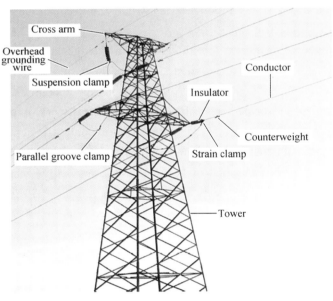

Fig. 7-3 Compositions of an overhead line

(6) Tower foundation: for stabilizing towers and prevent them from up-pull, sinking or tilting due to vertical loads, horizontal loads, emergency disconnection tension and external forces.

(7) Stay wire: for balancing lateral load and conductor tension acting on towers, reducing consumption of tower materials and reducing cost of the line.

(8) Grounding device: for reducing overhead potential and protecting line insulation from breakdown and flashover.

A power cable is composed of core, insulation layer, shielding layer and protective layer.

(1) As a conductive part of a power cable, the core is for transmitting electric energy and is the main part of a power cable.

(2) Insulation layer is an indispensable part in power cable structure, which electrically isolates cores from the ground and cores of different phases and ensures transmission of electric energy.

(3) Shielding layer is for equalizing conductor core and insulating electric field. 6kV and higher power cables usually have conductor shielding. Some LV cables do not have shielding layer. Shielding layers include semi-conductive shielding and metal shielding.

(4) Protective layer is for preventing external impurities and moisture and protecting power cables from damage by external force.

III. Safety requirements for electric power lines

(I) Basic conditions of operators

(1) They are free of disease that hinders their work (physical examination is required at least once every two years).

(2) They have necessary safety production knowledge, know methods of emergency rescue, especially first aid against electric shock.

(3) They must attend education on safe production knowledge and skill training so as to know well electrical knowledges and operating skills for power distribution; they must know well related sections of the Regulation depending on natures of their work and take up the job after exam.

(4) External organizations or personnel involved in electrical work on system of the company shall be familiar with this Regulation; They can attend work only after exam and recognized by equipment operation and maintenance management organization.

(5) The operators shall be informed of hazard factors, preventive measures and emergency accident settlement measures for the work site and operating posts. Before start of work, the equipment operation and maintenance management organization shall inform the workers of on-site electrical equipment wiring, dangerous points and safety precautions.

(6) Operators shall wear safety helmets before start of work on site, and on-site operators shall wear cotton long sleeved work clothes and insulating shoes.

(7) Doors shall be closed after the workers enter and leave power distribution substations and switching stations.

(8) Opening HV distribution equipment cabinet doors, box covers, sealing plates, etc. for enclosing live parts by operators without authorization is forbidden.

(9) Operators shall attend annual exam on this Regulations. Those who have suspended electrical work for three consecutive months or longer time for some reasons shall restudy this

Regulation and pass exams before return to work.

(10) New electrical workers, interns and temporary workers (management personnel, part-time workers, etc.) must attend education on knowledge of safety in production before starting work at designated work sites and they cannot work alone.

(Ⅱ) Basic conditions of distribution lines and equipment

(1) At HV system access point of a user line with multiple power supplies and self-contained power supply, obvious disconnection points are required.

(2) All power supply sides and appropriate positions (such as support points, tension rods, etc.) on insulated conductors for HV leads of pole-type transformer shall have electrical grounding rings or other electrical and grounding devices.

(3) HV power distribution equipment such as HV power distribution substations, switching stations, box-type substations and ring main units shall have anti-misoperation locking devices.

(4) Special screws and tools shall be used for sealing plate at busbar side of cabinet-type distribution equipment, and special tools shall be properly stored. Opening the cabinet when it is electrified is forbidden.

(5) Incoming and outgoing power supply sides of enclosed high-voltage distribution equipment shall have energized display device.

(6) There shall be operating instructions and status indications in Chinese on power distribution equipment operating mechanism.

(7) A distribution equipment shall have qualified grounding resistance.

(8) Electroscopes and grounding devices are advisable for ring main units and cable branch boxes and other box-type distribution equipment.

(9) There shall be mechanical indication of closing and opening positions on post mounted circuit breaker.

(10) Spare holes for cables or spare holes for busbar terminal in an enclosed composite apparatus shall be sealed with special tools.

(11) A standby compartment that has been connected to busbar shall have a name and number and shall be within jurisdiction of the dispatch and control center. Operating handles of isolating switches (knife switches) and net doors shall allow locking.

(12) After HV handcart switch is pulled out, the isolating baffle shall be reliably closed.

(Ⅲ) Safety requirements for overhead lines

1. Insulation strength

Overhead lines are installed outdoors and are greatly affected by the environment. For overhead lines, sufficient insulation strength is required to meet requirements for interphase insulation and ground insulation. In addition, requirements for operating overvoltage and grounding overvoltage shall be met, and atmospheric overvoltage shall be withstood. To meet the above requirements, overhead lines shall be erected using corresponding insulator strings considering different voltage classes, and sufficient distance between lines shall be kept.

In areas at 1,000 m or lower altitude, the minimum number of insulators required for operating overvoltage and lightning overvoltage at different voltage classes for suspension insulator strings is 2 for 10 kV voltage, 3 for 35 kV voltage, 7 for 110 kV voltage, 13 for 220 kV voltage, 17 for 330 kV voltage, 25 for 500 kV voltage and 32 for 750 kV voltage respectively.

2. Mechanical strength

Overhead lines shall have high mechanical strength. On one hand, they must bear tensile force generated by their own gravity, and on the other hand, they must withstand loads such as wind, rain, snow, and icing. Therefore, overhead conductors shall have adequate cross sectional areas, a conductor usually has no less than 2.5 safety factor of mechanical strength. That is to say, cross-sectional current of conductors shall be 2.5 times higher than load current.

3. Conductivity

The most important function of a conductor is conducting and transmitting electric energy. According to the requirement for conductivity, cross-sectional area of a conductor must accommodate operating heating and operating voltage loss. Heating during operation is controlled by the maximum continuous load current. When load current is excessively high and the conductor overheats, conductor fusing may occur which will lead to power outage or electric fire. Therefore, when the line is in operation, operating temperature shall be monitored to ensure that it does not exceed the specified value (no more than 70 °C for bare conductors and rubber insulated conductors and no more than 65 °C for plastic insulated conductors in general). Operating voltage loss means voltage drop on the line during operation. When there is excessive line voltage drop, electrical equipment cannot have qualified voltage and operate normally, which may also cause accident. In selecting cross sectional area of a conductor, economic current density, safe current-carrying capacity, voltage loss and mechanical strength shall be considered.

(Ⅳ) Safety requirements for cable lines

Buried cables shall meet the following requirements:

(1) When cable lines intersect one another, HV cable shall be below LV cable. If one of the cables is protected by conduit or separated by baffle plate within 1m in front of and behind the intersection, the minimum allowable distance is 0.15 m.

(2) When cables are close to or intersect thermal pipelines and insulation measures are taken, the minimum distances for parallel and cross laying are 0.5 m and 0.15 m respectively.

(3) Protective conduit is required at cable-railway or road crossings, and protective conduit shall run out the track or pavement by 2 m.

(4) The distance between a cable and a building foundation shall allow cable laying outside building apron; A cable leading into a building shall be protected by conduit, and the protective conduit shall run out building apron.

(5) The distance between cables directly buried underground and grounding of general

grounding devices shall be 0.15-0.5 m; Burial depth of a buried cable is generally no less than 0.7 m, and cables shall be buried below frozen soil layer.

And in order to ensure safe operation of cables, cable channels are to be provided as shown in Fig. 7-4. The supports shall be firm and free of rust; cables shall be neatly placed and firmly fixed; all cables shall be flame-retardant or have fireproof trays; holes where cables pass through must be sealed with fireproof blocking materials; fire isolation measures shall be taken at regular intervals along cable trenches; and firewall body shall be intact and well sealed.

Fig. 7-4 Cable channels

Ⅳ. Electric power line routine inspection

(Ⅰ) Power line routine inspection

(1) Regular routine inspection means inspection by line inspectors at certain time intervals, including routine inspection of line equipment (line itself and auxiliary equipment) and line protection areas (line channels).

(2) Fault inspection refers to line inspection organized by the operator to identify fault points, causes and faults on the line.

(3) Special routine inspection is line inspection by special routine inspection methods in special cases or as required. Special routine inspection includes night inspection, cross inspection, inspection on poles, routine inspection for preventing external force damage and aerial inspection by UAV.

(4) Supervisory inspection is generally organized by leaders of operation and inspection office or the technician for line works of the company for learning situations of the main body and channels, identifying problems and proposing countermeasures and for checking quality of inspection personnel. They are generally implemented through on-site and GPS positioning or patrol inspection system background.

(5) Night routine inspection is for identifying defects and abnormal situations that are non-detectable in daytime inspection. Generally, inspections focus on nodal "thermal" defects in line equipment and abnormal spark discharge, insulator discharge, conductor corona, etc. on the components.

(Ⅱ) Requirements for overhead lines routine inspection

Routine inspection of lines is for timely identification of equipment defects and fault points through careful routine inspection along the lines, and records shall be kept which will be basis for line maintenance.

In order to carry out periodic routine inspections and quickly identify problems, an entire line can be divided into a few inspection sections considering line length, administrative area where the line is located, the terrain along the line and the need for easy operation. Power supply department for the line will be responsible for inspection and maintenance and operation. Sections for routine inspection shall be clearly defined, and presence of unmanned blank spot is not allowed. Line routine inspection shall be carried out periodically by experienced personnel.

1. Line attendants shall meet the following requirements.

(1) They must know well equipment operating status and characteristics on the line for routine inspection.

(2) They must know well line design and construction so that abnormity can be identified.

(3) They must know well basic technical knowledge about the line and be familiar with relevant regulations.

(4) They must know well faults and abnormal conditions that occur during line operation and preventive measures to be taken.

(5) They must know well terrain, topography, transportation roads and villages and towns distribution in the areas along the line where towers are provided to allow passage of vehicles and arrival at the maintenance location for repair as soon as possible.

(6) They must learn rules of meteorological changes along the line so that preventive measures can be taken to prevent faults in different seasons.

2. As specified in *Electric Power Safety Working Regulations of State Grid Corporation of China,* line routine inspection shall meet the following requirements:

(1) Routine inspection shall be carried out by personnel experienced in power distribution. Routine inspection by a single person shall be approved and declared by the work area.

(2) Routine inspection in cable tunnels, remote mountainous areas, at night, accident conditions or adverse weather requires two persons as a group.

(3) Inspectors shall wear insulating shoes in normal routine inspection; Inspectors shall wear insulating boots or insulating shoes in routine inspection in rainy, snowy and windy days or in emergencies; Necessary protective equipment, self-rescue equipment and medicines shall be provided for line patrol in bad weather such as flood season, hot summer days and snowy days and in mountainous areas; Sufficient lighting equipment is required for line patrol at night.

(4) For line patrol in windy weather, the inspector shall move forward along windward side of the line to avoid touching broken conductors. In emergency routine inspection, the line shall always be considered alive, and safe distance must be kept. Line patrol at night shall be along outside of the line. Swimming is prohibited during line patrol.

(5) Line patrol during thunder and lightning is forbidden.

(6) In the event of an earthquake, typhoon, flood, mudslide, and other disasters, routine inspection of the affected site is forbidden.

(7) When distribution line and equipment inspection is required after a disaster, approval from equipment operating and management organization is required. Inspectors shall keep in touch with the dispatching department.

(8) Climbing up towers and distribution transformer racks for routine inspection by a single person is forbidden.

(9) When high-voltage distribution lines, equipment grounding or high-voltage conductors, cables are found to have fallen to the ground or suspended in the air during routine inspection, indoor personnel shall stay 4 m away from the fault point, and outdoor personnel shall stay 8 m away from the fault point. They shall report to the dispatch and control center and superiors for handling without delay. Before handling, efforts shall be made to prevent people from approaching the grounding or disconnection points to avoid damage by step voltage. Persons entering the above locations shall wear insulating boots, and those contact metal housing of the equipment shall wear insulating gloves.

(10) Inspectors are shall not remove or pass over barrier alone no matter if HV distribution lines and equipment are energized; when removing the barrier is required, someone shall be assigned for supervision, and safe distance specified in Tab. 7-1 shall be kept.

Tab. 7-1 Safe distance for HV lines and equipment at charged state

Voltage class/kV	Safe distance/m	Voltage class/kV	Safe distance/m
10 and lower	0.7	330	4.0
20, 35	1.0	500	5.0
66, 110	1.5	750	8.0
220	3.0	1,000	9.5

Note: when no voltage is listed in the table, safe distance for higher voltage class shall be applied. At 750 kV voltage, the data is to be calibrated based on 2,000 m altitude, while at other voltage classes, the data is to be calibrated based on 1,000 m altitude.

(11) Before entering SF_6 power distribution unit room, ventilation is required.

(12) Power distribution substations, switching stations, box-type substations, etc. shall have at least three keys, one for emergency use and one for use by O&M personnel. Other keys can be available for use by approved HV equipment inspectors and approved maintenance and construction team leaders, but registration and signature are required, and the keys shall be returned immediately after routine inspection or work is completed.

(13) In LV distribution network routine inspection, touching bare live parts is forbidden.

Ⅴ. Cut trees

Safe voltage for human body is 36 V, while voltage on HV lines in open field is at least 10 kV.

Bare aluminum wire used for HV lines at 35 kV and higher voltage class does not have insulation layer, and high voltage can cause ionization in the air. Therefore, not only HV line itself, but also the areas around the HV line are dangerous areas.

HV lines are mostly bare aluminum wires. When trees under a line touch the HV line, they will have discharge effect on the trees. Long-time friction between trees and HV lines will have impact on HV line strength and cause breakage. In windy and heavy rain days, trees swaying in the wind will cause HV lines grounding or trip, giving rise to large-area blackout.

When trees come into contact with or are very close to a HV line, the HV line will ignite and discharge electricity to the trees. In heavy wind and rainy days, this may cause trip and large-area short circuit and power failure.

As specified in Article 53 of *Electric Power Law* of China, power management department shall provide signs at power facility protection zones in accordance with relevant provisions of the State Council for power facility protection. Any organization or individual shall not construct buildings or structures within the legally designated power facility protection zones that may endanger the power facilities, cultivate plants that may endanger safety of the power facilities or stack items that may endanger safety of the power facilities. Plants cultivated before the power facility protection zone is designated that impede safety of power facilities shall be trimmed or felled.

According to Article 24 of the *Regulations on the Protection of Electric Power Facilities* in China, when construction, reconstruction or expansion of power facilities will inevitably cause damage to crops, require trees and bamboos cut down or dismantling buildings and other facilities, electric power facilities construction enterprises shall provide one-time compensation in accordance with relevant national regulations. Trees and bamboos cultivated or naturally growing in legally designated power facility protection zone that may endanger safety of power facilities shall be trimmed or cut down by electric power enterprises.

Power line operators must carry out regular line inspection to learn growth of trees and bamboo in line passages, and trees and bamboos that may impede safe operation of electric power lines shall be cut down. The following safety requirements shall be met during trees and bamboos cut down:

(1) Trees and bamboos shall be trimmed and cut down under supervision.

(2) Before cutting trees near a live line, the person in charge of work shall tell all workers that the electric power line is energized; Personnel, trees and ropes shall stay at a safe distance from the conductor as specified in Tab. 7-2.

(3) No one shall stay below the trees to be cut and within the range of fallen trees.

(4) To prevent trees (branches) from falling onto the line, insulating ropes shall be used to pull them in the opposite direction of the line. The ropes shall be of sufficient length and strength to protect personnel from injury by fallen trees.

(5) Measures shall be taken to prevent trees from bouncing down and approaching power lines during hillside trees cutting.

(6) During tree cutting, measures shall be taken to prevent personal injury by insects or animals such as wasps.

(7) When climbing trees, safety belts shall be used and shall not be fastened near or above fractures of a branch to be cut. Climbing and grabbing fragile and dead branches are not allowed; climbing and grabbing unbroken trees that have been sawn or cut are not allowed.

(8) Cutting trees high above or close to live lines when wind force is over 5 is forbidden.

(9) Operations using chainsaws and electric saws shall be carried out by personnel familiar with mechanical properties and operating methods. Before use, make sure that there is no metal objects such as iron nail within the range of sawing to prevent injury by the metal object.

Tab. 7-2 Safe distance for working nearby or across other HV power lines

Voltage class/kV	Safe distance/m	Voltage class/kV	Safe distance/m
10 and lower	1.0	330	5.0
20, 35	2.5	500	6.0
66, 110	3.0	750	9.0
220	4.0	1,000	10.5

VI. Line maintenance

(Ⅰ) Safety precautions for overhead line maintenance

For overhead line interruption maintenance, the following measures shall be taken to ensure safety of personnel, equipment and power grid:

(1) On-site survey system, work ticket system, work permit system, work supervision system, and work interruption, transfer and termination system shall be implemented in strict accordance with safety organization measures for line maintenance. For site survey, job task, starting and ending pole numbers in power cut area must be ascertained, live parts to be reserved and geographical situation of work site must be recorded, and organizational, technical, and safety measures must be prepared. Type 1 work ticket for work on electric power line is to be issued, and work permit is to be obtained. Work supervisors shall be always present at the work site and provide careful supervision over safety of the staff. To resume work after interruption, all safety measures shall be checked for completeness before start of work. When work is completed, completion procedures shall be completed. After receiving a completion report from the person in charge of work, the work permitter shall go through completion procedures, make sure that all operating staff have leave the line and all grounding wires have been removed and issue an instruction for removing power plant and safety measures at substation line side and restore power transmission through the line after keeping records and checking.

(2) Outage, electricity inspection, provision of grounding wire, sign boards and safety fences shall be completed as required by technical measures for safety. For installing grounding wires, be noted that grounding wires must be provided on the branch lines that may transmit power to the line

in outage so that sudden power-on will be prevented. Possibilities of situations such as reverse power transmission by users, crossing disconnection or induced electricity and misoperation by operators on duty shall be fully considered.

(3) Requirements in *Electric Power Safety Working Regulations* shall be met.

The following behaviors shall be prohibited during work on a tower:

① Climbing a newly erected tower with insecure foundation or without temporary bracing.

② Climbing or moving on towers while carrying equipment.

③ Climbing up/down towers or gliding along towers using ropes and stay wires.

The following precautions shall be taken for work on a tower:

① When operators are climbing towers, moving or working on towers, the members they hold shall be firm and provide safety protection, measures shall be taken to prevent the safety belt from detaching from tower top or damage by sharp objects.

② When working on a tower, dual-control safety belt with backup protection rope or speed differential self-locking device shall be shall be separately hung on fixed members of different parts of the tower.

③ Before climbing a cross arm, it shall be checked for corrosion and firm connection. During the inspection, safety belt (rope) shall be attached to the main pole or a secure member.

④ For working on towers in densely populated or passable areas, fences or other protective measures shall be provided below the working point considering falling radius.

⑤ When vertical crossing operations on and below a tower are unavoidable, measures shall be taken to prevent falling objects from injuring people. During operations, mutual care and close cooperation are required.

⑥ During work on a tower, activities unrelated to the work are not allowed.

When a ladder or temporary work platform is used on a tower, both ends shall be reliably connected to the fixtures, and the work is usually completed by one person. Work on line towers in thunderstorm days is forbidden.

(4) "Ten Don'ts" of State Grid Corporation of China for on-site operations.

① Don't start work without ticket.

For work on electrical equipment and in related places, correctly completing and using work tickets and operation tickets are basic organizational measures for protecting safety. Operation without tickets may lead to unclear responsibilities for safety, incomplete technical measures for safety and inadequate implementation of organizational measures, which may result in controllable management and accident. For switching operations, formal instructions from on-duty control personnel and person in charge of operation and maintenance are required and pre-approved qualified operation tickets shall be used; For work on electrical equipment, work ticket or emergency repair sheet are required, and procedures such as issuing permits shall be gone through, and work tickets shall be filled out for different work contents; For hot work, work ticket for hot work must be obtained, and procedures such as issuance and permission shall be gone through.

② Do not start work if job task and dangerous points are unclear.

For work (operation) on electrical equipment, work tasks and dangerous points shall be ascertained, which is a prerequisite for safety in work. Unclear work tasks and dangerous points will give rise to problems such as failure to properly fulfill responsibilities for safety, blind operation and insufficient risk control. Before a switching operation, operators (including supervisors) shall know purpose and sequence of the operation. If they have any question about command for operation, they shall inquire of the command issuer and make sure that the commands are correct before start of work. Before start of work holding a work ticket, the person in charge of work and the special supervisor must know clearly job content, scope of supervision, division of work, live parts, safety measures and technical measures, dangerous points and safety precautions and inform shift team members of the same. Before start of work, shift team members shall carefully listen to instructions from the person in charge of work and the special supervisor, familiarize themselves with job content and work process, know well safety measures, dangerous points in the work and go through confirmation procedures. Before start of maintenance, emergency repair, testing and other work, the person in charge of work shall explain in detail the safety precautions to all operators, tell them nearby live parts and nearby live parts during work and take safety measures.

③ Do not start work before dangerous point control measures are ascertained.

All-round effective risk control measures provide a fundamental guarantee for safety in on-site operation. Dangerous points identified and prevention and control measures are also key contents of "two tickets" and "three measures", which will effectively prevent foreseeable safety risks if the operating staff are informed before start of work. O&M personnel shall analyze dangerous points during switching operation based on work task, equipment condition and power grid operation mode, develop prevention and control measures and make sure that the measures are put into practice during operation. After completing work permit procedures, the person in charge of work shall organize the workers to enter the work site and inform them of the dangerous points and safety precautions, and then all operating staff shall affix signature for confirmation. During operation, all personnel shall know well existing hazards in all aspects and check that control measures for dangerous points are complete and fit for site situations at any time. If dangerous points control measures are not put into practice or are damaged in completeness, the operation shall be suspended immediately and shall not be resumed before improvements are made as specified.

④ Do not start work when range of work is not approved.

Working within range of work is a basic requirement for guaranteeing personnel and equipment safety. Unauthorized expansion of work scope and increase or change of work tasks will remove operators from range of protection by original safety measures, which results in high risk of electric shock and other safety misadventures. If increase of work tasks does not result in power outage and changes in safety measures, existing conditions can ensure safety in work, and with consent of the work ticket issuer and the work permitter, the original work ticket can be used, but the increased work items shall be indicated on the work ticket and shall be informed to the operators. If increase of work tasks results in changes in or need for additional safety measures, procedures for work permit termination shall be completed and new work ticket shall be issued. Work can be resumed only after procedures for issuance and permission are completed.

⑤ Do not start work out of range of grounding protection.

When working on electrical equipment, grounding can effectively prevent situations such as sudden power restoration on equipment or lines under maintenance. If equipment under maintenance is suddenly electrified or there is induced electricity on nearby HV live equipment during operation in the range of grounding protection, electric shock accident may occur. After the equipment under maintenance is powered off, operators must work within range of grounding protection. Moving or removing grounding wires by operators without authorization is forbidden. Before start of work on HV circuit, all or some grounding wires must be removed, and permit from O&M personnel is required (permit from control personnel is required for installing grounding wire as instructed by control personnel), and the grounding wires shall be restored to original status after work.

⑥ Do not start work if the on-site safety measures are not in place and safety tools and instruments are not qualified.

Providing sign boards and barriers (fences) is one of the technical measures for safety protection. Sign boards are for warning and reminding. Failure to provide sign boards or providing wrong sign boards will give rise to risks such as closing equipment by mistake or climbing or touching live equipment by mistake. Fences are for blocking and cutting off. Failure to provide fences leading to access around a work site and failure to provide fully enclosed fences or correctly provide fences around live equipment will give rise to risks of entering electrified compartments by mistake and considering electrified body as shut down equipment. Safety tools and instruments can effectively prevent electric shock, burn, falling, trip, etc. and protect safety of workers. Qualified safety tools and instruments are essential conditions for safety in site work. Before use, make sure that they are free of defect, qualified through test and are within test period, and use of unacceptable safety tools and instruments is not allowed.

⑦ Do not start work when tower root and foundation and stay wire are not firm.

In recent years, there are multiple casualty accidents due to tower collapses, and painful lessons are learnt from the accidents. Stability of towers is crucial for preventing accidents in which workers fall or suffer injuries due to tower toppling. Before climbing a tower, operators shall check that tower root, foundation and stay wire are firm, tower materials are complete and bolts are complete, matched and tightened. After a tower is erected, pad shall be provided for foundation bolts immediately and nuts shall be tightened and threads shall be roughened. Foundation of a new tower shall be checked. Before the foundation is not completely firm, backfilled earth or concrete strength conforms to standards or temporary bracing is provided, climbing is not allowed.

⑧ Dot not start work when protection against falling from height is incomplete.

Falling from height is the highest safety risk in work at height, and falling prevention measures can effectively protect safety of operators working at height. For work at height, scaffold shall be erected, overhead working truck, lifting platform or fall preventing means shall be used. When working without scaffold or working on a scaffold without handrails at over 1.5 m height, safety belt or other reliable safety precautions shall be used. During work at height, safety belt shall be frequently checked for tightness. When high-altitude workers are shifting to another operating location, safety protection shall always be available.

⑨ Do not start work if the confined space has not undergone gas content test or fails the test.

A confined space has narrow entrance and exit and poor natural ventilation, and toxic, harmful, flammable and explosive substances may gather or oxygen content may be insufficient. Operators entering a space without undergoing gas detection or entering a unqualified space may suffer poisoning, and explosion may occur in such spaces. Work environments in cable shafts, cable tunnels, foundation pits with more than 2 m depth and ditches (trenches) are complicated, and the spaces are enclosed spaces where flammable, explosive and toxic gases may gather. To avoid poisoning and oxygen deficit during work in the above spaces, foul smells shall be removed and work can be started only if the spaces are qualified through gas detection.

⑩ Do not start work the person in charge of work (special supervisor) is not on the site.

Work supervision is the most basic requirement in safety organization measures. The person in charge of work is the organizational commander and safety officer for the work task. The person in charge of work and the special supervisor shall always be on the site to provide strict supervision and correct unsafe behaviors in a timely manner. When the special supervisor temporarily leaves the site during work, the supervised personnel shall be instructed to suspend work or leave the work site; if the special supervisor has to leave the work site for a long time, a new special supervisor shall be assigned. If the person in charge of work temporarily leaves the work site for a reason during work, competent personnel shall be assigned to temporarily act on behalf, and shift team members shall be informed; if the person in charge of work has to leave the site for a long time, another person in charge of work shall be assigned, and all operators and work permitter shall be informed of the situation.

(Ⅱ) Safety measures for cable line maintenance

(1) For power cable interruption maintenance, operating tickets shall be filled out.

(2) Before start of work, cable designations and sign boards must be checked against the work ticket; and when multiple cables are laid together, maintenance cable must be identifiable.

(3) Cables maintenance (shifting, removing, retrofitting or replacing joints or cable sockets) must be carried out at power-off state. Considering the special nature of cables, in addition to general safety precaution, the cables shall be considered as being at charged state if it is not sure whether they are energized.

(4) Excavation for cables or junction boxes must be carried out step by step from the starting end as indicated by the marker pegs, and use of metal tools for excavation is not allowed, otherwise the steel armour or insulation may be damaged.

(5) Cables or joint boxes for which excavation has been completed must be hung for protection if the area below them is to be excavated to allow easy operation; Suspended cables shall be hung at intervals of 1.0–1.5 m and have bending radius within the allowable range. Suspended joint box

must be placed flat and firmly, and the joints must be free from tension. Before moving a cable joint box, power interruption is usually required. When moving a cable joint box at charged state is required, it shall be moved slowly to prevent insulation damage which may result in explosion or electric shock. Bent portion of a moving cable must have the allowable minimum bending radius.

(6) Before cutting a cable, cable list and cable laying plan shall be checked, and instruments shall be used to verify that there is no electricity. Switches at power supply side must have a warning sign and locked, and person shall be specially assigned for watching as required. The saw blade for cutting cables shall be reliably grounded and have insulating handle. Operators shall wear insulating gloves and stand on insulating cushion and wear safety glasses as required.

(7) Insulating gloves and insulating boots shall be worn in operations such as cable heads or cable joints maintenance.

(8) In fabricating epoxy resin cable heads and mixing epoxy resin, measures shall be taken to prevent poisoning and fire. Fall preventing measures shall be taken for fabrication on poles.

(9) For maintenance in cable tunnels or manholes, foul smells shall be removed first, and ventilation equipment shall be used if necessary.

(10) In determining whether a cable is at charged state and checking its specifications and models, records cannot be used as absolute basis, and actual measurement is required.

(11) When there is obvious damage to a cable during operation, the operator shall not touch the damaged area. Before start of handling, power outage shall be requested. Before formal power outage, cutting steel armor or using spray lamp to seal and repair lead sheath is not allowed.

(12) For cable maintenance, if conditions permit, a cable leakage detector shall be provided to accurately determine whether the cable is at charged state.

(13) After cable maintenance, it must undergo test before electrification and putting into operation. Before and after a test, inspection and discharge are required.

(14) After maintenance of cable trenches and cable tunnels, measures against water flooding and animal entry shall be taken, otherwise repair is required. Cable trench excavated in rainy season for maintenance shall be backfilled and compacted immediately after work is completed; measures shall be taken to prevent flood inflow when backfilling is not possible, and flood control measures shall be taken for cable tunnels.

(15) Safety measures for working with portable stoves or blowtorch:

① The safe distance between flame and live parts shall be more than 1.5 m at 10 kV and lower voltage and more than 3 m at 10 kV and higher voltage.

② Fuel filling or igniting stoves or blowtorches nearby live conductors, live equipment, transformers, oil circuit breakers (switches) or in cable mezzanines, tunnels or trenches is not allowed.

任务二　电力线路带电作业

一、带电作业及分类

带电作业是指在高压电气设备上不停电进行检修、测试的一种作业方法。电气设备在长期运行中需要经常测试、检查和维修。带电作业是避免检修停电，保证正常供电的有效措施。

带电作业的内容可分为带电测试、带电检查和带电维修等。带电作业的对象包括发电厂和变电所电气设备、架空输电线路、配电线路和配电设备。带电作业的主要项目包括带电更换线路杆塔绝缘子、清扫和更换绝缘子、水冲洗绝缘子、压接修补导线和架空地线、检测不良绝缘子、测试更换隔离开关和避雷器、测试变压器温升及介质损耗值。

带电作业根据人体与带电体之间的关系可分为三类：等电位作业、地电位作业和中间电位作业。

等电位作业时，人体直接接触高压带电部分。处在高压电场中的人体，会有危险电流流过，危及人身安全，因而所有进入高压电场的工作人员，均应穿合格的全套屏蔽服，包括衣裤、鞋袜、帽子和手套等。全套屏蔽服的各部件之间，须保证电气连接良好，最远端之间的电阻不能大于 20 Ω，使人体外表形成等电位体。

地电位作业时，人体处于接地的杆塔或构架上，通过绝缘工具带电作业，因而又称绝缘工具法。在不同电压等级电气设备上带电作业时，必须保持空气间隙的最小距离及绝缘工具的最小长度。在确定安全距离及绝缘长度时，应考虑系统操作过电压及远方落雷时的雷电过电压。

中间电位作业是通过绝缘棒等工具进入高压电场中某一区域，但还未直接接触高压带电体，是前两种作业的中间状况。因此，前两种作业时的基本安全要求，在中间电位作业时均须考虑。

按照作业人员是否直接接触带电导体，带电作业又可分为直接作业和间接作业两种。等电位作业属于直接作业，地电位和中间电位作业均属于间接作业范畴。

二、带电作业的一般规定

以下规定适用于在海拔 1 000 m 及以下交流 10 kV（20 kV）的高压配电线路上，采用绝缘杆作业法和绝缘手套作业法进行的带电作业。其他等级高压配电线路可参照执行。

（1）参加带电作业的人员，应经专门培训，考试合格取得资格、单位批准后，方可参加相应的作业。带电作业工作票签发人和工作负责人、专责监护人应由具有带电作业资格和实践经验的人员担任。

（2）带电作业应有人监护。监护人不得直接操作，监护的范围不得超过一个作业点。复杂或高杆塔作业，必要时应增设专责监护人。

（3）工作负责人在带电作业开始前，应与值班调控人员或运维人员联系。需要停用重合闸的作业和带电断、接引线工作应由值班调控人员履行许可手续。带电作业结束后，工作负责人应及时向值班调控人员或运维人员汇报。

（4）带电作业应在良好天气下进行，作业前须进行风速和湿度测量。风力大于 5 级，或湿度大于 80%时，不宜带电作业。若遇雷电、雪、雹、雨、雾等不良天气，禁止带电作业。带电作业过程中若遇天气突然变化，有可能危及人身及设备安全时，应立即停止工作，撤离人员，恢复设备正常状况，或采取临时安全措施。

（5）带电作业项目，应勘察配电线路是否符合带电作业条件、同杆（塔）架设线路及其方位和电气间距、作业现场条件和环境及其他影响作业的危险点，并根据勘察结果确定带电作业方法、所需工具以及应采取的措施。

（6）带电作业新项目和研制的新工具，应进行试验论证，确认安全可靠，并制定出相应的操作工艺方案和安全技术措施，经本单位批准后，方可使用。

三、带电作业安全技术措施

（1）高压配电线路不得进行等电位作业。

（2）在带电作业过程中，若线路突然停电，作业人员应视线路仍然带电。工作负责人应尽快与调度控制中心或设备运维管理单位联系，值班调控人员或运维人员未与工作负责人取得联系前不得强送电。

（3）在带电作业过程中，工作负责人发现或获知相关设备发生故障，应立即停止工作，撤离人员，并立即与值班调控人员或运维人员取得联系。值班调控人员或运维人员发现相关设备故障，应立即通知工作负责人。

（4）带电作业期间，与作业线路有联系的馈线需倒闸操作的，应征得工作负责人的同意，并待带电作业人员撤离带电部位后方可进行。

（5）带电作业有下列情况之一者，应停用重合闸，并不得强送电：

① 中性点有效接地的系统中有可能引起单相接地的作业。

② 中性点非有效接地的系统中有可能引起相间短路的作业。

③ 工作票签发人或工作负责人认为需要停用重合闸的作业。禁止约时停用或恢复重合闸。

（6）带电作业，应穿戴绝缘防护用具（绝缘服或绝缘披肩、绝缘袖套、绝缘手套、绝缘鞋、绝缘安全帽等）。带电断、接引线作业应戴护目镜，使用的安全带应有良好的绝缘性能。带电作业过程中，禁止摘下绝缘防护用具。

（7）对作业中可能触及的其他带电体及无法满足安全距离的接地体（导线支承件、金属紧固件、横担、拉线等）应采取绝缘遮蔽措施。

（8）作业区域带电体、绝缘子等应采取相间、相对地的绝缘隔离（遮蔽）措施。禁止同时接触两个非连通的带电体或同时接触带电体与接地体。

（9）在配电线路上采用绝缘杆作业法时，人体与带电体的最小距离不得小于表 7-3 的规定，此距离不包括人体活动范围。

表 7-3 带电作业时人体与带电体的安全距离

电压等级/kV	10	20	35	66	110	220	330	500	750	1 000
安全距离/m	0.4	0.5	0.6	0.7	1.0	1.8	2.6	3.4	5.2	6.8

（10）绝缘操作杆、绝缘承力工具和绝缘绳索的有效绝缘长度不得小于表 7-4 的规定。

表 7-4　绝缘工具最小有效绝缘长度

电压等级/kV	有效绝缘长度/m	
	绝缘操作杆	绝缘承力工具、绝缘绳索
10	0.7	0.4
20	0.8	0.5

（11）带电作业时不得使用非绝缘绳索（如棉纱绳、白棕绳、钢丝绳等）。

（12）更换绝缘子、移动或开断导线的作业，应有防止导线脱落的后备保护措施。开断导线时不得两相及以上同时进行，开断后应及时对开断的导线端部采取绝缘包裹等遮蔽措施。

（13）在跨越处下方或邻近有电线路或其他弱电线路的档内进行带电架、拆线的工作，应制定可靠的安全技术措施，经本单位批准后，方可进行。

（14）斗上双人带电作业，禁止同时在不同相或不同电位作业。

（15）禁止地电位作业人员直接向进入电场的作业人员传递非绝缘物件。上、下传递工具、材料均应使用绝缘绳绑扎，严禁抛掷。

（16）作业人员进行换相工作转移前，应得到监护人的同意。

（17）带电、停电配合作业的项目，当带电、停电作业工序转换时，双方工作负责人应进行安全技术交接，确认无误后，方可开始工作。

四、带电作业安全工器具

（一）常用配网带电作业安全工器具

1. 绝缘帽（图 7-5）

普通安全帽的绝缘特性很不稳定，一般不能在带电作业中使用，带电作业用绝缘安全帽，采用高密度复合聚酯材料，除具有符合安全帽检测标准的机械强度外，还完全符合相关配电带电作业电气检测标准，其介电质的强度通过 20 kV 检测试验。

2. 防护眼镜（图 7-6）

10 kV 带电作业通常以空中作业为主，因此眼部的保护十分重要。正确使用安全防护眼镜能够避免阳光刺激，有效地预防因铁屑、灰沙等物飞溅而击伤眼部的危险，还可防烟雾、化学物质对眼部的刺激，以及防止因水蒸气的凝聚而对视线的影响。

图 7-5　绝缘帽

图 7-6　防护眼镜

3. 绝缘衣、裤（图 7-7）

绝缘衣、裤采用乙烯-乙酸乙烯酯共聚物（EVA）材料，绝缘性能好，机械强度适中；柔软轻便，穿着舒适；每件产品出厂前均经过严格测试；能提供全面的绝缘保护。

4. 绝缘手套（图 7-8）

绝缘手套是带电作业中作业人员最重要的人身防护用具，只要接触带电体，不论其是否在带电状态，均必须戴手套作业。绝缘手套兼备高性能的电气绝缘强度和机械强度，同时具备良好的弹性和耐久性，具有柔软的服务性能，可将手部的不适感和疲劳降低到最小程度。

图 7-7 绝缘衣

图 7-8 绝缘手套

5. 保护手套（羊皮手套，图 7-9）

柔软的皮革手套作为绝缘手套的机械保护，可防止绝缘手套被割伤、撕裂或刺穿，不可单独用作防止电击的保护。皮革保护手套需专用皮革制作，在提供足够的机械强度保护的同时，具备良好的服务性能，尺寸与绝缘手套相符，其开口顶端与橡胶绝缘手套的开口顶端保持最小清除距离。

6. 绝缘靴（鞋）（图 7-10）

绝缘靴（鞋）是 10 kV 配电网带电作业时使用的辅助绝缘安全用具，除应具备良好的电气绝缘性外，还必须应具有一定的物理机械强度，防止刺穿或磨损。长筒绝缘靴配合绝缘裤使用，可提供全面的人身绝缘安全保护，由天然脱蛋白弹性橡胶制成，具有穿着舒适、穿脱容易的优点。

图 7-9 保护手套

图 7-10 绝缘靴

（二）保管、使用和试验

带电作业工具存放应符合《带电作业用工具库房》（DL/T 974—2018）的要求。

带电作业工具使用前应进行检查，要求绝缘良好、连接牢固、转动灵活。同时，应根据工作负荷校核机械强度，并满足规定的安全系数。

运输过程中，带电绝缘工具应装在专用工具袋、工具箱或专用工具车内，以防受潮和损伤。发现绝缘工具受潮或表面损伤、脏污时，应及时处理并经试验或检测合格后方可使用。

使用时应按厂家使用说明书和现场操作规程正确使用。进入作业现场应将使用的带电作业工具放置在防潮的帆布或绝缘垫上，以防脏污和受潮。禁止使用有损坏、受潮、变形或失灵的带电作业装备、工具。操作绝缘工具时应戴清洁、干燥的手套。

带电作业工器具试验应符合《带电作业工具、装置和设备预防性试验规程》（DL/T 976—2017）的要求。带电作业遮蔽和防护用具试验应符合《配电线路带电作业技术导则》（GB/T 18857—2019）的要求。

五、配电线路带电作业案例

工作内容：绝缘手套作业法更换避雷器。

引用规范：《10 kV 配网不停电作业规范》（Q/GDW 10520—2016）。

危险点分析：

（1）作业人员在接触带电导线和换相工作前应得到工作监护人的许可。

（2）在作业时，要注意避雷器引线与横担及邻相引线的安全距离。

（3）作业中及时恢复绝缘遮蔽隔离措施。

（4）拆避雷器引线应先从与主导线或其他搭接部位拆除，防止带电引线突然弹跳。

（5）上、下传递工具、材料均应使用绝缘传递绳，严禁抛掷。

试验应符合《带电作业工具、装置和设备预防性试验规程》（DL/T 976—2017）的要求，工器具及材料准备见表 7-5。

表 7-5 工器具及材料准备

名称	单位	数量	名称	单位	数量
绝缘斗臂车	辆	1	5 000 V 绝缘电阻表	支	1
绝缘防护用具	套	2	10 kV 验电器	支	1
绝缘毯	张	20	多功能环节测试仪	台	1
绝缘毯夹	个	40	防潮苫布	块	2
绝缘传递索	根	1	个人工具	套	1
避雷器	只	3	其他绝缘遮蔽用具	—	根据需要准备

带电作业流程：

（1）现场操作前的准备。

现场复勘及查看气象条件。天气应晴好，无雷、雨、雪、雾；风力不大于 5 级；相对湿度不大于 80%。

（2）检查工器具、车辆和避雷器，工器具如图 7-11 所示。各绝缘工具进行分段绝缘检测，绝缘电阻不得低于 700 MΩ。利用绝缘斗臂车查看绝缘臂、绝缘斗绝缘是否良好。绝缘臂的有效绝缘长度应大于 1.0 m（10 kV）、1.2 m（20 kV），下端应装设泄漏电流监测报警装置。车体

图 7-11　检查工器具

应使用截面积不小于 16 mm² 的软铜线良好接地，并进行空斗试操作，确认液压传动、回转、升降、伸缩系统工作正常、操作灵活，制动装置可靠。新避雷器需查验试验合格报告，绝缘电阻不得低于 1 000 MΩ。

（3）穿戴个人防护用具及斗内准备。检查绝缘手套、绝缘安全帽等防护用具。作业人员穿戴好全套个人绝缘防护用具。将工具搬运进斗内，作业人员进斗，如图 7-12 所示。禁止绝缘斗超载工作。绝缘斗臂车应选择适当的工作位置，支撑应稳固可靠；机身倾斜度不得超过制造厂的规定，必要时应有防倾覆措施。

图 7-12　作业人员进斗

（4）设置绝缘遮蔽。

操作绝缘斗臂车进入带电区域，如图 7-13、图 7-14 所示。绝缘斗臂车操作人员应服从工作负责人的指挥，作业时应注意周围环境及操作速度。在工作过程中，绝缘斗臂车的发动机不得熄火（电能驱动型除外）。接近和离开带电部位时，应由绝缘斗中人员操作，下部操作人员不得离开操作台。绝缘斗臂车的金属部分在仰起、回转运动中，与带电体间的安全距离不得小于 0.9 m（10 kV）或 1.0 m（20 kV）。

图 7-13　绝缘斗臂车上升

斗内作业人员使用验电器按照导线→绝缘子→避雷器→横担→电杆顺序进行验电，确认无漏电现象。

图 7-14　绝缘斗臂车进入带电区域

斗内作业人员应按照"从近到远、从下到上、先带电体后接地体"的遮蔽原则对作业范围内的所有带电体和接地体进行绝缘遮蔽，如图 7-15 所示。

图 7-15　所有带电体和接地体完成绝缘遮蔽

（5）更换避雷器。

拆除避雷器上引线。斗内作业人员将绝缘斗调整至避雷器横担下适当位置，使用断线剪将近边相避雷器引线从主导线（或其他搭接部位）拆除，将引线妥善固定。

作业人员打开待更换避雷器的绝缘遮蔽，对避雷器进行更换，连接上引线。其余两相避雷器的更换按相同方法进行。三相避雷器接线器的拆除，可按由简单到复杂、先易后难的原则进行，先近（内侧）后远（外侧），或根据现场情况先两边相、后中间相。

（6）拆除绝缘遮蔽。避雷器更换完毕后，作业人员按照"从远到近、从上到下、先接地体后带电体"的原则拆除绝缘遮蔽。

（7）工作终结。绝缘斗退出工作区域，作业人员返回地面，将工器具收拾整理好，做到"工完料尽场地清"，工作终结。

（8）其他注意事项。

绝缘斗臂车支腿打在夯实牢固的地面上并用垫块或枕木垫好，绝缘臂最小有效长度为 1.0 m 及以上；绝缘遮蔽要完全，防止单相接地或相间短路；作业人员应戴好安全帽，严禁在绝缘斗臂车绝缘斗臂下方逗留；作业人员在接触带电导线和进行换相作业转移前，应得到监护人的许可；工作时应认真执行监护制度，及时纠正不安全动作，作业人员注意力应高度集中；作业线路下层有低压线路同杆并架时，如妨碍作业，应对作业范围内的相关低压线路采取绝缘遮蔽措施；拆避雷器引线应先从与主导线或其他搭接部位拆除，防止带电引线突然弹跳；在同杆架设线路上工作，与上层线路小于安全距离规定且无法采取安全措施时，不得进行该项工作；作业过程中禁止摘下绝缘防护用具。

Task II Live-wire work on electric power line

I. Classification of live-wire work

Live-wire work is a method of maintenance and testing on HV electrical equipment without power outage. Electrical equipment requires frequent testing, inspection and maintenance during long-term operation. Live-wire work is an effective way of avoiding power outage for maintenance and ensuring normal power supply.

Live-wire work include live testing, live inspection and live repair. Objectives of live-wire work include electrical equipment of power plants and substations, overhead transmission lines, distribution lines and distribution equipment. Main items of live-wire work include replacing line tower insulators, cleaning and replacing insulators, washing insulators with water, crimping and repairing conductors and overhead grounding wires, detecting faulty insulators, testing and replacing isolating switches and lightning arresters and testing transformer temperature rise and dielectric loss.

Live-wire work includes equal potential working, earth potential working and medial potential working.

During equal potential working, human body directly contacts HV live parts. There may be dangerous current flowing through a human body in HV electric field which will endanger personal safety. Therefore, all workers entering a HV electric field shall wear a whole set of qualified screening clothing, including clothing, shoes, socks, hats and gloves. There must be good electrical connection between components of a screening clothing, and resistance between the farthest ends shall be no more than 20 Ω, otherwise equipotential body may be generated at human body surface.

During earth potential working, human body is on a grounded tower or structure, and the operator is carrying out live-wire work using insulation tools. Therefore, the work is also known as insulation tool method. During live-wire work on electrical equipment at different voltage classes, the minimum clearance between air gaps and the minimum length of insulation tools must be maintained. In determining safe distance and insulation length, consideration shall be given to system operating overvoltage and lightning overvoltage during remote thunderclap.

During medial potential working, the operator enters a certain area in a HV electric field using insulating rod and does not come into contact with HV charged body. Such work is the medial operation between the above-said two operations. Therefore, the basic safety requirements for the first two types of operations must be considered during medial potential working.

Live-wire work can be further classified into direct work and indirect work depending on whether the operator coming into contact with the charged conductor. Equal potential working is a direct work, and both earth potential working and medial potential working are indirect working.

II. General provisions for live-wire work

The following provisions apply to live-wire work on AC 10 kV (20 kV) HV distribution lines at 1,000 m and lower altitudes using insulating rod and insulating gloves. They are available for reference for HV distribution lines of other classes.

(1) Live-wire work staff shall attend special training, pass exams, obtain qualifications and be approved by the organization before participating in the corresponding work. Live-wire work ticket issuer and the person in charge of work and the special supervisor shall be personnel qualified for and experienced in live-wire work.

(2) Supervision is required during live-wire work. The supervisor shall not carry out operation directly, and scope of supervision shall not exceed one work point. For complicated work or work on high towers, additional special supervisor shall be assigned as required.

(3) Before start of live-wire work, the person in charge of work shall get in touch with the control staff on duty or O&M personnel. For operations that require reclosing, disconnecting and connecting leads at charged state, procedures for permission shall be completed by the control staff on duty. After live-wire work, the person in charge of work shall report to the control staff on duty or O&M personnel.

(4) Live-wire work shall be carried out in good weather, and wind speed and humidity must be measured before start of work. When wind is over force 5 or humidity is over 80%, live-wire work is not advisable. In bad weather such as lightning, snow, hail, rain, fog, etc., live-wire work is prohibited. If there is a sudden change in weather during live-wire work that may endanger personal and equipment safety, work shall be suspended immediately, personnel shall be evacuated, equipment shall be restored to normal condition, or temporary safety measures shall be taken.

(5) For live-wire work, distribution lines shall be checked for suitability for live-wire work, lines provided on a same pole (tower) and their orientations and electrical spacing, work site conditions and environment and other dangerous points with impact on work shall be determined, and method of work, tools required and measures to be taken shall be determined considering results of the check.

(6) New live-wire work items and new tools shall undergo experimental demonstration to make sure that they are safe and reliable, and operating process and safety measures shall be developed which can be put into practice only after being approved by the organization.

III. Technical measures for safety in live-wire work

(1) Equal potential working is not allowed on HV distribution lines.

(2) If there is sudden power failure during live-wire work, the operator shall still consider the line as live line. The person in charge of work shall get in touch with the dispatch and control center or equipment operation and maintenance management organization as soon as possible, and the control personnel on duty or O&M personnel shall not restore power supply without contacting the person in charge of the work.

(3) During live-wire work, if the person in charge of work discovers or learns that relevant equipment is in fault, they shall immediately stop working, evacuate the personnel and immediately contact the control personnel on duty or O&M personnel. Upon discovery of equipment fault, the control personnel on duty or O&M personnel shall inform the person in charge of work without delay.

(4) When switching operation on the feeder in connection to the line is required during live-wire work, consent from the person in charge of work is required, and the work can be started only after live-wire work staff leave the live parts.

(5) In any of the following situations during live-wire work, reclosing shall be stopped and power transmission is not allowed.

① Work in a system with neutral point earthed that may result in single-phase grounding.

② Work in a system in which the neutral point is not effectively earthed which may result in interphase short circuit.

③ Work in which reclosing deactivation is required in the opinion of the work ticket issuer or the person in charge of work. Disabling or resuming reclosing at scheduled time is forbidden.

(6) During live-wire work, insulation protective equipment (such as insulating clothing or insulation shawls, insulating sleeves, insulating gloves, insulating shoes, insulating safety helmets, etc.) shall be used. In lead disconnection or connection at charged state, safety glasses shall be used, and safety belts used shall be of high insulation performance. During live-wire work, removing the insulation and protection devices is forbidden.

(7) Insulation and shielding measures shall be taken for other live objects that may be touched during operation and grounding bodies that do not allow safe distance (conductor supports, metal fasteners, cross arms, stay wires, etc.).

(8) For charged bodies, insulators, etc. in work area, interphase and phase-ground insulation and isolation (shielding) measures shall be taken. Touching two non-connected charged bodies or a charged body and a grounding body simultaneously is forbidden.

(9) During work on a distribution line by insulating rod method, the minimum distance between human body and a charged body shall be no less than that specified in Tab. 7-3 and does not include range of human body movement.

Tab. 7-3 The safe distance between human body and a charged body during live-wire work

Voltage class/kV	10	20	35	66	110	220	330	500	750	1,000
Safe distance/m	0.4	0.5	0.6	0.7	1.0	1.8	2.6	3.4	5.2	6.8

(10) Effective length of insulation of insulating bar, insulation load-bearing tools and insulation ropes shall be no less than that specified in Tab. 7-4.

Tab. 7-4　The minimum effective length of insulation of insulation tool

Voltage class/kV	Effective insulation length /m	
	Insulating bar	Insulation load-bearing tools, insulation ropes
10	0.7	0.4
20	0.8	0.5

(11) During live-wire work, use of non-insulated ropes (e.g., cotton yarn ropes, white palm ropes, steel wire ropes, etc.) is not allowed.

(12) For operations such as replacing insulators, moving or disconnecting conductors, backup protection measures are required to prevent wire detachment. Disconnecting conductors at two and more phases simultaneously is not allowed, and ends of a disconnected conductor shall be wrapped with insulation for shielding.

(13) Reliable safety and technical measures shall be developed for removing charged racks and conductors below a crossing or in an area nearby live lines or other weak current lines, which can be put into practice only after approval by the organization.

(14) During live-wire work by two person on an insulated arm, work at different phases or different potentials simultaneously is forbidden.

(15) Passing non-insulated objects to an operator entering electric field by earth potential working staff is forbidden. Insulating rope is required for transmitting tools and materials upward and downward, and throwing is strictly prohibited.

(16) Before an operator shifts to another phase, consent from the supervisor is required.

(17) During work in which live-line work and black-out work are coordinated, for shifting between live-line work and black-out work, the person in charge of work shall provide technical and safety clarification and make sure that everything is going well before start of work.

Ⅳ. Safety tools and instruments for live-wire work

(Ⅰ) Commonly used safety tools and instruments for live-wire work on distribution networks

1. Insulating cap (Fig. 7-5)

As ordinary safety helmets have very unstable characteristics, they are used in live-wire work in general. Insulating safety helmets for live-wire work are made of high-density composite polyvinyl acetate materials shall be used. They shall not only have mechanical strength conforming to safety helmet test standards, but also have dielectric strength withstanding 20 kV inspection test.

2. Goggles (Fig. 7-6)

As 10 kV live-wire work is usually aerial work, eye protection is very important. Correct use of safety goggles can avoid sunlight stimulation, effectively prevent risk of eye injuries caused by splashes of iron filings, dust and other objects and prevent eyes irritation by smoke and chemicals and impact of steam condensation on vision.

Fig. 7-5　Insulating cap

Fig. 7-6　Goggles

3. Insulating clothes and trousers (Fig. 7-7)

EVA material with high insulation performance and moderate mechanical strength are to be used; soft and lightweight and comfortable clothing shall be worn; every product has undergone strict test before leaving the factory; they can provide full insulation protection.

4. Insulating gloves (Fig. 7-8)

Insulating gloves are the most important personal protective equipment for live-wire work staff. Whenever an operator comes into contact with a charged body, regardless of whether it is at charged state, insulating gloves must be used. Insulating gloves have high-performance and electrical insulation strength and mechanical strength as well as high elasticity and durability. They provide soft service performance and can minimize discomfort and fatigue in the hands.

Fig. 7-7　Insulating clothes

Fig. 7-8　Insulating gloves

5. Protective gloves (cape gloves) (Fig.7-9)

Soft leather gloves can provide mechanical protection for insulating gloves, preventing them from cut, tear or piercing. But they shall not be used alone for protection against electric shock. Leather protective gloves must be made of special leather, which not only provides sufficient mechanical strength, but also provides high service performance and have sizes matching insulating gloves, and top of the opening shall maintain the minimum clearance from top of rubber insulating gloves.

6. Insulating boots (shoes) (Fig. 7-10)

Insulating boots (shoes) are auxiliary insulation safety appliances used during live-wire work on 10 kV distribution networks. In addition to good electrical insulation, they must have certain physical and mechanical strength to prevent piercing or wear. Long insulating boots using with insulating pants can provide all-round insulation and safety protection. They are made of natural deproteinized elastic rubber and have advantages of comfortableness and are easy to put on and take off.

Fig. 7-9　Protective gloves

Fig. 7-10　Insulating boots

(Ⅱ) Storage, use and testing

Live-wire work tools shall be stored as required in *Depot of Tools for Live-working* (DL/T 974－2018).

Live-wire work tools shall be checked before use for intact insulation, firm connection and flexible rotation. And mechanical strength shall be checked according to working load and shall conform to specified safety factors.

During transportation, electrified insulation tools shall be packed in special tool bags, toolboxes or special tool cars to prevent moisture and damage. Insulation tools that are damped, damaged or have contaminated surface shall be promptly treated and tested or qualified before use.

The tools shall be used correctly according to instructions from the manufacturer and on-site operating procedures. Before entering the work site, live-wire work tools to be used shall be placed on moisture-proof canvas or insulation cushion to prevent contamination and damp. Use of live-wire work equipment and tools that are damaged, damp, deformed or malfunctioning is forbidden. Clean and dry gloves shall be worn for operating insulation tools.

Test of tools and instruments for live-wire work shall conform to *Preventive Test Code of Tools, Devices and Equipment for live-Working* (DL/T 976－2017). Test of shielding and protective equipment for live-wire work shall comply with *Technical Guide for Live Working on Distribution Line* (GB/T 18857－2019).

Ⅴ. Examples of live-wire work on distribution lines

Job content: replacing lightning arrester by insulating gloves method.

Code applied: *Criterion for Overhaul without Power Interruption in 10 kV Distribution Network* (Q/GDW 10520-2016).

Dangerous points analysis:

(1) Before coming into contact with a live conductor and phase shift, permit from the supervisor is required.

(2) During work, attention shall be paid to the safe distance between lightning arrester lead and cross arm.

(3) Insulation, shielding and isolation measures shall be restored during operation.

(4) Lightning arrester lead removal shall begin with the joints to the main conductor or other parts so that live conductor bouncing is prevented.

(5) Insulating rope is required for transmitting tools and materials upward and downward, and throwing is strictly prohibited.

Test shall conform to *Preventive Test Code of Tools, Devices and Equipment for Live-working* (DL/T 976-2017). See Tab. 7-5 for tools and instruments and materials preparation.

Tab. 7-5 Tools and instruments and materials preparation

Name	Unit	Quantity	Name	Unit	Quantity
Aerial lift device with insulated arm	Pcs.	1	5,000 V insulation resistance meter	Pcs.	1
Tools for insulation protection	Set	2	10 kV electroscope	Pcs.	1
Insulating mat	Pcs.	20	Multifunctional link tester	Pcs.	1
Insulating mat clamp	Pcs.	40	Moisture-proof tarpaulin	Pcs.	2
Insulated transmission cable	Pcs.	1	Personal tools	Set	1
Lightning arrester	Pcs.	3	Other tools for insulation shielding	—	To be prepared as required

Live-wire work process flow:

(1) Preparations before on-site operation.

On-site recheck and meteorological conditions check. The weather shall be sunny and there is no thunder, rain, snow or fog; No more than force 5 wind; No more than 80% relative humidity.

(2) Check tools, vehicles and lightning arresters. Tools and instruments are as shown in Fig. 7-11. The insulation tools shall undergo insulation testing on section basis, and insulation resistance shall be no less than 700 MΩ. Insulating arm and insulated arm of aerial lift device with insulated arm shall be check for good condition. Effective insulation length of an insulated arm shall be more

than 1.0 m (10 kV) and 1.2 m (20 kV), and leakage current monitoring and alarm devices shall be installed at the lower end. The body shall be well grounded with no less than 16mm^2 soft copper wire, and the insulated arm shall be tested at empty state to make sure that hydraulic transmission, rotation, lifting and telescopic systems are working properly, operating freely and the braking device is reliable. Conformity test report of new lightning arrester must be checked, and insulation resistance shall be no less than 1,000 MΩ.

Fig. 7-11　Checking tools and instruments

(3) Use personal protective equipment and make preparations in insulated arm. Check insulating gloves and insulating safety helmet and other protective devices. Operators shall use a complete set of insulation and protection devices. Tools shall be moved into the insulated arm, and operators shall enter the insulated arm, as shown in Fig. 7-12. Overload operation of insulated arm is forbidden. Appropriate working position shall be selected for aerial lift device with insulated arm, and the support shall be stable and reliable; body inclination shall be no more than that specified by the manufacturer, and anti-overturning measures shall be taken as required.

Fig. 7-12　Operators enter an insulated arm

(4) Insulation shielding is required.

Operate the aerial lift device with insulated arm to move it into a live-wire work area as shown in Fig. 7-13 and Fig. 7-14. Operator of aerial lift device with insulated arm shall obey command from the person in charge of work and pay attention to surrounding environment and operating speed during operation. During work, engine of aerial lift device with insulated arm shall not be shut down (except for electric powered ones). When approaching and leaving a live part, it shall be operated by personnel in the insulated arm, and operators at lower part are not allowed to leave the operating platform. Metal parts of an aerial lift device with insulated arm during lifting up and rotation shall be at no less than 0.9 m (10 kV) or 1.0 m (20 kV) safe distance from a charged body.

Fig. 7-13 Rising of aerial lift device with insulated arm

Operator in the insulated arm shall use electroscope for electricity inspection in the sequence of conductor-insulator-lightning arrester-cross arm and pole and confirms that there is no leakage.

Fig. 7-14 Aerial lift device with insulated arm entering a charged region

Operating staff in the arm shall provide insulation shielding for all charged bodies and grounding bodies in the range of work "from the near to the distant, from bottom to top and from charged body to grounding body" as shown in Fig. 7-15.

Fig. 7-15　Insulation shielding for all electrified bodies and grounding bodies

(5) Replace lightning arrester.

Remove the upper lead from lightning arrester. Operators in the insulated arm adjusts insulated arm to appropriate position below the lightning arrester cross arm; Lightning arrester lead at side phase is removed from the main lead (or other connection parts) using a conductor cutter; properly fix the leads.

The operator opens insulation shielding of the lightning arrester to be replaced to allow lightning arrester replacement and lead connection. Lightning arresters at the other two phases are to be replaced in the same way. In removing three-phase lightning arrester connectors, removal of the simple ones shall be followed by the complex ones, easier work shall be followed harder work and removal of the farther (inside) shall be followed by the nearer (outside), or removal of two edge phases shall be followed by removal of intermediate phase depending on the site situations.

(6) Remove Insulation shielding. After replacing lightning arrester, the operators shall remove insulation shielding in the sequence of "from far to near, from top to bottom and from grounding body to charged body".

(7) End of work. The insulation arm leaves the work area, the operators return to the ground, tidy up the tools and instruments to ensure that the site is cleared after work is completed, and the work is completed.

(8) Other precautions.

Outriggers of the aerial lift device with insulated arm shall be placed on compacted and firm ground and leveled with blocks or sleepers. The minimum effective length of an insulated arm shall be no less than 1.0 m; Insulation shielding shall be complete so that single-phase grounding or interphase short circuit can be avoided; the operators shall wear safety helmets, and staying under insulated arm of aerial lift device with insulated arm is strictly forbidden; Before coming into contact with live conductors and phase shift by operators, permit from the supervisor is required. During the work, the supervision system shall be conscientiously implemented, unsafe actions shall be corrected in a timely manner, and the operators shall be highly focused; When there are LV lines blow the line that share a same pole and work is impeded, insulation shielding shall be provided for the relevant LV lines within the work area; Lightning arrester lead removal shall begin with the joints to the main conductor or other parts so that live conductor bouncing is avoided. For work on lines sharing a same tower, if the safe distance from upper lines is less than the specified distance and safety measures cannot be taken, the work is not allowed; Removing insulation protection appliances during work is prohibited.

任务三　电力电缆绝缘测试实训

一、作业任务

3 人一小组，一人扮演工作负责人，两人扮演班组成员，按照现场安全要求，在布置好安全措施后，正确对电缆进行放电操作，对兆欧表进行检查，并使用兆欧表测量电缆的绝缘电阻，填写记录表，清理并结束现场工作。

二、引用标准及文件

（1）《国家电网公司电力安全工作规程（变电部分）》。
（2）《国家电网公司十八项电网重大反事故措施》。

三、作业条件

测量应在天气良好的情况下进行，且空气相对湿度不高于 80%；禁止在有雷电或邻近高压设备时使用绝缘电阻表，以免发生危险。在有接地系统的室内完成作业；作业人员精神状态良好，熟悉安全措施、技术措施以及现场工作危险点。

四、作业前准备

1. 现场试验的基本要求及条件

（1）着装准备。着全棉工作服，正确佩戴安全帽，穿绝缘软底鞋。
（2）检查安全工器具、设置安全围栏。用正确方法检查绝缘手套、10 kV 放电棒/接地线、绝缘靴等，并在工作区域设置好安全围栏，在唯一进出口悬挂"在此作业""由此进出"标示牌，向围栏外悬挂"止步，高压危险！"标示牌若干。
（3）兆欧表的选择和检查。根据电缆不同电压等级正确选择兆欧表：

① 500 V 及以下电缆，橡塑电缆的外护套及内衬层，采用 500 V 兆欧表。
② 0.6/1 kV 电缆，采用 1 000 V 兆欧表。
③ 0.6/1 kV 以上电缆，采用 2 500 V 兆欧表。
④ 6/6 kV 及以上电缆，采用 2 500 V 或 5 000 V 兆欧表。

图 7-16　兆欧表

对兆欧表进行外观检查：如图 7-16 所示，外观应良好，外壳完整，玻璃无破损，摇把灵活，指针无卡阻，接线端子应齐全完好，表线应是单芯软绝缘铜线且完好无损，其长度不应超过 5 m。

对兆欧表进行开路试验：两条线分开（L 和 E）处于绝缘状态，摇动兆欧表的手柄达 120 r/min 表针指向无限大（∞）为好。

对兆欧表进行短路试验：摇动兆欧表手柄到 120 r/min，将两只表笔瞬间搭接一下，表针指向"0"（零），说明兆欧表正常。

测试线绝缘应良好，禁止使用双股麻花线或平行线。

（4）电缆放电、装设接地线。正确选择接地点，用锉刀、砂布等工具将接地点打磨清洁，去除污迹，接线应牢靠。将电缆各相分别放电并接地。

（5）电缆支撑、清洁。按试验标准剥切电缆芯线，支撑电缆达到试验要求，擦去芯线绝缘上面的污迹。

2. 工器具及材料选择

电缆（10 kV）；兆欧表（500 V、1 000 V、2 500 V 供选择）；测量用绝缘线；10 kV 放电棒/接地线；绝缘手套；绝缘靴；安全围栏；安全标示牌"在此作业""由此进出"各一块，"止步，高压危险！"若干；绝缘垫；防水带；一字螺丝刀；计时器；温湿度计。

3. 危险点及预防措施

（1）仪表本身故障可能对试验造成安全隐患。试验前务必对仪表进行检测，确保仪器能够正常使用。

（2）若被测电缆带电，会对人身安全造成极大威胁。所以务必对电缆进行断电，试验前以及每次试验完一相，均须对电缆进行短路放电。

（3）现场测试电缆另一端情况不明，也会对试验安全造成隐患。整个试验过程中，在测试另一端增加看护人员和安全栏。

（4）作业现场为电气设备检修场地，有很多其他电气设备、设施，可能会对现场人员造成机械伤害。作业区域设置好安全围栏，现场人员着工装、戴好安全帽，按照工作负责人要求在指定范围内活动。

4. 作业人员分工

现场工作负责人（监护人）：×××

现场作业人员：×××

五、作业规范及要求

（1）非被测部位均应短路接地且接触良好；被试电缆无人工作，测试过程中禁止他人接近；派人到另一端看守或装好安全遮栏，防止有人接触被试电缆。

（2）测试前后均应对被试电缆进行充分放电；摇表停止转动前或被测电缆未放电前，严禁用手触及；拆线时，不要触及引线的金属部分。

（3）测量时摇动摇表手柄的速度均匀保持 120 r/min；读取数值后，应先断开兆欧表与被试品的连线，然后再将兆欧表停止转动，以免被试品的电容上所充的电荷经兆欧表放电而损坏兆欧表。

（4）摇测时，两测试线不能相交，红线必须架空；人体不能接触红线和接地线。

六、作业流程及标准（表7-6）

表7-6　电力电缆绝缘测试流程及评分标准

班级		姓名		学号		考评员		成绩		
序号	作业名称	质量标准			分值/分	扣分标准			扣分	得分
1			准备工作							
1.1	着装	按规程要求穿戴工作装			2	未按规程要求着装一项扣1分；着装不规范一处扣1分				
1.2	摇表、工器具、绝缘工具、安全用具	正确选择兆欧表、干湿温度计、接地线、放电棒、绝缘手套、绝缘靴、试验围栏、标示牌、绝缘垫、活动扳手、一字改刀、平口钳、榔头、钉子、防水带、试验记录表			6	选择错误每项扣2分；漏一项扣2分				
1.3	检查设备、仪表、安全用具	用正确方法检查仪表、绝缘工具、安全用具等是否有合格标示，且未过期；兆欧表检查是否能达无穷位及短路时指零			6	方法错误扣2分；漏一项扣1分				
1.4	做现场安全措施、主要危险点及安全注意事项	经许可，将测试现场按规程要求装设围栏，向外悬挂安全警示牌，接到监护人指令并大声复诵后方可开始工作；测试中主要的危险点及安全注意事项编写			10	未按照规程操作一项扣2分；未经许可、未复诵一项扣2分；未写分值全扣，未写全按每项分值扣分				
2	兆欧表的选择	根据电缆不同电压等级正确选择兆欧表：500 V 及以下电缆，橡塑电缆的外护套及内衬层，采用 500 V 兆欧表；0.6/1 kV 电缆，采用 1 000 V 兆欧表；0.6/1 kV 以上电缆，采用 2 500 V 兆欧表；6/6 kV 及以上电缆采用 2 500 V 或 5 000 V 兆欧表			6	选择错误每项扣2分				

续表

序号	作业名称	质量标准	分值/分	扣分标准	扣分	得分
3		绝缘电阻的测量				
3.1	接地线	正确选择接地点，用锉刀、砂布等工具将接地点打磨清洁，去除污迹，接线应牢靠	2	接地点不正确扣2分；未打磨、清洁、线松动扣2分		
3.2	电缆支撑、清洁	按试验标准剥切电缆芯线，支撑电缆达到试验要求，擦去芯线绝缘上面的污迹	5	未按要求剥切支撑电缆扣3分；不清洁有残留物扣2分		
3.3	电缆、表、放电棒接地	分别将电缆芯线、铠装、屏蔽层、表的"E"端、放电棒等可靠接地，根据现场情况在适当位置摆放兆欧表，正确接线	4	位置错误、接地不良扣2分；接地不正确、接线错误扣2分		
3.4	外护层绝缘电阻测量	选用500 V兆欧表；取下电缆铠装、屏蔽层的接地线；测试人员站在绝缘垫上，先将兆欧表转速摇至120 r/min，再用测试线接入铠装，读取60 s值或指针稳定值；为防止试品反充电，读完数后应先将测试线断开试品，再停止摇动兆欧表，然后对试品放电并接地（绝缘电阻不低于0.5 MΩ/km）	8	选表、接线错误、表的操作不规范扣3分；读数不正确扣3分；先停止摇动兆欧表，后将测试线断开试品，造成反充电者扣4分；放电接地操作不规范扣3分		
3.5	内衬层绝缘电阻测量	选用500 V兆欧表；取下电缆屏蔽层的接地线；测试人员站在绝缘垫上，先将兆欧表转速摇至120 r/min，再用测试线接入屏蔽层，读取60 s值或指针稳定值；为防止试品反充电，读完数后应先将测试线断开试品，再停止摇动兆欧表，然后对试品放电并接地（绝缘电阻不低于0.5 MΩ/km）	8	选表、接线错误、表的操作不规范扣3分；读数不正确扣3分；先停止摇动兆欧表，后将测试线断开试品，造成反充电者扣4分；放电接地操作不规范扣3分		

续表

序号	作业名称	质量标准	分值/分	扣分标准	扣分	得分
3.6	主绝缘电阻测量	选用 2 500 V 或 5 000 V 兆欧表；取下电缆被试芯线的接地线；测试人员站在绝缘垫上，先将兆欧表转速摇至 120 r/min，再用测试线接入被试芯线，读取 15 s 和 60 s 时的绝缘吸收比数值；为防止试品反充电，读完数后应先将测试线断开试品，再停止摇动兆欧表，然后对试品放电并接地；做其余两相测试重复以上步骤（绝缘电阻值与上次相比应无明显差别或耐压前后应无明显变化）	15	用绝缘垫、选表接线，错误扣 3 分；未达到转速或转速不稳定扣 3 分；读数不正确扣 5 分；先停止摇动兆欧表，后将测试线断开试品，造成反充电者扣 3 分；放电接地操作不规范扣 3 分；未作记录或记录不正确扣 5 分		
3.7	拆除接线	拆除所有因试验而接的引线及接地线、接地线最后拆除至接地点处	3	线未拆完扣 2 分；未按要求拆线一项扣 2 分		
3.8	将剥切的电缆作防水处理	用防水带 1/2 搭接包缠不少于来回 3 次；防水带使用时首先应去除隔离层，一般应拉伸 100%（一倍）后绕包，不能将带脱手落地，保持清洁，不松带、断带	5	未拉伸、搭接 1/2 不够各扣 3 分；跌落、层数不正确各扣 3 分；松带、断带各扣 3 分		
4			工作终结			
4.1	整理工具、设备	收拾工器具、安全用具、仪表、摆放整齐合理	2	摆放不合理扣 2 分		
4.2	清理现场	拆除现场安全遮栏、标示牌等，现场清理干净，不留任何物品，然后再向评委报告试验结束	3	有物品留下、不做清理各扣 2 分；不作结束报告扣 1 分		
4.3	试验记录	按要求作记录，字迹清楚、整洁	8	错一处扣 3 分；表达不清楚扣 3 分		

续表

序号	作业名称	质量标准	分值/分	扣分标准	扣分	得分
4.4	画兆欧表测量电缆绝缘电阻的示意图	标出各部的名称、各部绘制表达清楚，应简单清晰	7	示意图表达不清一处扣5分；名称错一处扣5分		
合计			100			

Task III　Practical training on power cable insulation test

I. Operating tasks

In a 3-person group, one person acts as work leader, and the other two persons act as group members. As required for safety on work site, when safety precautions are in place, they will start cable electro-discharge, use megohmmeter for check, use megohmmeter to measure cable insulation resistance, complete record forms and clear the site and finish site work.

II. Referenced standards and documents

(1) *Electric Power Safety Working Regulations (Power Transformation) of State Grid Corporation of China*;

(2) *18 Major Anti-accident Measures for Power Grid of State Grid Corporation of China*.

III. Operating conditions

Measurement shall be carried out in good weather, and relative humidity in the air shall be no more than 80%; Use of insulation resistance meter when there is lightning or HV equipment nearby is prohibited, otherwise danger may occur. Work shall be completed in a room with grounding system, and the operators shall be at good mental state and know well the safety measures, technical measures and dangerous points in on-site work.

IV. Preparation before operation

1. Basic requirements and conditions for on-site test

(1) Preparations for clothing. Operators shall wear cotton work clothes, wear safety helmets correctly and wear soft soled insulating shoes.

(2) Safety tools and instruments shall be checked, and safety fences shall be provided. Insulating gloves, 10 kV discharging rods/grounding wires, insulating boots, etc. are shall be checked properly, safety fences shall be provide in work area. "Work In Progress" and "Enter and exit from here" sign boards shall be provided at the only access, and "Stop! High voltage, danger!" sign boards shall be provided outside the fences.

(3) Selection and inspection of megohmmeters. Correct megohmmeter shall be selected based on different voltage classes of cables:

① For outer sheath and inner lining of 500 V and lower rubber and plastic cables, 500 V megohmmeter shall be used;

② For 0.6/1 kV cables, 1,000 V megohmmeter shall be used.

③ For 0.6/1 kV and higher cable, 2,500 V megohmmeter shall be used.

④ For 6/6 kV and higher cables, 2,500 V or 5,000 V megohmmeter shall be used.

Fig.7-16　Megohmmeter

Visual inspection on the megohmmeter: as shown in Fig. 7-16, the megohmmeter shall have good appearance, the housing shall be intact, the glass shall be free of damage, the rocker shall be flexible, the pointer shall be free of jamming, the connecting terminals shall be complete and intact, the meter wire shall be single-core soft insulated copper wire and shall be intact and has no more than 5 m length.

Open circuit test on megohmmeter: two wires are separated (L and E) and at insulated state, and setting megohmmeter handle to 120 r/min and having the pointer directing infinity (∞) are advisable.

Short circuit test on megohmmeter: turn megohmmeter handle to 120 r/min, momentarily contact the two test pens, and the megohmmeter shall be normal when the pointer directs "0" (zero).

The test wire shall be well insulated, and use of double-strand twisted wire or parallel wire is forbidden.

(4) Cable discharge and installation of grounding wires. Correct grounding points shall be selected, tools such as file or emery cloth shall be used to polish and clean the grounding points and remove stains, and the wires shall be firmly connected. Phases of the cable shall be discharged and grounded.

(5) Cable support and cleaning. Peel and cut cable cores according to test standards, provide cable support as required for the test, and wipe off any stain on core insulation.

2. Selection of tools and instruments and materials

Cable (10 kV); Megohmmeter (500 V, 1,000 V, 2,500 V available for selection); Insulated wire for measurement; 10 kV discharging rod/grounding wire; Insulating gloves; Insulating boots; safety fence; provide a "Work In Progress" and an "Enter and exit from here" safety sign board and a number of "Stop! High voltage, danger!" sign boards; insulation cushion; waterproof tape; slotted screwdriver; timer; hygrothermograph.

3. Dangerous points and preventive measures

(1) Instrument fault may result in safety hazard in the test. Before a test, the instruments shall

be check to make sure that they are available for normal use.

(2) If the cable to be tested is electrified, it will pose great threat on personal safety. Therefore, the cable must be powered off, and short-circuit discharge on the cable is required before and after test at each phase.

(3) Unclear state of another end of the cable for test will also cause hidden trouble in the test. Throughout the test, additional watchers and safety fences shall be provided at the other side.

(4) The work site is an electrical equipment maintenance site where there are many other electrical equipment and facilities that may cause mechanical injury to persons on the site. Safety fences shall be provided for the work area, and persons on the site shall wear work clothes and safety helmets and work within designated areas as instructed by the person in charge of work.

4. Division of labor among operators

Person in charge of on-site work (supervisor): ×××

On-site operator: ×××

Ⅴ. Operating specifications and requirements

(1) The parts that are not to be tested shall be short circuited to ground and have good contact; The cable for test is unmanned, and no one is allowed to approach it during the test; Person shall be assigned to the other end for watching or safety barrier shall be installed to prevent anyone from coming into contact with the test cable.

(2) The test cable shall undergo full discharge before and after the test; Touching with hands is strictly forbidden before the megameter stops rotation or the test cable is discharged; In disassembling wires, touch metal part of the lead is not allowed.

(3) During measurement, speed of megameter handle shaking shall be kept at 120 r/min; After reading the values, the wire between megohmmeter and the test object shall be disconnected and then the megohmmeter shall be shut down to prevent electric charge on capacitor of the test object from discharging through megohmmeter and cause damage to megohmmeter.

(4) During measurement with a tramegger, two test lines shall not intersect, and the red line must be overhead; Human body shall not come into contact with red wires and grounding wires.

Ⅵ. Operation processes and standards (see Tab. 7-6)

Tab. 7-6　Power cable insulation test processes and scoring standards

Class		Name		Student ID		Examiner		Score		
S/N	Operation name	Quality standard			Points	Deduction criteria		Deduction		Score
1	Preparations									
1.1	Clothing	Operators shall wear work clothing as required in the Regulation			2	Deduct 1 point for failure to dress as required Deduct 1 point for non-standard clothing				

Continued

S/N	Operation name	Quality standard	Points	Deduction criteria	Deduction	Score
1.2	Megameter, tools and instruments, insulation tools and safety appliances	Select correct megohmmeter, psychrometer, grounding wire, discharging rod, insulating gloves, insulating boots, test fence, sign board, insulation cushion, monkey spanner, slotted screwdriver, flat pliers, hammer, nail, waterproof tape and test record sheet	6	Deduct 2 points for each incorrect selection; Deduct 2 points for each omission		
1.3	Check equipment, instruments and safety appliances	Correctly check instruments, insulation tools and safety appliances for presence of acceptable labels and being within valid period; Make sure that the megohmmeter can be up to infinity and indicate zero during short circuit	6	Deduct 2 points for incorrect method; Deduct 1 point for each omission		
1.4	Ascertain on-site safety measures, major dangerous points and safety precautions	Provide fences on the test site, provide off-site warning signs and start work only after receiving command from the supervisor and repeating the command loudly; write down main dangerous points and precautions in the test	10	Deduct 2 points for failure to act as specified; Deduct 2 points for failure to get permit and repeat the instructions; Deduct all points for failure to keep records and deduct point(s) for incomplete record		
2	Megohmmeter selection	Correct megohmmeter shall be selected based on different voltage classes of cables: For outer sheath and inner lining of 500 V and lower rubber and plastic cables, 500 V megohmmeter shall be used; For 0.6/1 kV cable, 1,000 V megohmmeter shall be used; For 0.6/1 kV and higher cable, 2,500 V megohmmeter shall be used; For 6/6 kV and higher cable, 2,500 V or 5,000 V megohmmeter shall be used	6	Deduct 2 points for each incorrect selection		
3		Insulation resistance measurement				
3.1	Grounding wire	Correctly select grounding points, use tools such as file or emery cloth to polish and clean the grounding points, remove stains and firmly connect the wires	2	Deduct 2 points if grounding points are not incorrect; Deduct 2 points for failure to grind and clean grounding points and line loosening		

Continued

S/N	Operation name	Quality standard	Points	Deduction criteria	Deduction	Score
3.2	Cable support and cleaning	Peel and cut the cable cores according to test standards, provide cable support as required for the test, and wipe off any stain on core insulation	5	Deduct 3 points for failure to peel and cut support cable as required; Deduct 2 points if there is residue		
3.2	Grounding of cables, meters and discharging rods	Reliably ground "E" ends of cable cores, armors, shields and meters and discharging rods. Place the megohmmeter in an appropriate position considering site situations and connect it correctly	4	Deduct 2 points if positions are incorrect and grounding is poor; Deduct 2 points if grounding and wiring are incorrect;		
3.4	Measurement of outer sheath insulation resistance	Use 500 V megohmmeter; Remove cable armors and grounding wire of shielding layer; The tester shall stand on insulation cushion, set megohmmeter speed to 120 r/min, connect test wire into the armor and read values at 60 s or stable value indicated by the pointer; To prevent reverse charge by the test object, after reading the data, the test wire shall be disconnected from the test object, and then megohmmeter shaking shall be stopped, and the test object shall be discharged and grounded (Insulation resistance shall be no less than 0.5 MΩ/km)	8	Deduct 3 points for incorrect meter selection and wiring and nonstandard operation of meters; Deduct 3 points if readings are incorrect; Deduct 4 points if one stops shaking megohmmeter and then disconnect test wire from test object which results in reverse charge; Deduct 3 points for nonstandard discharge grounding		
3.5	Measurement of inner liner insulation resistance	Use 500 V megohmmeter; Remove grounding wire of cable shielding layer; The tester shall stand on insulation cushion, set megohmmeter speed to 120 r/min, connect test wire into shielding layer and read values at 60 s or stable value indicated by the pointer; To prevent reverse charge by the test object, after reading the data, the test wire shall be disconnected from the test object, and then megohmmeter shaking shall be stopped, and the test object shall be discharged and grounded (no less than 0.5 MΩ/km insulation resistance)	8	Deduct 3 points for incorrect meter selection and wiring and nonstandard operation of meters; Deduct 3 points if readings are incorrect; Deduct 4 points if one stops shaking megohmmeter and then disconnect test wire from test object which results in reverse charge; Deduct 3 points for nonstandard discharge grounding		

Continued

S/N	Operation name	Quality standard	Points	Deduction criteria	Deduction	Score
3.6	Main insulation resistance measurement	Use 2 500 V or 5 000 V megohmmeter; Remove grounding wire of core of the cable to be tested The tester shall stand on an insulation cushion, set megohmmeter speed to 120 r/min, connect test wire into the core for test and read insulation absorption ratio at 15 s and 60 s; To prevent reverse charge by the test object, after reading the data, the test wire shall be disconnected from the test object, and then megohmmeter shaking shall be stopped, and the test object shall be discharged and grounded; Repeat the above steps in test on the other phases (insulation resistance shall not differ greatly from the previous one, or withstand voltage shall not vary significantly)	15	Deduct 3 points for incorrect selection of insulation cushion and meter wiring; Deduct 3 points for failure to have rotating speed or unstable rotating speed; Deduct 5 points if readings are incorrect; Deduct 3 points if one stops shaking megohmmeter and then disconnect test wire from test object which results in reverse charge; Deduct 3 points for nonstandard discharge grounding Deduct 5 points for failure to keep records or incorrect records		
3.7	Remove wires	Remove all leads and grounding wires up to the grounding point	3	Deduct 2 points if wires are not completely removed; Deduct 2 points if wires are not removed as required		
3.8	Provide waterproof treatment on cables peeled and cut	Wrap it with waterproof tape overlapping by 1/2 for at least three turns back and forth. Before use of waterproof tape, remove the isolation layer, stretch the tape by 100% (one fold), hold the tape with hand to prevent it from falling to the ground, keep it clean and tight and avoid breakage	5	Deduct 3 points respectively for failure to stretch and overlap the tape by 1/2; Deduct 3 points respectively for dropping the tape to ground and incorrect number of layers; Deduct 3 points for loosening and breakage		
4			End of work			
4.1	Clean up tools and equipment	Clean up tools and instruments, safety appliances and instruments, and place them neatly and reasonably	2	Deduct 2 points for failure to put the tools reasonably		
4.2	Clear the site	Dismantle safety barriers, sign boards, etc. on the site, clean up the site, do not leave any object on the site, and then report the end of the test to the review committee	3	Deduct 2 points respectively for leaving objects on the site and failure to clean the site Deduct 1 point for failure to submit completion report		

模块七 电力线路安全运检（Module Ⅶ Safe operation and inspection of electric power lines）

Continued

S/N	Operation name	Quality standard	Points	Deduction criteria	Deduction	Score
4.3	Test records	Keep records as required, and handwriting must be clear and tidy	8	Deduct 3 points for an error; Deduct 3 points for unclear expression		
4.4	Draw schematic diagram of cable insulation resistance measurement using megohmmeter	Mark designations of the parts and render them clearly to allow easy understanding	7	Deduct 5 points for an unclear illustration on schematic diagram; Deduct 5 points for a wrong designation		
Total			100			

任务四 导线在绝缘子子上的绑扎操作实训

一、作业任务

单人操作,以"线路绝缘子"为工作对象,按照《国家电网公司电力安全工作规程(线路部分)》中的要求,完成绝缘子绑扎操作实训。

二、引用标准及文件

(1)《国家电网公司电力安全工作规程(线路部分)》。
(2)《国家电网公司十八项电网重大反事故措施》。

三、作业条件

应在良好的天气操作;作业人员精神状态良好,熟悉工作中的安全措施、技术措施以及现场工作危险点。

四、作业前准备

1. 现场操作的基本要求及条件

勘察现场设备情况,查阅相关技术资料,包括历史数据及相关规程。

2. 工器具及材料选择

个人工器具1套、抹布1只、扎线1把、安全带1套、瓷绝缘子1个、登高踩板1副。

3. 危险点及预防措施

(1)高压触电。
预防措施:正确进行模拟预演,做好个人防护。
(2)使用不合格的安全工器具。
预防措施:操作设备前,应正确选取安全工器具并检查合格。

4. 作业人员分工

现场工作负责人(监护人):×××
现场作业人员:×××

五、作业规范及要求

(1)安全施工,正确使用个人工器具。
(2)不得浪费材料。
(3)请做到工完料尽,场地干净。
(4)登杆前检查:线路名称和杆号、埋深杆身、杆根、拉线及安全带、脚扣(表7-7)。

表 7-7 登杆前检查工作流程和说明

序号	工作流程	说明
1	第一步	用盘好的绑线短头在绝缘子左侧导线上缠绕 3 圈，其方向是从导线外侧，经导线上方绕向导线内侧。绑扎起头位置应靠近绝缘子
2	第二步	用盘起的绑扎线在绝缘子颈外侧绕到绝缘子右侧导线下方，在导线顶打一交叉，左侧导线下方绕过绝缘子颈内侧，自绝缘子右侧导线下方在导线顶再打一交叉到绝缘子左侧导线下方，绝缘子顶部形成一个"X"形交叉线，要求交叉线紧固，不能松动
3	第三步	用盘起的绑线绕过绝缘子外侧到绝缘子右侧导线下方缠绕 3 圈；其方向是从导线下方经绝缘子外侧绕向上方； 或用盘起的绑线绕过绝缘子内侧导线下方在导线顶部再交叉，再经绝缘子右侧导线下方向绝缘子外侧
4	第四步	用盘起的绑线经绝缘子外侧到绝缘子左侧导线下方在导线顶端打回交叉，其方向从绝缘子右侧导线下方，用盘起的绑线经绝缘子内侧在绝缘子右侧导线下方缠绕 3 圈，再经绝缘子外侧在绝缘子右侧导线下方缠绕 3 圈缠绕完毕后，绝缘子顶部有两个 X 形交叉线，导线两侧缠绕各 6 圈
5	第五步	收尾时，将绑线头从导线下方经绝缘子内侧绕至绝缘子左侧短头处，并与短头拧小辫（3 个麻花），麻花大小均匀，辫头顺向导线外侧

（5）遵循安全操作规程，按照操作票的步骤正确操作。
（6）操作结束后，对操作质量进行检查。

六、作业流程及标准（表 7-8）

表 7-8 变压器分接开关的调整操作流程及评分标准

班级		姓名		学号		考评员		成绩	
序号	作业名称	质量标准		分值/分		扣分标准		扣分	得分
1	正确填写记录页	按照作业任务要求正确填写记录页		30		按照作业任务要求正确填写记录页，记录页填写不规范的，视情况扣 2~30 分			
2	安全意识	准备好该项操作所需的安全用具并进行检验； 做好个人防护，做好登杆前检查； 正确持操作票在模拟系统上模拟操作一次，核对设备的位置、名称、编号和运行方式		20		未能准备好该项操作所需的安全用具并进行检验，扣 2~5 分； 未做好个人防护，未做好登杆前检查，扣 2~5 分；			

续表

序号	作业名称	质量标准	分值/分	扣分标准	扣分	得分
2	安全意识			要持操作票在模拟系统上模拟操作一次，未核对设备的位置、名称、编号和运行方式，扣 5～10 分		
3	操作技能	能正确选择、检查、使用工器具及材料； 绑扎和导线应使用同种金属，绑线本身不能有接头，直径不得小于导线绞线单股直径但也不能太大，以免施工不便和不易绑紧； 10 kV 线路导线采用双十字绑扎，农网低压线路采用单十字绑扎； 导线绑扎应牢固； 扎线操作工作轻松流畅工艺美观； 作业前，应检查检查绝缘子的瓷质部分有无裂纹、硬伤、脱釉等现象，瓷质部分与金属部分的连接是否牢固可靠，金属部分有无严重锈蚀现象； 导线的绑扎必须牢固可靠，不得有松脱、空绑等现象 绑扎：对于直线杆，针式绝缘子的导线应安放在顶槽，顶槽应顺线路方向，水平瓷横担的导线应安放在端部的边槽上，对于转角杆，导线应固定在针式绝缘子转角外侧的脖子上	50	操作工艺不美观，扣 10 分； 导线绑扎流程出错，扣 10～40 分； 导线绑扎出线空绑，扣 10～40 分		

续表

序号	作业名称	质量标准	分值/分	扣分标准	扣分	得分
4	选择绑扎方法	选择绑扎方法错误	否定项	根据杆塔类型、选择用的导线绑扎方法不正确；根据电压等级选用的导线绑扎方法不正确		
合计			100			

Task IV Training on Tying Conductors on Insulators

I. Tasks

The operation should be completed by one person, who is required to tie a conductor on an insulator in accordance with the requirements in the *Electric Power Safety Working Regulations (Power Line) of State Grid Corporation of China.*

II. Reference standards and documents

(1) *Electric Power Safety Working Regulations (Power Line) of State Grid Corporation of China*;

(2) 18 *Major Anti-accident Measures for Power Grids of State Grid Corporation of China.*

III. Operating conditions

Operations should be carried out in good weather. Operators should be in a good state of mind, and be familiar with the safety measures, technical measures, and dangerous points of the work on the site.

IV. Preparation for work

1. Basic requirements and conditions for on-site operation

Operators should check the condition of on-site equipment, and understand relevant technical data, including historical data and relevant procedures.

2. Selection of tools and instruments and materials

1 set of personal tools and instruments, 1 rag, 1 bundle of cable ties, 1 set of safety belt, 1 porcelain insulator, and 1 pair of climb plank.

3. Dangerous points and preventive and control measures

(1) High voltage electric shock.

Prevention and control measures: Carry out simulation and take proper personal protection measures.

(2) Use of non-conforming safety tools and instruments.

Prevention and control measures: Before operating the equipment, operators should select and check the safety tools and instruments properly.

4. Division of labor among operators

Person in charge of on-site work (supervisor): ×××

On-site operator: ×××

V. Operating procedures and requirements

(1) Construction should be carried out safely and personal tools and instruments used properly.

(2) Materials should not be wasted.

(3) Upon completion of the work, the materials shall run out and the site shall be clean.

(4) Before climbing the pole, the operator should check the line name, pole number, buried depth of pole body and pole root, stay, safety belt, and grappler, as shown in Tab. 7-7.

Tab. 7-7 checking procedure before climbing the pole

S/N	Work process	Description
1	Step 1	Wind the short end of the coiled cable tie around the conductor on the left side of the insulator three times, in the direction from the outside to the inside of the conductor via the top. The starting position of the tie should be close to the insulator
2	Step 2	Wind a coiled cable tie around the outer side of the insulator neck to the bottom of the conductor on the right side of the insulator, and make a cross on the top of the conductor; wind it around the inside of the insulator neck via the bottom of the conductor on the left to the bottom of that on the right, and make another cross on the top of the conductor; wind it to the bottom of the conductor on the left, then make another cross on the top of the conductor. The crosses must be tightened and not be loose
3	Step 3	Wind a coiled cable tie around the outside of the insulator to the bottom of the conductor on the right side of the insulator three times, in the direction from the bottom of the conductor to the top via the outside of the insulator; or wind a coiled cable tie around the bottom of the conductor on the inside of the insulator, make a cross on the top of the conductor, and then wind it via conductor on the right side of the insulator to the outside of the insulator
4	Step 4	Wind a coiled cable tie around the outside of the insulator to the bottom of the left conductor of the insulator, and make a cross on the top of the conductor, and wind it towards the bottom of the conductor on the right side of the insulator. Wind a coiled cable tie around the inside of the insulator to the bottom of the conductor on the right side of the insulator three times, and then wind it around the outside of the insulator to the bottom of the conductor on the right side of the insulator three times. Now there are two crosses on the top of the insulator, and the conductors are wound by six times on each side
5	Step 5	When finishing, wind a cable tie end around the bottom of the conductor via the inside of the insulator to the short end on the left side of the insulator, and twist it with the short end make a small braid (3 twists). The twist should be uniform in size, with the end pointing to the outside of the conductor.

(5) Follow the safety operating procedures and operate correctly according to the steps in the operation ticket.

(6) Check the quality of operation after the operation.

VI. Operating procedures and standards (Tab. 7-8)

Tab. 7-8 Operation process and scoring criteria for adjustment of transformer tap changer

Class		Name		Student ID		Examiner		Score	
S/N	Description	Quality standard			Score (points)	Standard for deduction (points)		Deduction	Score
1	Fill out the record page correctly	Fill out the record page correctly according to the requirements of the operation task			30	Fill out the record page correctly according to the requirements of the operation task; otherwise, 2–30 points will be deducted as appropriate.			
2	Safety consciousness	Prepare and check the safety appliances required for the operation; Take proper personal protection measures and check the specified items; Correctly hold the operation ticket to simulate the operation in the simulation system, and check the location, name, number and operation mode of the equipment			20	Failure to prepare and check the safety appliances required for the operation will result in a deduction of 2–5 points; Take proper personal protection measures and check the specified items; otherwise, 2–5 points will be deducted; Correctly hold the operation ticket to simulate the operation in the simulation system. Failure to check the location, name, number and operation mode of the equipment will result in a deduction of 5–10 points;			
3	Operational skills	Operators should be able to select, check, and use tools and instruments and materials correctly; The cable tie and conductor should be made of the same metal. The cable tie itself should be free of joints, the diameter should not be smaller than that of a single strand of the conductor; however, it should not be too large, so as to avoid construction inconvenience and difficulty in tying. Double-cross should be used for 10 kV line conductors and single-cross for low voltage lines of rural power grids. The conductors should be tied firmly, The tying operation is easy and smooth, presenting aesthetic technique;			50	10 points will be deducted for unaesthetic operation technique; 10–40 points will be deducted for errors in the conductor tying process; 10–40 points will be deducted for loose turns			

Continued

S/N	Description	Quality standard	Score (points)	Standard for deduction (points)	Deduction	Score
3	Operational skills	Before operation, check the porcelain part of the insulator for cracks, flaws, deglazing, etc., the connection between the porcelain part and the metal part for firmness and reliability, and the metal part for serious corrosion; The conductors must be tied firmly and reliably, and there must be no loosened ties or loose turns. Tying: For intermediate supports, the conductor of the pin insulator should be placed in the top groove, which should be in the direction of the line. The conductor of the horizontal porcelain cross arm should be placed in the side groove of the end. For angle supports, the conductor should be fixed on the neck of the outside of the angle of the pin insulator				
4	Selection of tying method	An incorrect tying method is selected	Negative	According to tower type, an incorrect conductor tying method is selected; According to voltage class, an incorrect conductor tying method is selected		
Total			100			

模块 八　现场风险辨识与现场急救

在电力生产使用维护过程中，存在着人、机、料、法、环、测等方面的风险因素以及职业病危害因素。在作业前分析出可能引发事故的风险因素及职业病危害因素，再采取针对性的预防控制措施，在作业过程中实施控制，可预防人身、设备、电网事故发生，远离职业病，从而实现安全生产"可控、能控、在控"的目标。但电力作业过程中各种不规范、不负责行为导致风险分析预控不到位，人身触电、机械伤人等事故仍有发生，严重威胁着作业人员的人身安全。因此，作业人员除了需学会现场风险分析预控，还需学会紧急救护的基本知识及技能，尤其应学会触电急救，事故发生时能及时救护，避免可能发生的伤亡，达到现场自救、互救的目的。

学习目标：

（1）会正确进行作业现场风险辨识。

（2）会正确进行作业现场风险预控。

（3）能正确分析出作业现场的职业病危害因素并加以预控。

（4）会正确、成功地进行触电急救。

（5）会进行简单的现场外伤急救。

Module VIII Risk identification and first aid on the site

In electric power production, use and maintenance, there are risk factors with respect to human, machine, material, law, environment and measurement and occupational disease hazards. Before start of work, risk factors and occupational disease hazards that may cause accident shall be analyzed, and targeted prevention and control measures shall be taken in the process of work to prevent personal injury and equipment damage, power grid accident and occupational diseases so that safety in production is "controllable and controlled". However, non-standard and irresponsible behaviors in work on electric power system have led to inadequate risk analysis and control and accidents such as electric shock and mechanical injury, posing serious threat to safety of the operators. Therefore, the operators shall not only be capable of site risk analysis and prevention and control, but also learn basic knowledge and skills for emergency rescue, especially first aid against electric shock. When an accident occurs, they shall be able to provide timely rescue to avoid possible casualty and achieve the goal of self-rescue and mutual rescue on the site.

Learning objectives:

(1) Be able to identify risks on work site.

(2) Be able to carry out risk prevention and control on work site.

(3) Be able to correctly analyze and prevent and control occupational disease hazards on work site.

(4) Be able to correctly provide first aid against electric shock.

(5) Be able to provide simple off-site first aid.

任务一 作业现场风险辨识及职业病

一、作业现场风险辨识

人类从事的每一项生产活动都存在着包括劳动者本身、工具设备、劳动对象、作业环境等方面不同程度的危险性和不安全因素。所谓风险，就是指可能发生的危险或灾祸，因此风险辨识又称为危险点辨识。通过作业前对作业现场人的不安全行为、物的不安全状态和安全管理的缺失进行辨识，分析出可能引发事故的风险因素，再采取针对性的预防控制措施，在作业过程中实施控制，预防人身事故、人为责任事故的发生，实现安全生产的目标。

（一）作业现场风险因素

在实际生产活动中，风险是一种诱发事故的隐患。事先对风险进行分析辨识并采取措施加以防范，就会化险为夷，确保安全。电力作业中的风险因素，是指在作业中有可能发生危险的地点、部位、场所、设备、工器具和行为动作等，主要包括以下几个方面：

（1）作业人员不严格遵守安全工作规程。如雷雨天气巡视户外高压设备时，不穿绝缘靴，致使跨步电压触电等。

（2）不安全的机器设备等物体。如设备高速旋转的部分裸露在外，人不小心与之碰触，造成机械伤害；设备的外壳没有接地或接地电阻不符合要求，当绝缘损坏时，就可能造成触电事故。

（3）不安全的作业环境。环境中的有害粉尘、有毒气体、高温、噪声、辐射等都是不安全因素。如作业环境中存在的有害粉尘，可能会造成尘肺病；作业环境中存在的生产性毒物，可能会造成职业中毒等。

作业现场风险因素的特点：

（1）具有客观实在性。作业现场的风险因素客观实在存在，不以人的主观意识为转移。

（2）具有复杂多变性。作业现场风险因素的复杂性是由作业情况的复杂程度决定的。即使每次作业任务相同，若参加作业的人员、作业的场合地点、使用的工具以及所采取的作业方式不同，存在的风险点也可能会不同。即使是相同情况的作业，存在的风险点也不是固定不变的，旧的风险消除了，新的风险又会出现，所以分析预控风险点不能一劳永逸，一定要具体情况具体分析，按照实际情况决定所应采取的方法。

（3）具有潜在性。作业现场的风险因素存在于即将进行的作业过程，虽然客观存在，但还没有转变为现实的危害。

（4）具有可知可防性。电力作业现场的风险因素具有潜在性、隐蔽性，做好风险点可防可控可管理具有一定的难度，但既然风险点是客观存在的事物，就一定能被我们发现、认知，从而预防控制，只要思想重视，举措得力，一定能够把风险扼杀在萌芽状态。

（二）作业现场风险分析

风险分析，即危险点分析，是指在一项作业或工程开工前，对该作业项目（工程）所存在的危险类别、发生条件、可能产生的情况和后果等进行判断和推测，找出风险点。其目的是控制事故发生。

作业中的风险点有的是显现的易发觉的,比如电气作业中的触电危险、机械伤人危险、登高作业中的坠落危险。对于显现的危险,我们可以事先采取足够的安全措施,比如操作过程中戴好绝缘手套、戴好安全帽、穿好绝缘靴、与带电体保持足够的安全距离、系好安全带等。作业中的风险点有的是潜在的不易发现的,这就需要我们事先进行科学的分析,对潜在的风险点做出准确的判断。

1. 风险分析的要点

(1)作业场地:作业场地不同(如高空、带电、停电、井下、箱柜内等不同的作业场地,道路、交通状况的不同),可能给作业人员、设备、电网带来的安全危害也不同。

(2)作业环境:作业环境不同(如高温、高压、高海拔、易燃、易爆、辐射、有毒有害气体、缺氧等不同的作业环境),给作业人员、设备、电网带来的安全危害可能也不同。

(3)设备情况:作业过程使用的机械、设备、工具等可能给作业人员、设备、电网带来的危害。

(4)操作规范程度:操作流程不规范、操作方法不当、不严格遵守安全规程规定等可能给作业人员、设备、电网带来危害。

(5)作业人员状况:作业人员思想意识不清醒(如酒后、使用某些药品后、连续熬夜后)、身体健康状况不佳、不安全的操作行为、技术技能水平不足等可能给作业人员、设备、电网带来的危害。

(6)其他因素:其他可能给作业人员、设备、电网带来危害的不安全因素。

2. 风险分析的步骤

(1)资料分析:收集本单位过去同类作业或其他单位同类作业的有关资料,对同类作业发生的安全事故进行深入分析,找出事故发生的根本原因及应对措施。

(2)全面勘察:组织全面现场勘察,从将要从事的作业范围、作业环境、作业设备的状态等方面入手,对作业中可能出现的风险源头与隐患之处进行评测,以此来作为制定具体的风险控制与防范措施的基础依据。

(3)综合分析:相关技术人员要以现场作业内容及方法为根据,结合作业所涉及的环境条件、设备状况、工器具状况、相关技术人员的业务素质、身体素质及作业习惯等方面,仔细查找对人身、设备、电网造成安全危害的关键点。作业人员的经验越丰富,对此次作业情况越熟悉,所分析的风险点就越透彻,对可能出现的问题估计得越充分。综合分析阶段的具体做法是:

① 工作负责人召集工作班成员对此次作业进行全方位的分析预测。对于复杂大型作业,应邀请经验丰富、技术精湛的员工参会指导。

② 把作业过程分解为若干阶段,结合过去同类作业的经验教训,逐个阶段分析有可能存在的风险点。参会成员采用头脑风暴法,各抒己见,广泛讨论,科学分析,尽可能地找出存在的所有风险点。作业阶段越短,预测出的风险点越可靠。

③ 研究风险点是如何演变为事故的,即风险点变为事故的条件是什么。阻断了风险点演变为事故的条件,也就阻止了事故的发生。

④ 对分析出的风险点进行评估,根据风险点危险程度的大小、轻重划分出危险等级,制

定各风险点的防控措施，对危险等级高的风险点应重点加以防范。

⑤ 在作业过程中，应根据作业过程及气候、环境等的变化不断辨识、查找可能的风险点。

（三）风险预控

风险预控就是对电力生产中的每项工作，根据作业内容、工作方法、环境、人员状况、设备实际等去分析、查找可能导致事故的风险因素，再依据规程制度，制定防范措施的一种安全管理方法。它的基础工作是风险点识别与查找，控制的关键在于生产现场实施程序化、规范化作业。实践证明，它是预防事故的有效方法。

1. 风险预控类别

（1）宏观控制。以整个系统为对象，对风险点进行控制。采用的手段主要有法制手段（政策、法令、规章）、经济和行政手段（奖、罚、惩、补）以及教育手段（安全教育培训）。

（2）微观控制。以具体的风险点为控制对象，对风险点进行控制。所采用的主要手段是整改措施、组织措施、安全技术措施、预警提醒和监护。宏观控制和微观控制互相结合、互相补充、互相制约，可以建立安全事故的多层次、立体的控制系统。

2. 风险预控手段

（1）必须从源头抓起。设备固有缺陷是风险点的源头，要通过消缺、检修、改造，及时消除缺陷，从根本上消除潜在的风险。对一时无法消除的缺陷，或技术上及经济上难以进行根本性的治理时，可采用其他预控措施和应急方案，但应视风险点的性质得到有关部门或上级的批准。

（2）要从安全教育培训抓起。实际上，《国家电网公司电力安全工作规程》中"不得""严禁""防止"等词汇的表述，已明确了作业过程中应该怎么做、不应该怎么做，已指明了各类作业中具体存在的风险点，对同一类作业具有普遍的适用性和可操作性。残酷的经验教训证明，风险点的生成、扩大、突变以致最终酿成事故，皆是因为有关人员不熟悉或不能严格遵守安全工作规程。因此，作业人员必须加强安全工作规程学习，熟练掌握《国家电网公司电力安全工作规程》，严格按照规程规定执行工作任务。

（3）要突出作业和操作的过程预控。风险点预控重在作业过程控制，要加强现场安全措施的执行落实，保证现场每一位作业人员清楚本次作业任务、所存在的风险点和应采取的安全措施，如在进行检修工作时需填写工作票，履行保证安全的组织措施，以书面的形式使风险点预控措施得以确认，对设备检修的每一环节、每一工作步骤进行风险点确认、控制，从而达到对作业全过程的控制。

（4）开展标准化、规范化、程序化作业。为加强对作业人员行为风险点预控，除开展安全教育培训、加强防护外，还可以开展标准化、规范化、程序化作业。如制定倒闸操作标准化操作流程，所有作业人员都必须严格按照标准化流程开展工作，规范作业人员个体行为，防止漏项、顺序颠倒，防止误操作。

（四）风险分析、预控的作用

国内外资料统计表明，90%以上的事故是由于当事人对有可能造成伤害的风险点缺乏事先预想或虽然预想到却缺乏有效的防范措施。因此，做好风险点的分析预控工作，就能使有可能诱发事故的人为因素得以避免，把事故遏制在萌芽状态。具体来说，做好风险点分析、预控工作，有如下作用：

（1）增强作业人员对风险的认识，克服麻痹思想，防止冒险行为。一些事故的发生，与当事人对作业中可能存在的风险点及其危害认识不足、有险不知险、知险无措施有直接关系。

（2）能够防止由于仓促上阵而导致的危险。准备不充分，安排不周、忙乱无序或图方便简化、颠倒作业步骤，这本身就埋藏着事故隐患。

（3）能够防止由于技术业务不熟而诱发的事故。在作业前，开展风险点分析预控活动，实际上就是对安全工作重要性的再认识，对有关作业的工艺、技术业务的再学习。

（4）做好风险点分析预控工作，能够使安全措施更具针对性和实效性，确实起到预防事故的作用。以往的教训是：作业人员对作业中存在的风险点心中无数，工作票中提出的安全措施缺乏针对性和可操作性，从而导致事故的发生。

（5）做好风险点分析预控工作，能够减少以致杜绝由于指挥不力而造成的事故。指挥人员由于不熟悉作业中存在的险情或凭主观臆断进行指挥，极有可能造成事故，甚至会造成群死群伤。

二、职业危害及职业病

（一）职业病

职业病是指企业、事业单位和个体经济组织（统称用人单位）的劳动者在职业活动中，因接触粉尘、放射性物质和其他有毒、有害物质等因素而引起的疾病。各国法律都有对于职业病预防方面的规定，一般来说，只有符合法律规定的疾病才能称为职业病。

1. 职业病危害因素的分类

职业病危害因素按其来源可以分为以下三类：

（1）与生产过程有关的职业病危害因素。

① 物理因素，如高温、低温、辐射、生产性噪声、振动等。

② 化学因素，包括：生产性粉尘，如煤粉尘、硅尘、石棉尘、电焊尘等；铅、锰、汞、苯等有毒物质。

③ 生物因素，主要指某些病原微生物或致病寄生虫等，如炭疽杆菌、布氏杆菌、森林脑炎病毒及蔗渣上的霉菌等，医务工作者接触的传染性病源如 SARS 病毒、新冠病毒。

（2）与劳动过程有关的职业病危害因素。

作业时间过长、作业强度过大、劳动制度与劳动组织不合理、长时间强迫体位劳动、个别器官和系统的过度紧张、长时间处于不良体位或使用不合理的工具等，均可造成对劳动者健康的损害。

（3）与作业环境有关的职业病危害因素。

作业环境中的危害因素主要是自然环境中的有害因素，如高温、高压、潮湿、低温、不符合卫生要求或卫生标准、缺乏必要的卫生工程技术设施、安全防护设备和个人防护用品有缺陷等。

2. 职业病危害因素的预防

职业病的预防遵循三级预防原则，即：① 一级预防：从根本上着手使劳动者尽可能不接触职业病有害因素，或控制作业场所有害因素水平在卫生标准允许限度内。② 二级预防：对作业工人实施健康监护，早期发现职业损害，及时处理、有效治疗，防止病情进一步发展。③ 三级预防：对已患职业病的患者积极治疗，促进健康。

落实三级预防的基本措施有：① 实施劳动卫生监督，包括预防性和经常性卫生监督，以及事故性处理。② 降低有害因素浓（强）度，常见的卫生技术措施有从工艺上改进，防止有害因素逸散，推广运用低毒、无毒的材料或技术，配置个人防护用品、通风防尘等。③ 职业性健康筛检，如入职体检、定期体检和离退休体检等。

电力企业职工在生产、施工过程中接触的有害因素主要有生产性粉尘、噪声、高温、有毒物质、射线、微波辐射、低频辐射等。

（二）生产性粉尘与尘肺病

生产性粉尘是指在生产过程中形成，并能长时间悬浮在空气中的固体微粒。生产性粉尘来源于固体物质的机构加工、物质蒸气冷凝、物质的不完全燃烧等。在我国分布较广，对职业人群健康影响较大的生产性粉尘主要有硅尘、煤尘和石棉尘。

1. 生产性粉尘的来源

生产性粉尘来源十分广泛，如固体物质的机械加工、粉碎，金属的研磨、切削，矿石的粉碎、筛分、配料或岩石的钻孔、爆破和破碎等，耐火材料、玻璃、水泥和陶瓷等工业中原料加工，皮毛、纺织物等原料处理，化学工业中固体原料加工处理，物质加热时产生的蒸气、有机物质燃烧不完全所产生的烟，等。

2. 生产性粉尘的分类

根据生产性粉尘的性质可分为三类：① 无机性粉尘，包括矿物性粉尘，如石英、石棉、煤等；金属性粉尘，如铁、锡、铝等及其化合物；人工无机粉尘，如水泥、金刚砂等。② 有机性粉尘，包括植物性粉尘，如棉、麻、面粉、木材；动物性粉尘，如皮毛、丝尘；人工合成的有机染料、农药、合成树脂、炸药和人造纤维等。③ 混合性粉尘，指上述各种粉尘的混合存在形式，一般是两种以上粉尘的混合。生产环境中最常见的就是混合性粉尘。

3. 生产性粉尘引起的职业危害

粉尘对人体的危害程度取决于其化学成分和浓度。粉尘中游离二氧化硅含量愈高，对人体危害愈大。同一种粉尘，浓度愈高，对人体危害愈严重；另外，粉尘对人体的危害还与其被粉碎的程度即分散度有关，粒径较小和颗粒愈多，分散度愈高在空气中浮游的时间愈长，被人体吸入的机会就愈多，其危害也就愈大。粉尘引起的职业危害有全身中毒性、局部刺激性、变态反应性、致癌性、尘肺。其中以尘肺的危害最为严重。尘肺是目前我国工业生产中最严重的职业危害之一。

患者早期可无临床症状，部分患者有胸闷、咳嗽、咳痰等，随上述症状加重并有气紧气喘、呼吸困难，晚期可并发肺气肿及肺心病。

在尘肺治疗上目前尚无根治的药物，主要采取对症治疗和支持治疗控制病情的进一步发展。目前，国外已将重点转移到尘肺的预防上，主要措施有配置除尘设备，加强个人防护，接尘岗位职工定期到防保机构检查身体等。

（三）生产性毒物与职业病

生产过程中使用或产生的有毒物质，统称为生产性毒物。生产性毒物在一定条件下可通过不同途径进入人体而引起中毒，或引起免疫功能或其他生理功能改变，因而使人易患病或促使原有疾病的病情加重，病程延长。有的毒物具有局部刺激，致敏及腐蚀作用，有的还可致肿瘤、致畸胎及诱发遗传变异等。

劳动者在生产过程中由于接触毒物所发生的中毒称为职业中毒。

1. 生产性毒物进入人体的途径

了解生产性毒物进入人体的途径，对于采取相应的防治措施具有重要意义。毒物可通过呼吸道、皮肤和消化道侵入人体。

（1）通过呼吸道进入人体。整个呼吸道的黏膜和肺泡都能不同程度地吸收有毒气体、蒸气及烟尘，但主要的部位是支气管和肺泡，尤以肺泡为主。肺泡接触面积大，周围又布满毛细血管，有毒物质能很快地经过毛细血管进入血液循环系统，从而分布到全身。这一途径是不经过肝脏解毒的，因而具有较大的危险性。在石油化工企业中发生的职业中毒，大多数是经呼吸道吸入体内而导致中毒的。

（2）通过皮肤进入人体。皮肤脂溶性毒物，如苯胺、丙烯腈等，可以通过人体完整的皮肤，经毛囊空间到达皮脂腺及腺体细胞而被吸收，一小部分则通过汗腺进入人体。毒物进入人体的这一途径也不经肝脏转化，直接进入血液系统而散布全身，危险性也较大。

（3）消化道毒物由消化道进入人体的机会很少，多由不良卫生习惯造成误食或由呼吸道侵入人体，一部分沾附在鼻咽部混于其分泌物中，无意被吞入。毒物进入消化道后，大多随粪便排出，其中一部分在小肠内被吸收，经肝脏解毒转化后被排出，只有一小部分进入血液循环系统。

2. 常见的职业中毒

（1）金属及类金属中毒。

金属有多种分类方法，按照理化特性可分为重金属、轻金属、类金属三类。金属中毒有多种，如铅中毒、四乙基铅中毒、锰中毒、铍中毒、镉中毒。类金属中毒有砷中毒和磷中毒等。铅中毒者口内有金属味，出现恶心、呕吐、腹胀、阵发性腹绞痛、便秘或腹泻等症状，严重者出现抽搐、瘫痪、昏迷、循环衰竭、中毒性肝病、中毒性肾病、贫血、中毒性脑病等；四乙基铅中毒可产生严重神经系统症状，部分患者出现全身皮疹，可有呼吸道刺激症状；铍化合物的皮肤损害主要表现为皮炎、铍溃疡和皮肤肉芽肿；铬对皮肤的损害较明显；磷早期中毒症状一般为神经系统和消化系统症状等。

（2）有机溶剂中毒。

有机溶剂中毒引起的职业危害问题，目前在全国也是非常突出的。例如生产酚、硝基苯、橡胶、合成纤维、塑料、香料以及制药、喷漆、印刷、橡胶加工、有机合成等工作常与苯接触，可引起苯中毒；还有甲苯、汽油、四氯化碳、甲醇和正乙烷中毒等。苯中毒主要影响造血系统和中枢神经系统。甲苯与苯大体相同，但毒性略轻些。汽油主要经呼吸道吸入，急性中毒时，轻者由头痛、头晕、无力，呈"汽油醉态"。高浓度吸入还可以引起化学性肺炎、肺水肿，严重者出现中毒性脑病等。

（3）刺激性气体中毒。

工业生产中常遇到的一类气体主要有氯气、光气、氮氧化物及氨气等。刺激性气体对呼吸道有明显的损害，轻者为上呼吸道刺激症状，重者可产生喉头水肿、喉痉挛、中毒性肺炎，可导致肺水肿。刺激性气体大多是化学工业的原料和副产品，此外在医药、冶金等行业也经常接触到。刺激性气体多有腐蚀性，生产过程中常因设备被腐蚀而跑、冒、滴、漏现象，或因管道、容器内压力增高而致刺激性气体大量外溢造成中毒事故。刺激性气体中毒主要是眼、上呼吸道均有刺激征等，严重时，可发生黏膜坏死、脱落，引起突发性呼吸道阻塞而窒息。

（4）窒息性气体中毒。

常见的是二氧化碳中毒。不通风的发酵池、地窖、矿井、下水道、粮仓等处，可由较高浓度的二氧化碳蓄积。二氧化碳中毒常为急性中毒，患者进入高浓度的二氧化碳环境后，几秒钟内即迅速昏迷，如不能及时地救出可致死亡。另外还有氰化物及甲烷中毒等。

（5）高分子化合物中毒。

高分子化合物的生产包括有化学原料合成单体，单体经过聚合或缩聚成聚合物，聚合物的加工、塑制等。在整个合成、加工及使用过程中均可产生一些有害因素。

（四）高温作业的危害与预防

1. 高温作业的定义

高温作业是指有高气温，或有强烈的热辐射，或伴有高气湿（相对湿度≥80%）相结合的异常作业条件，湿球黑球温度指数（WBGT 指数）超过规定限值的作业，包括高温天气作业和工作场所高温作业。

高温天气是指地市级以上气象主管部门所属气象台站向公众发布的日最高气温 35 ℃ 以上的天气。高温天气作业是指用人单位在高温天气期间安排劳动者在高温自然气象环境下进行的作业。工作场所高温作业是指在生产劳动过程中，工作地点平均 WBGT 指数≥25 ℃ 的作业。

2. 高温对健康的影响

高温可使作业人员感到热、头晕、心慌、烦、渴、无力、疲倦等，可出现一系列生理功能的改变，主要表现如下：

（1）体温调节障碍，由于体内蓄热，体温升高。

（2）大量水盐丧失，可引起水盐代谢平衡紊乱，导致体内酸碱平衡和渗透压失调。

（3）心律脉搏加快，皮肤血管扩张及血管紧张度增加，加重心脏负担，血压下降；但重体力劳动时，血压也可能增加。

（4）消化道贫血，唾液、胃液分泌减少，胃液酸度降低，淀粉活性下降，胃肠蠕动减慢，造成消化不良和其他胃肠道疾病。

（5）高温条件下若水盐供应不足可使尿浓缩，增加肾脏负担，有时可导致肾功能不全，尿中出现蛋白、红细胞等。

（6）神经系统可出现中枢神经系统抑制，注意力和肌肉的工作能力、动作的准确性和协调性及反应速度的降低等。

高温环境下发生的急性疾病多为中暑，按发病机理可分为热射病、日射病、热衰竭和热痉挛。为使企业在职业病登记和报告中易于识别，在《防暑降温措施暂行办法》中将中暑分为如下三种：① 先兆中暑。在高温作业过程中出现头晕、头痛、眼花、耳鸣、心悸、恶心、四肢无力、注意力不集中、动作不协调等症状，体温正常或略有升高，但尚能坚持工作。② 轻症中暑。具有前述症状，而一度被迫停止工作，但经短时休息，症状消失，并能恢复工作。③ 重症中暑。具有前述中暑症状，被迫停止工作，或在工作中突然晕倒，皮肤干燥无汗，体温在 40 ℃ 以上或发生热痉挛。

3. 防暑降温措施

（1）改善作业环境。预防中暑的关键在于改善高温作业环境，使作业场所的气象条件符合国家规定的卫生标准。在高温班组内合理布置热源，避免作业人员周围受到热源作用。尽可能把各种加热设备置于班组之外。温度很高的产品应尽快运出班组，如果热源不能移动，应采取隔热措施。通风是防暑降温的重要措施，应加强自然通风，使班组内高温从高窗或气孔排出。班组屋顶可安装风帽，墙角可开窗，加强通风。当自然通风不能将余热全部排出时，应采用机械通风。

（2）加强个体防护。高温作业人员应穿耐热、坚固、导热系数小、透气功能好的浅色工作服，根据防护需要，穿戴手套、鞋套、护腿、眼镜、面罩、工作帽等。

（3）采取必要的组织措施和保健措施。制定合理的劳动和休息制度，调整作息时间，采取多班次工作办法；合理布置工间休息地点；加强宣传教育，使职工自觉遵守高温作业安全卫生规程；定期检测作业场所的气象条件；实行医务监督，对高温作业人员定期进行体检；为高温作业人员提供清凉饮料。

（五）噪声的危害与预防

1. 噪声的定义

噪声是指由不同频率和不同强度的音响，无规律地组合在一起所形成的声音。它是人们不需要的声音，或者是指那些听起来使人厌烦的声音，表示噪声强弱的物理量称为噪声级。

2. 噪声的来源

噪声的来源是多种多样的，工业生产过程中产生的噪声大体上可以分为三大类。

（1）空气动力性噪声，如鼓风机、空压机、汽轮机、风动工具、汽笛等产生的噪声。

（2）机械性噪声，如冲床、球磨机、车床、电锯、滚筒、剪板机、织布机等产生的噪声。

（3）电磁性噪声，如发电机、变压器、电力继电器等电气设备运转时所产生的噪声。

3. 噪声的危害

噪声对人体的危害是多方面的，不同的噪声级对人体造成的主观感受如表 8-1 所示。

表 8-1　不同的噪声级对人体造成的主观感受

声音	噪声级/dB	主观感受	声音	噪声级/dB	主观感受
刚刚引起听觉	0	极静	一般车辆行驶	80	较吵
风吹落叶沙沙声	20	极静	很嘈杂的马路	90	较吵 很吵
轻声耳语	20~30	极静 安静	拖拉机开动	100	很吵
卧室	20~30	安静	电锯工作	110	很吵 感到疼痛
图书馆阅览室	30~40	安静	球磨机工作	120	感到疼痛
办公室	40~50	安静 较静	螺旋桨飞机起飞	130	感到疼痛 无法忍受
一般说话	30~45	较静	喷气式飞机起飞	140	无法忍受
大声说话	45~60	较静 较吵	火箭、导弹发射	150	无法忍受

当噪声突然≥150 dB 时，会引起鼓膜破裂。为保护听力，声音不能超过 90 dB；为保证学习和工作，声音不能超过 70 dB；为保证休息和睡眠，声音不能超过 50 dB。

噪声的危害可分为听觉系统危害和听觉外系统危害。

（1）听觉系统危害。

长期接触强烈的噪声，听觉系统首先受损，听力的损伤有一个从生理改变到病理改变的循序渐进过程。

① 暂时性听阈位移。

暂时性听阈位移是指人或动物接触噪声后引起听阈变化，脱离噪声环境后经过一段时间听力可恢复到原来水平，根据变化程度不同分为听觉适应和听觉疲劳。

听觉适应：指短时间暴露在强烈噪声环境中，感觉声音刺耳、不适，停止接触后，听觉器官敏感性下降，脱离接触后对外界的声音有"小"或"远"的感觉，听力检查听阈可提高 10~15 dB（A），离开噪声环境 1 min 之内可以恢复。

听觉疲劳：指较长时间停留在强烈噪声环境中，引起听力明显下降，离开噪声环境后，听阈提高超过 15~30 dB（A），需要数小时甚至数十小时听力才能恢复。

② 永久性听阈位移。

永久性听阈位移是指噪声引起的不能恢复到正常水平的听阈升高。根据损伤的程度永久性听阈位移又分为听力损伤、噪声性耳聋及爆震性耳聋。

听力损伤：一般表现为听力曲线在 3 000~6 000 Hz 出现"V"形下陷，此时患者主观无耳聋感觉，交谈和社交活动能够正常进行。

噪声性耳聋：人们在工作过程中，由于长期接触噪声而发生的一种进行性的感音性听觉损伤。随着损伤程度加重，高频听力下降明显，同时语言频率（500~2 000 Hz）的听力也受到影响，语言交谈能力出现障碍。

③ 爆震性耳聋。

在某些生产条件下，如进行爆破，由于防护不当或缺乏必要的防护设备，可因强烈爆炸所产生的振动波造成急性听觉系统的严重外伤，引起听力丧失，称为爆震性耳聋。根据损伤程度不同可出现鼓膜破裂、听骨破坏、内耳组织出血，甚至同时伴有脑震荡。患者主诉耳鸣、耳痛、恶心、呕吐、眩晕，听力检查严重障碍或完全丧失。

（2）听觉外系统危害。

噪声还可引起听觉外系统的损害，主要表现在神经系统、心血管系统等，如易疲劳、头痛、头晕、睡眠障碍、注意力不集中、记忆力减退等一系列神经症状；高频噪声可引起血管痉挛、心率加快、血压增高等心血管系统的变化；长期接触噪声还可引起食欲不振、胃液分泌减少、肠蠕动减慢等胃肠功能紊乱的症状。

4. 噪声的预防

（1）噪声源的控制。

合理选择材料和改进机械设计来降低噪声；改进工艺和操作方法降低噪声；减少激振力来降低噪声；提高运动零部件间的接触性能；降低机械设备系统的噪声辐射部件对激振力的响应；使用低噪声设备。

（2）传播途径的控制。

在传播途径上控制噪声是目前噪声控制中采用最普遍的技术，按其工作原理可分为吸声、隔声和消声。

（3）个人防护。

进行反复的宣传教育，使广大职工认识到噪声的危害和防噪的重要性，自觉做好个人防护。常用的个人防护用品分为内用和外用两种：外用的有将耳部全部覆盖起来的耳罩或帽盔；内用的有插入外耳道中的耳塞。

（六）电离辐射的危害与预防

1. 概念及分类

辐射是电辐射源发射出的电磁波和微粒子流的总称。它包括电离辐射和非电离辐射。

电离辐射对人体的照射方式分为外照射和内照射。外照射是体外辐射源对人体的照射，外照射主要来源于 α、β、γ、X 等射线。内照射是指进入体内的放射性核素对人体造成的照射。人体摄入放射性核素的途径有吸入、食入和通过皮肤、毛孔或伤口吸收进入，以及医疗诊治等，如人体吸入空气中的放射性气体和颗粒，食入食物和水中的放射性物质等。

α 射线：带正电荷的粒子流，其电离作用很强，穿透能力差，在空气中的射程只有 3~8 cm，一张纸或健康皮肤便可将其挡住。

β 射线：带负电的高速电子源，电离能力较 α 射线小得多，穿透能力较强，在空气中的射程有几米至十几米。

γ 射线：波长较短，不带电的电离辐射（统称为光子），以光速向四周扩散，穿透能力极强，能穿透金属，但电离能力很弱。

X 射线：由原子核外的电子壳层中发射出来的，性质与射线大致相同。

2. 电离辐射对人体的危害

电力生产、电力建设中射线用于金属探伤，作业时由于不了解放射线特性，反复受超过允许剂量的体外照射，当累积量（一次连续照射或多次反复照射所受到的总剂量）达到一定量时，可能引起慢性放射病。如在短时间内，接受大剂量电离辐射，可能引起急性放射病。

电离辐射对人体的危害主要是慢、急性放射病，包括躯体效应和遗传效应。

慢性放射病的临床表现为头痛、乏力、易激动、睡眠不良、食欲降低、心悸、多汗等，进而出现顽固性头晕、头痛、记忆力减退，常伴有恶心、腹胀、鼻和牙龈出血、低烧、毛发脱落、贫血等症状，最后导致极度疲乏、血压降低、胃肠功能紊乱等，甚至发展为急性白血病。在血象方面，早期白细胞波动不定，以后逐渐下降，白细胞分类主要是中性粒细胞减少，淋巴细胞、单核细胞和嗜酸性细胞相对增加，较晚则有血小板、红细胞、血红蛋白的降低，以及白细胞形态的改变。

急性放射病的临床表现为神经系统过度兴奋、剧烈的头痛、头晕，甚至丧失知觉，同时伴有恶心、烦躁、食欲降低、腹泻、血便、脱水等症状。病情轻重与接受剂量有关，剂量当量在 $0.2\sim1\mathrm{Sv}$ 时，被照射者会短暂出现轻微淋巴细胞和中性粒细胞减少，在 $1\sim2\mathrm{Sv}$ 时，会出现疲乏、呕吐、淋巴细胞和中性粒细胞减少，而且恢复很慢，可能出现远期效应；当剂量更高时，会出现恶心、呕吐、周身不适，甚至死亡。病情发展一般要经过初期（$1\sim3\mathrm{d}$）、假愈期（$3\sim10\mathrm{d}$）、极期（$30\sim35\mathrm{d}$）和恢复期（$3\sim4$ 个月）四个阶段。电离辐射的远期效应潜伏期很长。远期效应主要有白血病、再生障碍性贫血、恶性肿瘤、白内障以及对早期胎儿的影响等。遗传效应是指电离辐射对生殖细胞的损伤并在后代身上显示出来的效应。

3. 电离辐射的防护

（1）外照射的防护方法。

① 时间防护：在不影响工作的前提下，尽可能减少对人体的受照时间，培训工作人员熟悉和掌握操作技巧，达到操作敏捷准确，以减少照射时间，也可采用多人轮换操作的方法。

② 距离保护：在保证效果的前提下，操作人员应尽量远离放射源，如利用镊子、钳子机械手等操作，增加人体和放射源之间的距离。

③ 屏蔽防护：外防护应用最多的方法。即在人与放射源之间放置屏蔽材料，以减少射线强度。

（2）内照射的防护方法。

内照射主要是放射性核素的吸入和食入，因此外照射防护三种方法同样适用于内照射。比如减少在辐射工作环境中的停留时间，远离辐射工作环境，佩戴口罩，穿着防护服，不要在工作区域饮水和饮食，不要使用受到辐射污染的食品；室内应用湿拖布擦拭、防止灰尘飞扬；放射性同位素储存室应安装排气管；所有操作要在隔离屏蔽通风小室内进行。操作人员操作时要非常小心，不要让沾有放射性物质的器物割破皮肤，手部有伤口，必须停止工作；不要用有机溶剂洗手。

（3）个体防护。

从事放射性工作的人员要配备个人防护用具，如工作服、帽子、橡皮手套、鞋子以及口罩、面具等。工作服和手套要经常清洗。

（4）卫生保健。

对从事放射线的工作人员，应进行就业前健康检查，凡血红蛋白低于 11%（男）或 10%（女），红细胞数低于 4 000/mm³，血小板持续低于 10 万/mm³ 者，均不能从事放射性工作。

对在职放射性人员，如受照范围接近年最大允许剂量当量水平者，每年体检一次；低于 30%者，每 2～3 年体检一次。如有条件，在职放射性人员，应佩戴剂量仪，工作场所剂量测量结果及个人每日所受剂量应记录备查。

4. 非电离辐射的危害与防护

非电离辐射是指量子能量在 12 eV 以下的粒子，这些粒子不引起物质产生电离。它包括紫外线、红外线、可见光、射频辐射（微波、高频电磁场）等。电力生产、建设中受影响的主要是紫外线（电焊工）和射频辐射。紫外线对人体的影响主要是造成皮肤和眼睛的损害；射频辐射主要是引起中枢神经和植物神经系统机能障碍，临床表现主要为神经衰弱综合征。因此，也应采取相应的防护措施。

Task Ⅰ On-site risk identification and occupational disease

Ⅰ. On-site risk identification

Every production activity of human beings involves dangers of different degrees and unsafe factors, e.g., workers themselves, tools and equipment, subjects of labour and working environment. Risks are dangers or disasters that may occur. Risk identification is also known as dangerous points identification. By identifying unsafe behaviors of the people, unsafe state of objects and lack of safety management on the work site, analyzing risk factors that may cause accident and then taking targeted prevention and control measures before start of work, the process of work is controlled, and personal accidents and injury and human liability accidents are prevented so that the objective of safe production is achieved.

(Ⅰ) Risk factors on the site

In actual production activities, risk is a hidden danger that gives rise to accident. By analyzing and identifying risks in advance and taking measures to prevent the risks, we can come safely out of danger. Risk factors in work on electric power system refer to locations, positions, places, equipment, tools and instruments and behaviors that may cause danger during work, including the following ones:

(1) The operating personnel fail to abide by regulations for safety in work. Outdoor high-voltage equipment inspection in thunderstorm days without wearing insulating boots may cause electric shock by step voltage.

(2) Unsafe objects such as mechanical equipment. Touching an exposed high-speed equipment rotating part will result in mechanical injury. When equipment has no grounded housing or has unacceptable grounding resistance, insulation damage may cause electric shock.

(3) Unsafe working environment. Harmful dusts, toxic gases, high temperature, noise, radiation, etc. in the environment are all unsafe factors. Harmful dust in working environment (if any) may cause pneumoconiosis; Occupational toxicants in working environment may cause occupational poisoning, etc.

Characteristics of risk factors on the site:

(1) Objective reality. Presence of risk factors on work site is independent of human subjective consciousness.

(2) Complexity and variability. Complexity of risk factors on work site depends on complexity of the work. Dangerous points may be different if operating personnel, operating sites, tools and methods used for work are different even if the job tasks are identical. Even in identical tasks, existing risk points are variable. After existing risks are eliminated, new risks will occur. Therefore,

analysis of risk points for prevention and control is not efficacious forever, and the specific situations shall be analyzed and actions to be taken shall be determined considering the practical situations.

(3) Be potential. Risk factors on a work site exist in upcoming work process, although they are in objective existence, they are not yet true hazards.

(4) Be knowable and preventable. Risks factors on electric power work site are potential and hidden and are not easily preventable, controllable and manageable. But as risk points are in objective existence, they are definitely identifiable, recognizable and preventable and controllable by us. So long as we attach importance to our thinking and take effective measures, we can definitely nip the risks in the bud.

(Ⅱ) Work site risk analysis

Risk analysis, also known as dangerous points analysis, means identification and inference of types of hazards in a job item (project), conditions for occurrence, possible resultant situations and consequences before start of a job or commencement of a project to find out risk points. The purpose is controlling occurrence of accident.

Some risk points during work are obvious and easily identifiable, e.g., danger of electric shock and mechanical injury in electrical work, risk of falling from height during working at height. To prevent apparent risks, we can take sufficient safety measures in advance, e.g., wearing insulating gloves, safety helmets, insulating boots, maintaining sufficient safe distance from charged objects and fasten safety belts during work. As some risk points during work are potential and not detectable, we must accurately identify potential risk points through analysis in a scientific manner.

1. Key points in risk analysis

(1) Work sites: work at different sites (for example, locations of work at height, live-wire work, work during outage, downhole operation, work in cabinets have different roads and traffic conditions) may impose different safety hazards on workers, equipment and power grids.

(2) Working environment: different working environments (e.g., high temperature, high pressure, high altitude, flammable, explosive, radiation, toxic and harmful gases, oxygen deficiency, etc.) may cause different safety hazards to operators, equipment and power grids.

(3) Equipment state: hazards on operators, equipment and power grid from use of machinery, equipment, tools, etc. used for work.

(4) Normalization of work: nonstandard operating procedures, improper operating methods and failure to strictly comply with safety specifications may cause harm to operators, equipment and power grids.

(5) Operator state: possible hazards to operators, equipment and power grids due to unconsciousness of operators (e.g., after drinking, use of medicine and staying up late), poor physical health, unsafe operating behaviors, insufficient technical skills, etc.

(6) Other factors: other unsafe factors that may do harm to operators, equipment and power grid.

2. Risk analysis steps

(1) Data analysis: collect relevant data about similar operations by the organization or other organizations in the past, carry out in-depth analysis of safety accidents that have occurred in similar operations, identify root causes of the accidents and take corresponding measures.

(2) Comprehensive investigation: organize comprehensive on-site investigation, begin with scope of the work to be carried out, working environment, status of working equipment and other aspects; evaluate potential sources of risks and hidden dangers that may occur during work, and develop specific risk control and prevention measures on this basis.

(3) Comprehensive analysis: relevant technical personnel shall carefully search for key points that result in hazards to personnel, equipment and power grid considering contents and methods of on-site operations, environmental conditions, equipment conditions, tools and instruments conditions, operational skills, physical fitness and work habits of relevant technicians involved in the operation. A more experienced operator knows more about the work and can provide thorough risk points analysis and adequate estimation of problems that may occur. Specific practices in comprehensive analysis stage:

① The person in charge of work organizes shift team members to carry out all-round analysis and prediction of the job task. For complex and large-scale work, experienced and skilled employees shall be invited to attend the meeting for providing guidance.

② Break down work process into a few phases and analyze possible risk points in all phases based on experiences in similar work. The members express their opinions, attend extensive discussions, conduct scientific analysis and make all efforts to identify all existing risk points through brainstorming. When work period is shorter, risk points predicted will be more reliable.

③ Learn how risk points evolve into accidents, i.e., what are the conditions for risk points to evolve into accident. By blocking conditions for risk points to evolve into accident, occurrence of accidents is prevented.

④ Evaluate risk points identified, determine risk level considering degree and severity of the risk points, develop prevention and control measures for each risk point and focus on prevention of high risk points.

⑤ During the work, potential risk points shall be continuously identified and searched for considering changes in work process, climate, environment, etc.

(Ⅲ) Risk prevention and control

Risk control is a safety management method for analyzing and identifying potential risk factors that may result in accidents in each work on electric power system considering content, work methods, environment, personnel conditions, equipment conditions, etc. so that preventive measures can be developed according to the regulations and systems. The basic work is to identify and search risk points, and procedural and standardized operations on work site are critical for risk control. Facts have proven that it is an effective method for preventing accidents.

1. Risk prevention and control include:

(1) Macro control. Risk point are controlled with a whole system as the object. Main methods include legal means (policies, laws and regulations), economic and administrative means (rewards, punishments, compensations) and educational means (safety education and training).

(2) Micro control. Risk point are controlled with specific risk points as control objects. The main means used are rectification measures, organizational measures, safety technical measures, early warning and supervision. Through combination, complementation and mutual restriction of macro and micro controls, a multi-level tri-dimensional control system for safety accident control can be established.

2. For risk prevention and control, the following things must be done.

(1) Start from the source. Inherent equipment defects are sources of risk points, and they shall be eliminated in a timely manner through defect elimination, maintenance and transformation to fundamentally eliminate potential risks. Other prevention and control measures and emergency plan can be used when there is defect non-eliminable at present or fundamental control and management are technically and economically infeasible, but approval from relevant authorities or higher levels is required considering nature of the risk points.

(2) Education and training on safety. In fact, expressions such as "not allowed", "strictly prohibited", and "prevented" in *Electric Power Safety Working Regulations of State Grid Corporation of China* tell us what can be done and what cannot be done in our work and indicate specific risk points in operations and are of universal applicability and operability for similar work. Experiences show that generation, expansion and sudden change of risk points leading to accident are as a result of unfamiliarity with or inability to strictly comply with safe work regulations. Therefore, operators must know well safe work regulations and *Electric Power Safety Working Regulations of State Grid Corporation of China* and complete work task by strictly following the regulations.

(3) Emphasize process control in work and operation. Risk point control shall focus on work process control, and every operator on the site shall be clearly informed of the job task, existing risk points and safety precautions to be taken. For example, work ticket is required for maintenance, organizational measures for safety protection must be taken, risk point prevention and control measures shall be confirmed in writing, and every stage and work step in equipment maintenance shall be confirmed and controlled so that the whole process of work is under control.

(4) Carry out standardized, normalized and routinized operations. To strengthening prevention and control of risk points in behaviors of operators, education and training on safety must be provided, protection must be strengthened, and standardized, normalized and routinized operations are required. If a standardized operation process for switching operation is established, all operators must strictly follow the standardized process to carry out their work, standardize their behaviors and avoid omission and reverse order and incorrect operation.

(Ⅳ) Purposes of risk analysis and prevention and control

Domestic and foreign data statistics show that over 90% of accidents are due to lack of prior anticipation of potential risk points that may cause damage or lack of effective preventive measures even if the risk points are anticipated. Therefore, effective analysis, prevention and control of risk points can prevent human factors that may trigger accidents and will nip the accidents in the bud. Specifically, risk point analysis and prevention and control have the following functions:

(1) Enhance awareness of risks among operators, overcome mental paralysis and prevent risky behaviors. Occurrence of some accidents is directly related to insufficient understanding of potential risk points and hazards of the risk points in work, lack of awareness of risks and lack of measures to prevent risks.

(2) Be able to prevent danger caused by hastiness. Inadequate preparation, inadequate plan, disorder or simplifying or reversing work steps are hidden dangers.

(3) Be able to prevent accidents due to inadequate operating skills. Before start of work, risk points analysis and prevention and control are for recognition of importance of safety and relearning processes and technical skills for the work. Relearn processes and technical skills for the work.

(4) Risk points analysis and control can make safety measures more targeted and effective and prevent accidents. Experiences tell us that if operators have no idea about risks in the operation and safety measures indicated in work ticket are not targeted and operable, accident may occur.

(5) Risk points analysis and control can reduce and even eliminate accidents due to poor command. Commanders who are unfamiliar with dangerous situations in the operation or provide commands based on subjective assumption are most likely to cause accident and even mass death and casualty.

Ⅱ. Occupational hazards and occupational diseases

(Ⅰ) Occupational diseases

Occupational diseases are diseases suffered by labors in enterprises, public institutions and individual economic organizations (collectively referred to as employers) who come into contact with dusts, radioactive substances and other toxic and harmful substances in their occupational activities. There are provisions on occupational disease prevention in laws of the countries. Generally, any disease specified in the laws can be referred to as occupational disease.

1. Classification of hazards from occupational diseases

Hazards from occupational diseases can be classified into the following three categories on source basis:

(1) Occupational disease hazards in relation to production process.

① Physical factors, e.g., high temperature, low temperature, radiation, productive noise, vibration, etc.

② Chemical factors, including industrial dusts such as coal dust, silicon dust, asbestos dust, welding dust and toxic substances such as lead, manganese, mercury, benzene, etc.

③ Biological factors, including pathogenic microorganisms or parasites such as bacillus anthracis, brucella, TBE and mold on bagasse; infectious disease sources contacted by medical workers, e.g., SARS virus and COVID-19.

(2) Occupational disease hazards in relation to labor process.

Excessively long working time, excessively high work intensity, unreasonable labor systems and organizations, prolonged compulsive position labor, excessive stress on individual organs and systems, prolonged poor posture or use of unreasonable tools can do harm to health of laborers.

(3) Occupational disease hazards in relation to working environment.

Main hazardous factors in working environment are harmful factors in natural environment, e.g., high temperature, high pressure, humidity, low temperature, non-compliance with hygiene requirements or standards, lack of necessary hygiene engineering and technical facilities, defective safety protection equipment and personal protective equipment, etc.

2. Prevention of occupational disease hazards

Occupational diseases shall be prevented by following the three-level prevention rule.

① Level-1 prevent: protect laborers from exposure to harmful factors of occupational disease or control harmful factor in work site within the limit permissible by hygienic standards.

② Level-2 prevention: monitor health of workers, take treatment measures and effective medical care to prevent progression of diseases upon identification of occupational hazards.

③ Level-three prevention: provide active medical care to occupational disease sufferer to promote health.

Basic measures for three-level prevention include:

① providing labor hygiene supervision, including preventive and regular hygiene supervision and emergency treatment.

② Reducing concentration (intensity) of harmful factors: common hygiene measures to reduce concentration (intensity) of harmful factors include improving process, preventing escape of harmful factors, promoting use of low-toxic and non-toxic materials or technologies, providing personal protective equipment, ventilation and dust prevention, etc.

③ Occupational health screening, e.g., occupational physical examination, regular physical examination and physical examination before retirement.

Harmful factors that power enterprise employees are exposed to during production and construction include industrial dust, noise, high temperature, toxic substances, rays, microwave radiation, low-frequency radiation, etc.

(Ⅱ) Industrial dusts and pneumoconiosis

Industrial dusts are solid particles generated in production process that suspend in the air for a long time. Industrial dusts are from mechanical processing of solid substances, condensation of substance vapors and incomplete combustion of substances. Industrial dusts that are widely distributed in China and have significant impact on health of occupational population include silicious dust, coal dust and asbestos dust.

1. Sources of industrial dusts

There are many sources of industrial dusts, e.g., mechanical processing and smashing of solid substances; metal grinding and cutting; ore crushing, screening and batching or rock drilling, blasting and crushing; processing of raw materials in industries such as refractory materials, glass, cement and ceramics; Processing of raw materials such as fur and textiles; vapor from solid materials processing and substances heating in chemical industry, smoke from incomplete combustion of organic substances, etc.

2. Classification of industrial dusts

Industrial dusts can be classified into three types:

① inorganic dust, including mineral dust such as quartz, asbestos, coal, etc.; Metallic dusts such as iron, tin, aluminum and their compounds; Artificial inorganic dusts such as cement, diamond sand, etc.

② Organic dusts, including plant dusts such as cotton, hemp, flour and wood; Animal dusts such as fur and silk dust; Synthetic organic dyes, pesticides, synthetic resins, explosives and artificial fibers.

③ Mixed dust means mixed form of various dusts, mixture of more than two above-said dusts in general. Mixed dust is the most common one in production environment.

3. Occupational hazard caused by industrial dusts

Harm of dusts to human body depends on chemical compositions and concentration. The higher the content of free silica in dust, the greater the harm to human health. The higher the concentration of a same dust, the more serious the harm to human health; In addition, extent of human body injury by dusts depends on degree of crushing or dispersity. When particle size is small and quantity of particles is large and dispersity is high and the particles suspend in the air for a longer time, there is higher probability of inhalation by human body and hazards are more serious. Occupational hazards caused by dusts include systemic toxicity, local irritation, allergic reaction, carcinogenicity and pneumoconiosis. Pneumoconiosis in particular poses the most serious threat. Pneumoconiosis is one of the most serious occupational hazards in industrial production in China at present.

The sufferers may have no clinical symptoms in the early stage, and some sufferers have chest tightness, cough, expectoration, etc. With aggravation of the above symptoms, they may have asthma, expiratory dyspnea and even emphysema pulmonum and pulmonary heart disease in later stage.

At present, there is no medicine for radical cure of pneumoconiosis, and symptomatic treatment and supportive treatment are provided for controlling progression of disease. At present, foreign countries have shifted their focus to prevention of pneumoconiosis, and measures taken include providing dust removal equipment, strengthening personal protection and attending regular physical examination by dust exposed workers at prevention and protection institutions.

(Ⅲ) Occupational toxicants and occupational diseases

Toxicants used or produced in production process are called occupational toxicants. Occupational toxicants can enter human body through different paths in certain conditions and cause poisoning, or give rise to changes in immune or other physiological functions, making people susceptible to illness or worsen conditions of the sufferers and prolonging course of the disease. Some toxicants cause local irritation, sensitization and corrosion, and some toxicants will cause tumors, teratogens and genetic mutations.

Occupational poisoning is a poisoning suffered by workers when they come into contact with toxic substances during production process.

1. Paths for occupational toxicants to enter human body

Knowing the path through which occupational toxicants enter human body is of great significance for taking corresponding prevention and control measures. Toxicants can enter human body through respiratory passage, skin and digestive tract.

(1) Enter human body through respiratory passage. Mucosa and alveoli of the entire respiratory tract can absorb toxic gases, vapors and smoke at different degrees, but the main parts are the bronchi and alveoli, especially the alveoli. Pulmonary alveolus has large contact area and is surrounded by capillaries. Toxic substances can quickly pass through capillaries and enter bloodstream and thus distribute throughout the body. As this path does not undergo detoxification by liver, there is high risk. Most of occupational poisoning in petrochemical enterprises is caused by inhalation through respiratory tract.

(2) Enter human body through skin. Skin fat soluble toxicants such as aniline and acrylonitrile can be absorbed through intact skin of human body which spread from hair follicle spaces to glandular cell and are absorbed by human body, and a mall part of toxicants enter human body through sweat glands. The path through which toxicants enter human body does not undergo liver transformation, and the toxicants enter bloodstream and spread throughout the body and give rise to high risk.

(3) There is minimum chance for digestive tract toxicants to enter human body. Such toxicants usually intrude into human body due to eating by mistake or through respiratory tract, some of which adhere to pharynx nasalis and mix with secretions and are unintentionally ingested. After going into digestive tract, the toxicants are usually excreted with feces, and some of them are absorbed in small intestine and excreted after detoxification by liver, and only a small portion enters blood circulation system.

2. Common occupational poisoning

(1) Metal and metalloid poisoning.

Metals can be classified into three categories: heavy metal, light metal and metalloid based on physical and chemical properties. There are many kinds of metal poisoning such as lead poisoning, tetraethyl lead poisoning, manganese poisoning, berylliosis poisoning and cadmium poisoning. Metalloid poisoning includes arsenic poisoning and phosphorus poisoning. Lead poisoning sufferer has metallic flavor in the mouth, and symptoms include nausea, vomiting, abdominal distention, paroxysmal abdominal angina, constipation or diarrhea; in severe cases, symptoms include convulsions, paralysis, coma, circulatory failure, toxic hepatic disease, toxic kidney disease, anemia, toxic encephalopathy, etc. Tetraethyl lead poisoning can cause serious nervous system symptoms, and some victims may have systemic rash and respiratory tract irritation; Skin damage by beryllium compounds mainly manifested as dermatitis, beryllium ulcer and skin granuloma. Chromium causes significant damage to skin; Early symptoms of phosphorus poisoning include neurological and digestive systems symptoms.

(2) Organic solvent poisoning.

Occupational hazards caused by organic solvent poisoning are also very prominent in our country at present. For example, production of phenol, nitrobenzene, rubber, synthetic fiber, plastics, flavors, pharmacy, painting, printing, rubber processing, organic synthesis and other work process exposure to benzene can cause benzene poisoning; There is also poisoning caused by toluene, gasoline, carbon tetrachloride, methanol and n-ethane. Benzene poisoning has impact on hematopoietic system and central nervous system. Toluene is basically the same as benzene, but it has slightly lower toxicity. Gasoline is inhaled through respiratory tract. An acute poisoning sufferer will feel headache, dizziness and weakness and will be at a "gasoline drunk state". One who inhales high-concentration gasoline may suffer chemical pneumonia, pulmonary edema and even toxic encephalopathy.

(3) Irritant gas poisoning.

Common gases in industrial production include chlorine, phosgene, nitrogen oxides and ammonia. Irritant gases have obvious damage to respiratory tract. In mild case, symptoms include upper respiratory tract irritation; and severe case, symptoms include laryngeal edema, laryngospasm, toxic pneumonia and even pulmonary edema. Most of irritant gases are raw materials and by-products in chemical industry, and they are also common in industries such as pharmacy and metallurgy. Most of irritant gases are corrosive. In production processes, irritant gases overflow, emit, drip and leak from eroded equipment or overfall from pipelines and pressure vessels due to pressure rise and cause poisoning. Irritant gas poisoning causes irritation in eyes and upper respiratory tract. In severe cases, mucosal necrosis and detachment may occur, resulting in sudden respiratory obstruction and asphyxia.

(4) Asphyxiating gas poisoning.

Carbon dioxide poisoning is very common. High-concentration carbon dioxide may gather in poorly ventilated fermentation tanks, cellars, mines, sewers, granaries and other spaces. Carbon dioxide poisoning is often acute poisoning. One who enters a high-concentration carbon dioxide environment will become unconscious within a few seconds and even die if there is no timely rescue. And there is cyanide and methane poisoning.

(5) polymer compound poisoning.

Production of polymer compounds includes monomer synthesis using chemical raw materials, monomer polymerization or condensation into polymers and polymers processing and molding. Harmful factors can occur throughout the entire process of synthesis, processing and use.

(IV) Hazards in work in the heat and prevention of the hazards

1. Definition of work in the heat

Work in the heat means work at high temperature, in abnormal work environment where there is intense heat radiation or high air humidity (relative humidity $\geq 80\%$RH), wet bulb and black bulb temperature index (WBGT index) exceeds specified limit value, including work in high temperature weather and high temperature workplace.

High temperature weather is the weather with daily maximum temperature over 35 °C as reported by meteorological stations affiliated with meteorological authorities at prefecture level or higher levels. Work in high temperature weather means work by laborers in high-humidity natural environment in high temperature weather as scheduled by the employer. Work in the heat at workplace means work at a site where average WBGT index is ≥ 25 °C in the process of work.

2. Impact of high temperature on human body

In high temperature environment, workers will feel hot, dizzy, flustered, annoyed, thirsty, weak and tired, and the following physiological function changes may occur:

(1) Thermoregulation disorder: heat accumulation in the body will give rise to body temperature rise.

(2) Loss of a large amount of water and salt may result in imbalance of water and salt metabolism, leading to acid-base imbalance and osmotic pressure imbalance in the body.

(3) Increased heart rate and pulse, vasodilatation and vasotonia will impose additional burden on the heart and cause blood pressure decline. But heavy physical labor may also give rise to elevation of blood pressure.

(4) Digestive anemia, decreased saliva and gastric secretion, decreased gastric acidity, decreased starch activity and decreased gastrointestinal motility will cause indigestion and other gastrointestinal diseases.

(5) At high temperature, insufficient water and salt supply will cause urine concentration, imposing additional burden on kidneys and sometimes leading to renal dysfunction, e.g., protein and red blood cells in the urine.

(6) The sufferer may experience central nervous system suppression, aprosexia and decreased muscle work ability and accuracy and coordination of actions and slow reaction.

Acute diseases occurring in high-temperature environment are often heatstroke which can be classified into heat apoplexy, asphyxia solaris, heat prostration and heat ulcer contracture based on pathogenesis. To allow easy identification in occupational disease registration and reporting by enterprises, in *"Interim Measures for Heatstroke Prevention and Cooling"*, heatstroke is classified into the following three types:

① premonitory heatstroke. During work in the heat, symptoms such as dizziness, headache, giddiness, tinnitus, palpitations, nausea, general fatigue, inattention and exercise not harmony may occur. Body temperature may be normal or slightly elevated, but the sufferer is still able to work.

② Light heat stroke. One who has to suspend work due to the above symptoms can self-healed and return to work after a short rest.

③ Severe heat stroke. A worker suffering the above-said heat stroke has to suspend work or suddenly fall in a faint during work, has dry skin without sweat, body temperature above 40T or heat cramp.

3. Measures for heatstroke prevention and cooling

(1) Improve working environment. To prevent heatstroke, improving high-temperature working environment and making sure that meteorological conditions in workplace meet national hygiene standards are crucial. Reasonably distributing heat sources at high-temperature work shift can prevent impact of heat sources in the areas around the operators. Heating devices shall be provided outside of workplace in so far as possible. Hot products shall be moved out of the workplace as soon as possible. If heat sources cannot be moved, insulation measures shall be taken. Ventilation is an important measure for heatstroke prevention and cooling, and natural ventilation shall be strengthened to ensure that high temperature is discharged from high windows or air holes in the workplace. A hood can be installed on the roof, and windows can be provided in the corners to improve ventilation. When natural ventilation is inadequate to remove all waste heat, mechanical ventilation is required.

(2) Use personal protection. High-temperature operators shall wear light colored work clothes that are heat-resistant, sturdy, have low thermal conductivity and high air permeability. As required for protection, they can wear gloves, shoe covers, leg guards, glasses, masks, work hats, etc.

(3) Take necessary organizational and health protection measures. Develop reasonable labor and rest systems, adjust work and rest schedules, and apply multi-shift work methods; Rationally provide rest areas in workplace; Strengthen publicity and education to ensure that employees consciously abide by the safety and hygiene regulations for work in the heat; Regularly monitor meteorological conditions in the workplace; Provide medical supervision and conduct regular physical examination for those working in the heat; Provide cool beverages for workers that work in the heat.

(Ⅴ) Hazard and prevention of noise

1. Definition of noise

Noise is a sound as a result of random combination of various sounds of different frequencies and intensities. It is not a sound that the people would like to hear, or it is an annoying sound. The physical quantity indicating noise strength is called noise level.

2. Sources of noise

There is a variety of sources of noise, and noise emitted in industrial production process can be generally classified into three categories.

(1) Aerodynamic noise: noise generated by blowers, air compressors, steam turbines, pneumatic tools, sirens, etc.

(2) Mechanical noise: noise from punch, ball mill, lathe, electric saw, drum, plate shear, loom, etc.

(3) Electromagnetic noise: noise emitted from electrical equipment such as generators, transformers and power relays during operation.

3. Hazards of noise

Harms of noise to human body are varied, and subjective feelings of human body caused by different levels of noise are as shown in Tab. 8-1.

Tab. 8-1 Subjective feelings caused by different levels of noise on human bodies

Sound	dB	Subjective feeling	Sound	dB	Subjective feeling
Just trigger auditory sense	0	Extremely quiet	General vehicle driving	80	Noisy
Rustling of fallen leaves in the wind	20	Extremely quiet	Very noisy road	90	Noisy Very Noisy
Whisper	20-30	Extremely quiet Quite	Tractor startup	100	Very Noisy
Bedroom	20-30	Quite	Electric saw operation	110	Very Noisy Feel pain
Reading rooms in a library	30-40	Quite	Ball mill operation	120	Feel pain
Office	40-50	Quite Quite	Propeller aircraft takeoff	130	Feel unbearable pain
Talking	30-45	Quite	Jet aircraft takeoff	140	Unbearable
Speak loudly	45-60	Quite Noisy	Rocket and missile launching	150	Unbearable

When noise is ≥150 dB suddenly, it may cause tympanic membrane rupture. To protect hearing, a sound shall be no more than 90 dB. To allow study and work, a sound shall be no more than 70 dB. To allow rest and sleep, a sound shall be no more than 50 dB.

Noise hazards can be classified into auditory system damage and external auditory system damage.

(1) Auditory system damage.

Long-time exposure to high noise will result in damage to auditory system, and hearing impairment is a result of a progressive process extending from physiological changes to pathological changes.

① Temporary hearing threshold shift.

Temporary hearing threshold shift means change in hearing threshold after exposure to noise by human or animals, and they can return to the original status after leaving the noisy environment for a period of time. It can be classified into auditory adaptation and auditory fatigue considering the extent of change.

Auditory adaptation: it means decreased sensitivity of auditory organ after brief exposure to highly noisy environment where the people feel harsh noise and discomfort; after such exposure, sounds may seem to be "low" or "far away" for the people, and auditory threshold may be increased by 10–15 dB (A), and auditory threshold may return to normal state within 1 min after leaving the noisy environment.

Auditory fatigue: it means significant hearing loss as a result of prolonged stay in a highly noisy environment. After leaving the noisy environment, auditory threshold will be increased by more than 15–30 dB (A), and auditory rehabilitation will occur after a few and even tens of hours.

② Permanent hearing threshold shift.

Permanent hearing threshold shift caused by noise that cannot be restored to normal level. Permanent threshold shift can be further classified into hearing loss, noise-induced deafness and explosive deafness.

Hearing loss is characterized by a "V"-shaped depression when auditory curve is at 3,000–6,000 Hz, at this moment the sufferer has no sense of deafness and is capable of talking and social activities.

Noise induced deafness is a progressive auditory impairment caused by long-term exposure to noise during work. As condition of the sufferer becomes worse, significant high-frequency hearing loss may occur, and hearing at language frequency (500–2,000 Hz) will also be impaired, resulting in speech disorder.

③ Explosive deafness.

In some production processes such as blasting, poor protection or lack of required protection equipment against blasting, vibration waves generated by violent explosion may cause severe acute trauma to auditory system, resulting in hearing loss, which is known as explosive deafness. The victim may suffer tympanic membrane rupture, ear bones damage, inner ear tissue bleeding and even cerebral concussion depending on degree of injury. The sufferer complains of tinnitus, earache, nausea, vomiting, dizziness, serious dysaudia or complete hearing loss.

(2) Damage to systems other than auditory system.

Noise can also cause damage to systems other than auditory system. The noise has impact on nervous system and cardiovascular system, and symptoms include fatigue, headache, dizziness, sleep disorder, inattention, memory loss and other neurological symptoms; High frequency noise can cause changes in cardiovascular system such as vasospasm, increased heart rate and elevation of blood pressure; Long-time exposure to noise can also cause gastrointestinal dysfunctions such as lack of appetite, decreased gastric secretion and slow intestinal peristalsis.

4. Noise prevention

(1) Noise source control.

Noise can be reduced by using proper materials and improving mechanical design; Improve technologies and operating methods to reduce noise; reducing noise by reducing excited force; improving contact performance between moving parts; reducing response of noise radiation components of mechanical equipment system to excited force; using low-noise equipment.

(2) Control route of transmission.

Transmission route control is the most commonly used technique for noise control, which can be classified into sound absorption, sound insulation and noise suppression on working principle basis.

(3) Personal protection.

Repeated publicity and education are required to enable the employees to recognize hazards of noise and importance of noise prevention, and personal protection is also required. Commonly used personal protective equipment include those for internal use and external use: devices for external use include ear muffs or helmets that cover the ears completely; devices for internal use include earplugs for inserting into auricular tubes.

(Ⅵ) Hazard and prevention of ionizing radiation

1. Concept and classification

Radiation is the general term for electromagnetic waves and microparticle flow emitted from electric radiation sources. It includes ionizing radiation and non-ionizing radiation.

Human body exposure to ionizing radiation include external exposure and internal exposure. External exposure is human body exposure to external radiation sources such as α, β, γ, X-rays. Internal exposure means human body exposure caused by radioactive nuclides entering human body. Ways of radioactive nuclides ingress into human body include inhalation, ingestion, absorption through skin, pores or wounds as well as medical diagnosis and treatment, e.g., inhalation of radioactive gases and particles in the air into human body and ingestion of radioactive substances into human body from foods and water.

α ray: a positively charged particle stream with high ionizing power and poor penetrating power, with 3-8 cm range in the air, be maskable by a piece of paper or health skin.

β ray: a high-speed electron source with negative charge, with ionizing power much lower than α rays, with high penetrating power and range in the air up to a few meters and even more than ten meters.

γ ray: uncharged ionizing radiation with short wavelength (collectively referred to as photons) which spreads around at speed of light. They have high penetrating power and can penetrate metals, but it has low ionization power.

X-rays are emitted from electron shell outside atomic nucleus and are roughly the same as rays in terms of properties.

2. Harm of ionizing radiation to human body

In power generation and power facilities construction, rays are used for metal flaw detection. Due to poor knowledge about characteristics of rays, operators are repeatedly exposed to external radiation at doses higher than the allowable dose during work; when cumulative exposure (total dose from a continuous exposure or repeated exposures) is up to a certain dose, chronic radiation sickness may occur. Exposure to high-dose ionizing radiation in a short period of time may cause acute radiation sickness.

Hazards of ionizing radiation to human body include chronic and acute radiation sickness, e.g., somatic and genetic effects.

Clinical manifestations of chronic radiation sickness include headache, lack of power, emotional excitement, dyssomnia, inappetence, palpitations, hidrosis, etc. which are followed by refractory dizziness, headache and memory loss and simultaneous phenomenons such as symptoms such as nausea, abdominal distension, nose and gum bleeding, low fever, trichomadesis, hypohemia, etc., and extreme fatigue, hypotension, gastrointestinal disorders and eve acute leukemia may occur. As shown on the hemogram, number of white blood cells fluctuated in early stage and then gradually decrease; in terms of white blood cells, neutrophile granulocyte decreases, lymphocytes, monocyte and eosinophils increase; in later stage, platelets, red blood cells and hemoglobin decrease and white blood cell morphology changes.

Clinical manifestations of acute radiation sickness include hyperactivity of nervous system, severe headache, dizziness and even loss of consciousness; Accompanied symptoms include nausea, fidget, loss of appetite, diarrhea, bloody stool, dehydration, etc. Severity of illness depends on accepted dose. When a dose equivalent is 0.2–1 Sv, the sufferer will experience slight momentary decrease in lymphocytes and neutrophils. When dose equivalent is 1–2 Sv, fatigue, vomiting and lymphopenia and neutropenia will occur, and period of recovery is very long and there may be long-term effects; Higher dose may cause nausea, vomiting, overall discomfort and even death. Progression of disease usually include four stages: initial phase (1–3 days), latent phase (3–10 days), stadium acmes (30–35 days) and recovery phase (3–4 months). Long-term effects of ionizing radiation have long latent period. Long-term effects include leukemia, aplastic anemia, malignant tumors, cataracts and impact on early fetuses. Hereditary effect refers to germ cell damage by ionizing radiation and its effect on later generations.

3. Protection against ionizing radiation

(1) Protection against external exposure.

① Time protection: minimize period of human body exposure without impact on work, provide training to operators to enable them to know well the operating skills and work quickly and accurately so as to reduce irradiation time; and alternate operations by multiple operators are advised.

② Distance protection: when effectiveness is ensured, operators shall stay away from radioactive sources in so far as possible. For example, using tweezers, pliers and manipulators for operation and increasing the distance between human body and radiation sources.

③ Shielding protection: it is the most commonly used method of external protection. Shielding materials are provided between the people and radioactive sources to reduce radiation intensity.

(2) Protection against internal radiation.

As internal irradiation involves inhalation and ingestion of radioactive nuclides, the three methods of external radiation protection are also applicable to internal irradiation protection, e.g., reducing work time in radiation environment, staying away from radiation environments, wearing masks, wearing protective suits, avoiding drinking or eating in work area and using food contaminated by radiation; using wet mops to wipe off dusts and prevent dust rising; Radioactive isotope storage room shall have exhaust pipes; All operations shall be completed in isolated, shielded and well-ventilated compartments. Operators shall be take great care during work to prevent objects contaminated by radioactive substances from cutting their skin, and an operator with hand injury must stop working; Do not wash hands with organic solvent.

(3) Personal protection.

Radioactive work staff shall have personal protective equipment such as work clothes, hats, rubber gloves, shoes, masks, etc. Work clothes and gloves shall be cleaned regularly.

(4) Health care.

Radioactive work staff shall undergo pre-job health check. Those with less than 11% (male) or 10% (female) hemoglobin, less than 4000/mm^3 RBC, less than 100,000/mm^3 blood cells shall be excluded from radioactive work.

In-service radioactive work staff shall attend annual health check if extent of exposure is up to the maximum allowable yearly dose equivalent; 2-3 physical examinations is required in a year if it is less than 30%. If conditions permit, in-service radioactive work personnel shall wear dosimeter, and doses measured at workplace and individual daily exposure dose shall be recorded for reference.

4. Hazards of non-ionizing radiation and protection

Non-ionizing radiation means particles with less than 12eV quantum energy which do not cause ionization. It includes ultraviolet, infrared, visible light, radio frequency radiation (microwave, high-frequency electromagnetic field), etc. The main impacts from power generation and construction are ultraviolet (on welders) and radio frequency radiation. Main impact of ultraviolet radiation on human body is damage to skin and eyes; RF radiation will cause dysfunctions of central nervous system and antagonistic system, and the main clinical manifestation is neurasthenia syndrome. Therefore, corresponding protective measures shall also be taken.

任务二 触电急救

一、现场抢救的基本原则

现场抢救必须做到迅速、就地、准确、坚持四项原则。

(一) 迅速

迅速就是要争分夺秒、千方百计地使触电者脱离电源,并将受害者放到安全的地方。这是现场抢救的关键。

(二) 就地

就地就是争取时间,在现场或附近(安全地方)就地救治触电者。

(三) 准确

准确就是抢救的方法和施行的动作姿势要合适得当。

(四) 坚持

坚持就是抢救必须坚持到底,直至医务人员判定触电者已经死亡,已无法抢救时,才可停止抢救。

二、影响触电损伤程度的因素

人体触及带电体并形成电流通路,致使组织损伤或功能障碍甚至死亡称为触电。

大量研究表明,电对人体的伤害主要来自于电流。电流流过人体时,电流的热效应会引起肌体烧伤、炭化或在某些器官上产生损坏其正常功能的高温;肌体内的体液或其他组织会发生分解作用,从而使各种组织的结构和成分遭到严重破坏;神经组织或其他组织因受到刺激而兴奋,内分泌失调,使人体内部的生物电破坏;产生一定的机械外力引起肌体的机械性损伤。

影响触电损伤程度的因素主要有以下几个方面:

1. 通过人体电流的大小

一般情况下,通过人体的电流越大、人体的生理反应越明显、越强烈,生命的危险性也就越大,见表 8-2。而通过人体的电流大小则主要取决于施加于人体的电压,电压越高,通过人体的电流越大。人体电阻的大小与皮肤干燥、完整、接触电极的面积以及人体的接触电压有关。一般情况下,人体电阻可按 $1\,000 \sim 2\,000\,\Omega$ 估算,而潮湿条件下的人体电阻约为干燥条件下的 1/2,人体电阻越小,危险性越大。

表 8-2 电流大小对人体的影响

电流强度/mA	对人体的影响	
	交流电/50 Hz	直流电
0.6~1.5	开始感觉，手指麻刺	无感觉
2~3	手指强烈麻刺、颤抖	无感觉
5~7	手部痉挛	热感
8~10	手部巨痛，勉强可以摆脱电源	热感加强
20~25	手迅速麻痹，不能自立，呼吸困难	手部轻微痉挛
50~80	呼吸麻痹，心室开始颤动	手部痉挛，呼吸困难
90~100	呼吸麻痹，心室经 3 s 停止跳动	呼吸麻痹

2. 电流通过人体的持续时间

通电时间越长，电击伤害程度越严重。通电时间短于一个心脏周期时（人的心脏周期约为 75 ms），一般不至于发生生命威胁。但若触电正好开始于心脏周期的易损伤期，仍会发生心室颤动，一旦发生心室颤动，如无及时地抢救，数秒钟至数分钟之内即可导致不可挽回的生物性死亡。

3. 电流通过人体的途径

电流通过人体不存在不危险的途径，以途径短而且经过心脏的途径的危险性最大，电流流经心脏会引起心室颤动而致死，较大电流还会使心脏立刻停止跳动。在通电途径中，从左手至胸部的通路为最危险。人体触电路径见图 8-1。

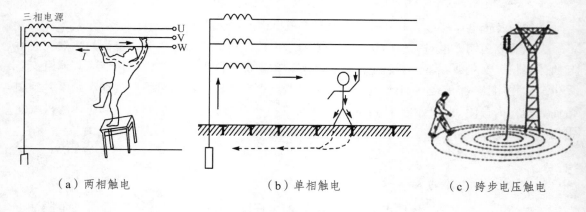

(a) 两相触电　　　　　(b) 单相触电　　　　　(c) 跨步电压触电

图 8-1 触电路径

4. 通过电流的种类

人体对不同频率电流的生理敏感性是不同的，因而不同种类的电流对人体的伤害程度也就有所区别。工频电流对人体伤害最为严重；直流电流对人体的伤害则较轻；高频电流对人体的伤害程度远不及工频交流电严重。

5. 人体状况

电对人体的伤害程度与人体本身的状况有密切关系。人体状况除人体电阻外，还与性别、健康状况和年龄等因素有关。主要表现为儿童、妇女和患有心脏病或患有中枢神经系统疾病的人，瘦小的人遭受电击后的危险性则会较大。

6. 人体电阻

人体电阻是不确定的电阻，它取决于多种因素，如接触电压、电流途径、持续时间、接触面积、温度、压力、皮肤厚薄及完好程度、潮湿、脏污等，一般认为人体电阻为 $1\,000 \sim 2\,000\,\Omega$，安全程度要求较高的场合按不受外界影响的 $500\,\Omega$ 来考虑，见表 8-3。

表 8-3 不同条件下的人体电阻

作用于人体的电压/V	人体电阻/Ω			
	皮肤干燥	皮肤干燥	皮肤干燥	皮肤浸入水中
10	7 000	3 500	1 200	600
25	5 000	2 500	1 000	500
50	4 000	2 000	875	440
100	3 000	1 500	770	375
250	2 000	1 000	650	325

7. 作用于人体的电压

触电伤亡的直接原因在于电流在人体内引起的生理病变。显然，此电流的大小与作用于人体的电压高低有关，电压越高，电流越大。人体电阻将随着作用于人体电压的升高而呈非线性急剧下降，致使通过人体的电流显著增大，使得电流对人体的伤害更加严重。

三、触电急救的步骤

触电急救包括迅速脱离电源、伤员脱离电源后的处理、心肺复苏（CPR）、抢救过程中的再判定 4 个步骤，其关键是迅速脱离电源及正确的现场救护。经验证明，在触电 1 min 内急救，有 60%~90% 救活的可能；在 1~2 min 内急救，有 45% 救活的可能；如果经过 6 min 才进行急救，只有 10%~20% 救活的可能；超过 6 min，救活的可能就更小了，但是仍有救活的可能。

CPR 是触电急救的关键步骤，美国心脏协会公布的《2020 年美国心脏协会心肺复苏及心血管急救指南》指出："尽管近年有所进展，仍只有不到 40% 的成人接受非专业人员启动的心肺复苏，而仅有不足 12% 的成人在急救医疗服务（EMS）到达之前接受了自动体外除颤器（AED）急救。"我国心肺复苏术的普及率不到 1%，心脏骤停抢救成功率更是不到万分之一。

电力行业是高危行业，触电急救是每一位电力员工必备技能之一，下面我们以院外成人触电、非专业急救为例来介绍如何实施触电急救。

（一）迅速脱离电源

电流作用的时间越长，对触电者的伤害就越重。因此，触电急救，首先要使触电者迅速脱离电源，越快越好。脱离电源，就是要把触电者接触的那一部分带电设备的所有断路器（开关）、隔离开关（刀闸）或其他断路设备断开，或设法将触电者与带电设备脱离开。在脱离电源过程中，救护人员也要注意保护自身的安全。如触电者处于高处，因采取相应措施，防止该伤员脱离电源后自高处坠落形成复合伤。

如果一旦触电，附近又无人救援时，务必镇静自救。人体一般在触电后的最初几秒内，人的意识并未完全丧失，触电者可用手抓住电线的绝缘处，把电线拉出，摆脱触电状态。如果触电时电线或电器固定在墙上，可用脚猛蹬墙壁，同时身体往后倒，借助身体重量甩开电源。触电时更多依靠互救脱离危险状态，下面介绍互救的措施。

1. 脱离低压电源

（1）如果接触电器触电，应立即断开近处的电源，可就近拔掉插头，断开开关或打开保险盒，如图8-2（a）所示。

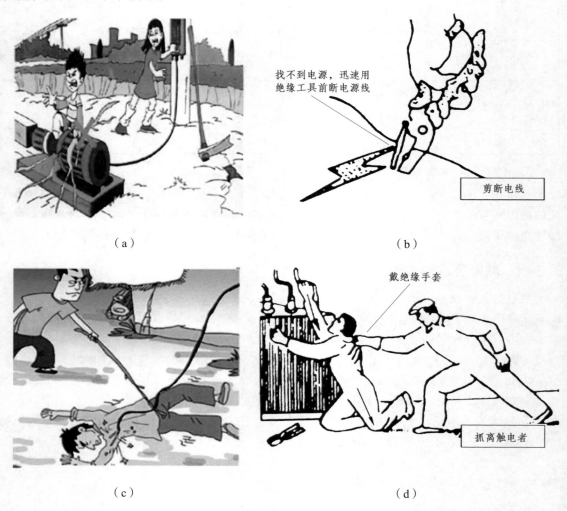

图8-2 脱离电源的方法

（2）如果电源开关或插座距离较远，可用有绝缘手柄的电工钳或干燥木柄的斧头、铁锹等利器切断电源，如图 8-2（b）所示。

注意：切断点应选择导线在电源侧有支持物处，防止带电导线断落触及其他人体。剪断电线要分相，一根一根地剪断，人尽可能站在绝缘物体或木板上。

（3）如果导线搭落在触电者身上或压在身下，可用干燥的木棒、竹竿等绝缘物品把触电者拉脱电源，如图 8-2（c）所示。

（4）如果电流通过触电者入地，并且触电者紧握导线，可设法用干燥的木板塞进其身下使其与地绝缘而切断电流，然后采取其他方法切断电源。

（5）如果触电者衣服是干燥的，又没有紧缠在身上，不至于使救护人员直接触及触电者的身体时，救护人员可直接用一只手抓住触电者不贴身的衣服，将触电者拉脱电源；也可站在干燥的木板、木桌椅或橡胶垫等绝缘物品上，用一只手把触电者拉脱电源；否则，须戴绝缘手套才能抓触电者，如图 8-2（d）所示。

注意：不得接触触电者的皮肤，也不能抓他的鞋。

（6）若触电发生在低压带电的架空线路上或配电台架、进户线上，对可立即切断电源的，则应迅速断开电源，救护者迅速登杆或登至可靠地方，并做好自身防触电、防坠落安全措施，用带有绝缘胶柄的钢丝钳、绝缘物体或干燥不导电物体等工具将触电者脱离电源。

2. 脱离高压电源

抢救高压触电者脱离电源与低压触电者脱离电源的方法大为不同，因为电压等级高，一般绝缘物对抢救者不能保证安全，电源开关距离远，不易切断电源，电源保护装置比低压灵敏度高等。为使高压触电者脱离电源，用如下方法：

（1）尽快与有关部门联系，停电。

（2）戴上绝缘手套，穿上绝缘鞋，拉开高压断路器或用相应电压等级的绝缘工具拉开高压跌落式熔断器，切断电源。

（3）如触电者触及高压带电线路，又不可能迅速切断电源开关时，可采用抛挂足够截面、适当长度的金属短路线的方法，迫使电源开关跳闸。

注意：抛挂前，将短路线的一端固定在铁塔或接地引下线上，另一端系重物。但是在抛掷短路线时，应注意防止电弧伤人或断线危及人员安全。

（4）如果触电者触及断落在地上的带电高压导线，救护人员应穿绝缘鞋或临时双脚并紧跳跃接近触电者，否则不能接近断线点 8 m 以内，以防跨步电压伤人。

3. 脱离电源的注意事项

（1）救护人员不得采用金属和其他潮湿的物品作为救护工具。

（2）未采取任何绝缘措施，救护人员不得直接触及触电者的皮肤和潮湿衣服。

（3）在使触电者脱离电源的过程中，救护人员最好使用一只手操作，并站在干燥的木凳、木板或绝缘垫上，以防触电。

（4）当触电者站立或位于高处时，应采取措施防止脱离电源后触电者摔跌。

（5）夜晚发生触电事故时，应考虑切断电源后的临时照明问题，以便急救。

（二）触电者脱离电源后的处理

触电者脱离电源以后，现场救护人员应迅速对触电者的伤情进行判断，对症抢救。同时设法联系医疗急救中心（医疗部门）的医生到现场接替救治。触电者伤情的判断方法如下：

1. 判断触电者意识、呼救

（1）轻轻拍打触电者肩部，高声喊叫："喂！你怎么啦？醒一醒。"如图 8-3 所示。

（2）呼救：一旦确定触电者神志不清，应立即大叫："来人啊，救命啊，有人触电了，快打 120 啊！"并招呼周围的人前来协助抢救。

注意：以上动作应在 10 s 以内完成，不可太长；拍打肩部不可用力太重，以防加重可能存在的骨折等损伤；一定要呼叫其他人来帮忙，因为一个人做心肺复苏术不可能坚持较长时间，而且劳累后动作容易走样，叫来的人除协助做心肺复苏外，还应立即打电话给救护站或呼叫受过救护训练的人前来帮忙。

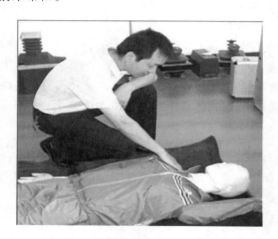

图 8-3　轻拍肩膀、高声喊叫

2. 摆平体位

触电者触电时可能有各种姿势，正确的抢救体位应是仰卧位：患者头、颈、躯干平卧无扭曲，双手放于两侧躯干旁，如图 8-4 所示。

如伤员摔倒时面部向下，应在呼救的同时小心将其转动，使伤员处于仰卧位。转动时要注意保护颈部，可以一手托住颈部，另一手扶着肩部，使伤员头、颈、胸平稳地直线转至仰卧，在坚实的平面上，四肢平放。

摆平体位后，解开伤员上衣，暴露胸部（或仅留内衣），便于观察其胸部起伏及体外按压，冷天要注意使其保暖。

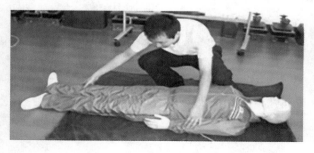 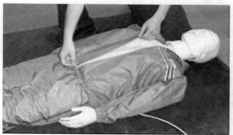

图 8-4　平整触电者肢体，解开上衣，暴露胸部

3. 判断触电者呼吸

触电者如意识丧失，应在 10 s 内用看、听、试的方法判定伤员有无呼吸。救护者跪于伤员一侧，用耳贴近伤员口鼻，头部侧向伤员胸部，见图 8-5。

看：看触电者的胸、腹壁有无呼吸起伏。

听：听触电者口鼻处有无呼气的声音。

试：用面部感觉触电者口鼻部有无呼气气流排出。

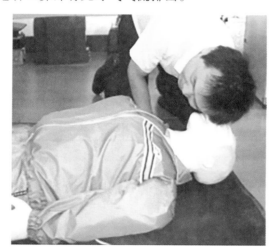

图 8-5　看、听、试的方法判断触电者有无呼吸

4. 判断触电者有无脉搏

在检查触电者的意识、呼吸之后，应对触电者的脉搏进行检查，以判断触电者的心脏跳动情况。具体方法如下：一只手置于触电者前额，使其头部保持后仰，另一只手的食指及中指指尖先触及气管正中部位（男性可先触及喉结），然后向两侧滑移 2~3 cm 至凹陷处，轻轻触摸颈动脉有无搏动，见图 8-6。

本步骤仅限于医务人员操作，非专业人员可在无意识和呼吸的前提下直接开始胸外按压操作。

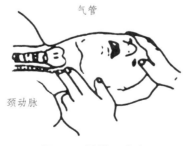

图 8-6　触摸颈动脉

注意：① 触摸颈动脉不能用力过大，以免推移颈动脉，妨碍触及；② 不要同时触摸两侧颈动脉，造成头部供血中断；③ 不要压迫气管，造成呼吸道阻塞；检查时间不要超过 10 s；④ 未触及搏动：心跳已停止，或触摸位置有错误；⑤ 触及搏动：有脉搏、心跳，或触摸感觉

错误（可能将自己手指的搏动感觉为伤员脉搏）；⑥判断应综合审定：如无意识，无呼吸，瞳孔散大，面色紫绀或苍白，再加上触不到脉搏，可以判定心跳已经停止；⑦婴、幼儿因颈部肥胖，颈动脉不易触及，可检查肱动脉，肱动脉位于上臂内侧腋窝和肘关节之间的中点，用食指和中指轻压在内侧，即可感觉到脉搏。

根据以上对意识、呼吸、心跳的判定综合审定，确定下一步的急救方案，如表8-4所示。

表8-4 不同状态下触电者的急救措施

神志	心跳	呼吸	对症救治措施
清醒	存在	存在	静卧、保暖、严密观察
昏迷	停止	存在	胸外心脏按压
昏迷	存在	停止	口对口（鼻）人工呼吸
昏迷	停止	停止	实施心肺复苏术

（三）心肺复苏（CPR）

若触电者已无意识，呼吸和心跳均已停止时，则立即按心肺复苏法帮助其恢复生命基本特征，心肺复苏包括胸外按压（人工循环）和口对口（鼻）人工呼吸。

1. 胸外按压（人工循环）

（1）按压位置：正确的按压位置是保证胸外按压效果的重要前提。右手的食指和中指沿触电伤员的右侧肋弓下缘向上，找到肋骨和胸骨接合处的中点；两手指并齐，中指放在肋骨和胸骨接合处的中点，见图8-7（a）；另一只手的掌根紧挨食指上缘置于胸骨上，即为正确的按压位置，见图8-7（b）。

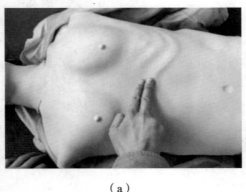 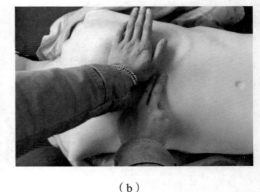

(a)　　　　　　　　　　　　　(b)

图8-7 找准按压点

（2）按压姿势：正确的按压姿势是达到胸外按压效果的基本保证。按压时掌根重叠，两手手指交叉翘起；按压与放松时间相同；两臂绷直，靠自身重量垂直向下按压；按压应平稳，有节律不间断地进行，避免在按压间隙依靠在触电伤员胸上，以便每次按压后使胸廓充分回弹，见图8-8。

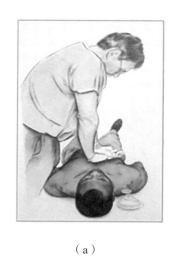
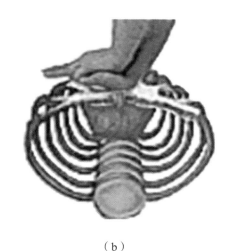

(a)   (b)

图 8-8　按压手势

（3）体位：触电伤员应仰卧于硬板床或地上。如为弹簧床，则应在其背部垫一硬板。硬板长度及宽度应足够大，以保证按压胸骨时，触电伤员身体不会移动。但不可因找寻垫板而延误开始按压时间。

（4）按压深度：通常成人伤员为 5～6 cm；青少年应采用成人的按压深度，即 5～6 cm；儿童及婴儿按压深度为胸部前后径的三分之一，大约相当于儿童 5 cm，婴儿 4 cm。

（5）按压频率：100～120 次/min。按压与人工呼吸比例：成人 30∶2（每按压 30 次吹气 2 次）；婴儿、儿童两人以上施救者为 15∶2（每按压 15 次吹气 2 次），一人施救为 30∶2（每按压 30 次后吹气 2 次）。

（6）按压间隙：为保证每次按压后使胸廓充分回弹，施救者在按压间隙，双手应离开患者胸壁，如果在两次按压之间，施救者依靠在触电伤员胸壁上，会妨碍患者的胸壁回弹。

（7）胸外按压常见的错误：① 按压除掌根部贴在胸骨外，手指也压在胸壁上，这容易引起骨折；② 按压定位不正确，向下易使剑突受压折断而致肝破裂，向两侧易致肋骨或肋软骨骨折导致气胸、血胸；③ 按压用力不垂直，导致按压无效或肋软骨骨折，特别是摇摆式按压更易出现严重并发症；④ 抢救者按压时肘部弯曲，因而用力不够，按压深度达不到；⑤ 按压冲击式，猛压，其效果差，且易导致骨折；⑥ 双手掌不是重叠放置，而是交叉放置；⑦ 放松时未能使胸部充分松弛，胸部仍承受压力，使血液难以回到心脏；⑧ 按压速度不自主地加快或减慢，影响按压效果；⑨ 按压间隙，双手未离开胸壁。

2. 畅通气道

有些触电伤员口腔、鼻腔内可能会有血液、呕吐物或假牙等异物阻塞气道，应将触电者身体侧向一侧，迅速将异物用手指抠出或是用纱布蘸出。

人受伤时，舌比较软，堆放在气道上，使呼吸不畅。因此，在人工呼吸之前应先畅通气道。通畅气道主要采用仰头抬颌法，用一只手放在触电者前额，另一只手的食指与中指置于下颌骨近下颌角处，两手协同将头部推向后仰，舌根随之抬起，见图 8-9。

注意：① 严禁用枕头或其他物品垫在触电者头下，头部抬高前倾，会加重气道阻塞，且使胸外按压时流向脑部的血流减少，甚至消失；②仰头抬颌时，成人头部后仰程度应为90°，儿童头部后仰程度应为60°，婴儿头部后仰程度应为30°；③抠出口腔或鼻腔异物时，要防止将异物推到咽喉深部。

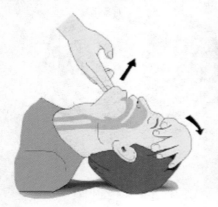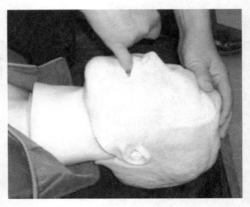

图 8-9　畅通气道

3. 口对口（鼻）人工呼吸

（1）保持气道畅通，用按于前额一手的拇指与食指捏住伤员鼻翼，以防止气体从口腔内经鼻孔逸出，深吸一口气屏住，并用自己的嘴唇包住伤员微张的嘴，用力快而深地向伤员口中吹气，同时仔细观察伤员胸部有无起伏。

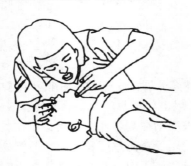

图 8-10　口对口人工呼吸

（2）一次吹气完毕后，脱离伤员口部，并放松捏鼻的手，看到伤员胸部回弹。

（3）每个吹气循环需连续吹气两次，每次 1~1.5 s，5 s 内完成。

（4）吹气时，胸廓隆起者，人工呼吸有效；吹气无起伏者，则可能是气道畅通不够，或鼻孔处漏气，或吹气不足，或气道有梗阻，应及时纠正，如图 8-10 所示。

注意：①每次吹气量不要过大，约 600 mL（6~7 mL/kg）即可，大于 1 200 mL，会造成胃扩张；②儿童伤员需视年龄不同而异，其吹气量约为 500 mL，以胸廓能上抬时为宜；③有脉搏无呼吸的伤员，则每 5 s 吹一口气，每分钟吹气 12 次；④口对鼻的人工呼吸，适用于有严重的下颌及嘴唇外伤、牙关紧闭、下额骨骨折等情况的伤员，难以采用口对口吹气法；⑤婴、幼儿急救操作时要注意，因婴、幼儿韧带、肌肉松弛，故头不可过度后仰，以免气管受压，影响气道通畅，可用一手托颈，以保持气道平直，另外，婴、幼儿口鼻开口均较小，位置又很靠近，抢救者可用口贴住婴、幼儿口与鼻的开口处，施行口对口鼻呼吸。

（四）抢救过程中的再判定

（1）实行心肺复苏法抢救触电者时，要随时注意发生的变化。按压吹气 2 min 后（相当于单人抢救时做了 5 个 30∶2 压吹循环），应用看、听、试方法在 5～10 s 内完成对触电者呼吸和心跳是否恢复的再判定。

（2）若判定颈动脉已有搏动但无呼吸，则暂停胸外按压而再连续大口吹气 2 次（每次 1～1.5 s），接着可每 4～5 s 吹气一次（即 12～16 次/min）。如脉搏和呼吸均未恢复，则应继续坚持心肺复苏法抢救。

（3）心肺复苏抢救成功的表现为瞳孔由大变小、面色转为红润、可见眼球活动、停止按压后颈动脉仍有跳动、有自主呼吸，昏迷逐渐变浅或出现挣扎。

（4）在整个抢救过程中，要每隔数分钟就进行一次再判定，判定时间均不得超过 5～10 s。在医务人员未接替抢救前，不得放弃现场抢救。

五、触电急救综述

（一）触电急救操作流程

综上所述，对于无意识、无呼吸、无心跳的触电伤员，触电急救操作步骤流程如图 8-11 所示。

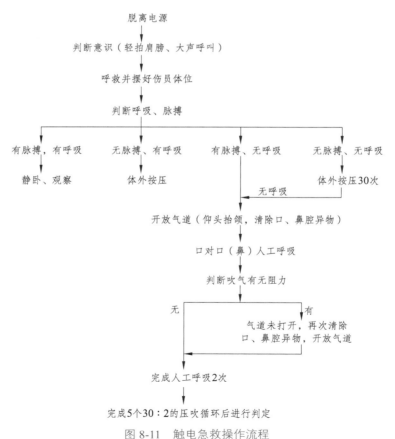

图 8-11　触电急救操作流程

（二）双人复苏操作要求

（1）两人应协调配合，吹气应在胸外按压的松弛时间内完成。

（2）按压频率为 100～120 次/min。

（3）成人按压与吹气比例为 30∶2，即 30 次心脏按压后，进行 2 次人工呼吸。

（4）为达到配合默契，可由按压者数口诀"1、2、3、4…29，吹"，当吹气者听到"29"时，做好准备，听到"吹"后，即向伤员嘴里吹气，按压者继而重数口诀"1、2、3、4…29，吹"，如此周而复始循环进行。

（5）施救者除做人工呼吸外，还应对触电者手试面部温度和观察瞳孔等，随时掌握伤员情况。

（三）抢救过程中触电伤员的移动与转院

（1）心肺复苏应在现场就地坚持进行，不要为方便而随意移动伤员，如确实需要移动时，抢救中断时间不应超过 30 s。

（2）移动伤员或将伤员送往医院时，应使伤员平躺在担架上，并在其背部垫以平硬阔木板。移动或送医院过程中应继续抢救，心跳呼吸停止者要继续用心肺复苏法抢救，在医务人员未接替救治前不能中止。

（3）应创造条件，用塑料袋装入砸碎了的冰屑做成帽状包绕在触电者头部，露出眼睛，使脑部温度降低，争取心、肺、脑完全复苏。

（四）触电者好转后的处理

（1）如触电者的心跳和呼吸经抢救后均已恢复，可暂停心肺复苏法操作，但心跳呼吸恢复的早期有可能再次骤停，应严密监护，不能麻痹，要随时准备再次抢救。

（2）初期恢复后，触电者可能神志不清或精神恍惚、躁动，应设法使触电者安静。

Task Ⅱ First aid for electric shock sufferers

Ⅰ. Basic rules of rescue on the site

On-site rescue requires quick, on-spot, accurate and persistent action.

(Ⅰ) Quickness

Quickness means seizing every minute and second and making every attempt to separate the electric shock sufferer from power source and move the sufferer to a safe place. This is crucial for rescue on the site.

(Ⅱ) On-the-spot action

On-the-spot action means buying time to rescue the sufferer on the spot or nearby (safe place).

(Ⅲ) Accuracy

Accuracy means suitable method of rescue and actions.

(Ⅳ) Persistence

Persistence means persisting in rescue until declaration of death of the sufferer by medical personnel.

Ⅱ. Factors affecting degree of electric shock injury

When a human body touches a charged body and a current path is generated, tissue damage, dysfunction or even death may occur. This situation is called electric shock.

Numerous studies have shown that harm of electricity to human body mainly comes from current. When current flows through a human body, thermal effect of the current can cause burn, carbonization or high temperature that damages normal function of some organs; Body fluid or other tissues in human body will decompose, resulting in serious damage to structures and components of the tissues; Nervous tissue or other tissues are excited due to stimulation, endocrine disorders will occur, and internal bioelectricity of human body is destroyed; Certain resultant mechanical external forces will cause mechanical damage to human body.

The factors that affect degree of electric shock damage include the following aspects:

1. The magnitude of the current flowing through human body

In general, higher current passing through human body gives rise to more significant and intense physiological response in human body and higher risk to life, as shown in Tab. 8-2. Magnitude of the current passing through human body depends on the voltage applied to human body. The higher the voltage, the higher the current passing through human body. Human body resistance is related to dryness and integrity of skin, area of contact electrodes and voltage coming into contact with human body. In general, human body resistance can be 1,000−2,000 Ω, and human body resistance in humid conditions is about 1/2 of that in dry conditions. Therefore, the smaller the human body resistance, the higher the danger.

Tab. 8-2　Impact of current on human body

Current intensity/mA	Impact on human body	
	AC/50 Hz	DC
0.6−1.5	Feel tingling in fingers	Insentience
2−3	Strong tingling and trembling of fingers	Insentience
5−7	Hand spasm	Hotness
8−10	Huge pain in the hands, just barely get away from power supply	Stronger hotness
20−25	Paralysis of hands, inability to stand on their own, expiratory dyspnea	Slight hands spasms
50−80	Respiratory paralysis, ventricular fibrillation	Hands spasm, expiratory dyspnea
90−100	Respiratory paralysis, ventricular arrest after 3 s	Respiratory paralysis

2. Duration of current flowing through human body

Longer duration of current flow causes more serious electric shock. When power-on time is shorter than a heart cycle (a human heart cycle is about 75 ms), there is no life threat in general. But if an electric shock occurs in the vulnerable phase of a cardiac cycle, ventricular fibrillation will still occur. In such cases, irreparable biological death may occur within a few seconds and a few minutes if there is no timely rescue.

3. Path of current flowing through human body

There is no safe path of current flowing through human body, and the path that is short and passes through the heart is the most dangerous one. Current flowing through the heart will cause ventricular fibrillation and death, and higher current will result in cardiac arrest. In current pathway, the path between left hand and chest is the most dangerous one. Path of electric shock on human body is as shown in Fig. 8-1.

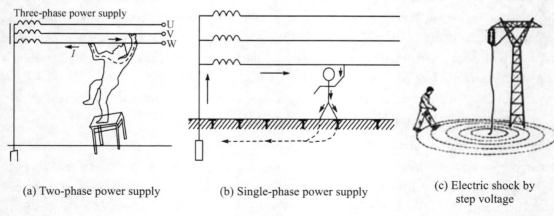

(a) Two-phase power supply　　(b) Single-phase power supply　　(c) Electric shock by step voltage

Fig. 8-1　Path of electric shock

4. Types of current

As physiological sensitivity of the human body to current of different frequencies varies, damage to human body varies with types of current. Power current causes the most serious harm to human body; DC current does slighter harm to human body; Harm of high-frequency current to human body is far less severe than that of power frequency AC.

5. Human body condition

Damage to the human body caused by electricity is closely related to human body condition. Human condition is not only related to human body resistance, but also related to factors such as gender, health status and age. Children, women and heart patients or central nervous system disease sufferers and skinny persons in particular are more dangerous after electric shock.

6. Human body resistance

Human body resistance is an uncertain resistance that depends on various factors such as contact voltage, current path, duration, contact area, temperature, pressure, skin thickness and intactness, humidity, dirt, etc. It is generally believed that human body resistance is 1,000–2,000 Ω. For situations requiring higher safety, 500 Ω free from external impact is considered, See Tab. 8-3.

Tab. 8-3 Human body resistance in different conditions

Voltage on human body /V	Human body resistance /Ω			
	Dry skin	Dry skin	Dry skin	Skin immersed in water
10	7,000	3,500	1,200	600
25	5,000	2,500	1,000	500
50	4,000	2,000	875	440
100	3,000	1,500	770	375
250	2,000	1,000	650	325

7. Voltage on human body

The direct causes of electric shock injury are physiological changes caused by electric current in human body. Obviously, magnitude of current is related to the voltage acting on human body. Higher voltage and current will result in significant increase in current flowing through human body and more serious damage to human body with non-linear sharp decrease in human body resistance as the voltage acting on human body increases.

III. Steps of electric shock sufferer rescue

First aid against electric shock includes four steps: quick separation from power supply, treatment of the wounded after separation from the power supply, cardiopulmonary resuscitation (CPR) and reassessment in the process of rescue. Quickly getting away from power supply and correct on-site rescue are critical. Experiences show that first aid within 1 min since electric shock will provide 60%–90% chance of survival; first aid within 1–2 min since electric shock will provide

45% chance of survival; first aid within 6 min since electric shock will provide only 10%-20% chance of survival; and first aid after 6 min since electric shock will provide minimum chance of survival.

CPR is a critical step in first aid against electric shock. The *2020 American Heart Association Guidelines for Cardiopulmonary Resuscitation and Emergency Cardiovascular Care* states that "Despite of the progress in recent years, less than 40% adults receive cardiopulmonary resuscitation by non professionals, and less than 12% adults receive first aid using automatic external defibrillator (AED) before emergency medical services (EMS) are available." In China, prevalence of cardiopulmonary resuscitation is less than 1%, and rate of success in rescue against cardiac arrest is less than 1/10,000.

Electric power industry is a high-risk industry, and first aid against electric shock is one of the essential skills for every employee. Now, we would like to present emergency responses to electric shock using adult sufferers and non-professional first aids as examples.

(Ⅰ) Quickly get away from power supply

Longer duration of current flow will cause more serious damage to electric shock sufferers. Therefore, to provide first aid against electric shock, the first step is to move the sufferer from power supply as soon as possible. To get away from power supply, all circuit breakers (switches), isolating switches (knife switches) or other disconnecting devices of the live equipment in contact with the sufferer must be disconnected, or efforts can make to move the sufferer away from the live equipment. In disconnecting from power supply, rescuers shall also take care of themselves. If the electric shock sufferer is at a high position, measures shall be taken to protect the injured from falling from height which gives rise to complex injury after separating from power supply.

Once an electric shock occurs and there is no one nearby, be sure to calm down and start self-rescue. In the first few seconds after electric shock, the sufferer is usually not completely unconscious, and the sufferer may catch wire insulation and pull out the wire to extricate himself from electric shock. If the wire or appliance is fixed on the wall when an electric shock occurs, the sufferer can forcefully push against the wall with the feet and lean backwards to get rid of the power source using his body weight. Electric sufferers are usually helped out of dangerous state through mutual aid. Mutual aid measures are given below.

1. Get away from LV power supply.

(1) If electric shock is caused by contacting with electric appliances, nearby power supply shall be disconnected, and plugs nearby can be pulled out and switches can be disconnected or fuse box can be opened, as shown in Fig. 8-2 (a).

(2) If power switch or socket is far away, it can be cut off with sharp tools such as electrician's pliers with insulated handle or axe and shovel with dry wooden handles, as shown in Fig. 8-2 (b).

Note: cut-off points shall be at a location where the conductor is supported on power supply side to prevent live conductor from falling and touching other human bodies. Wire cutting shall be on phase basis, and wires shall be cut off one by one, and the operator shall stand on insulating object or wooden boards in so far as possible.

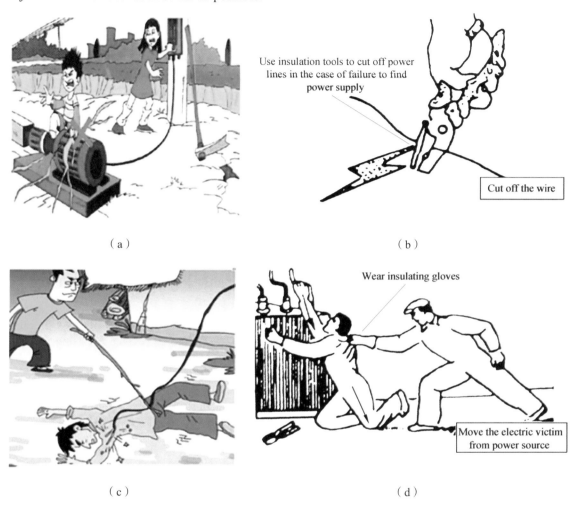

Fig. 8-2　Method of disconnecting from power supply

(3) If a conductor falls on or is held down by the electric sufferer, dry wooden sticks, bamboo poles and other insulating materials can be used to pull the person off the power source, as shown in Fig. 8-2 (c).

(4) If there is current flowing to the ground through the electric shock sufferer and the sufferer holds the conductor tightly, try to insert a dry wooden board under the body to isolate it from the ground and cut off current, and then cut off power supply by other means.

(5) If clothing of the sufferer is dry and not tightly wrapped around the body and the rescuers can easily touch the body, the rescuer can grasp clothing of the sufferer with one hand and pull him off the power supply; The rescuer can also stand on insulating items such as dry wooden boards, wooden tables and chairs or rubber pads and pull the electric shock victim off the power supply

with one hand; Otherwise, use the hand wearing insulating gloves to catch the electric shock sufferer, as shown in Fig. 8-2 (d).

Note: do not touch skin of the electric shock sufferer or catch shoes of the electric shock sufferer.

(6) If an electric shock occurs on a LV live overhead line, distribution rack or incoming line and the power source can be immediately cut off, immediate power cut is required. The rescuers shall quickly climb onto a pole or a safe place and protect themselves from electric shock and falling and use tools such as wire pliers with insulating rubber handle, insulating objects or dry non-conductive objects to remove the person from power source.

2. Get away from high voltage power supply

The methods of disengaging the victim from high voltage power supply are very different from those of disengaging the victim from low voltage power supply, since the voltage class is high, the general insulation cannot guarantee the safety of rescuers; the power switch is far away, making it hard to cut off the power, and the power protection device is more sensitive than the low voltage one. Please use the following methods to disengage the high voltage electric shock victim from the power supply:

(1) Contact the authorities as soon as possible for power outage.

(2) Put on insulating gloves and shoes, close the high voltage circuit breaker or close the high voltage drop-out fuse using the insulation tools of appropriate voltage class to cut off the power supply.

(3) If the victim touches the high voltage live line, and it is not possible to disconnect the power switch quickly, throw the metal short-circuit wire of sufficient cross section and appropriate length to force the power switch to trip.

Note: Before throwing, please secure one end of the short-circuit wire to the tower or down conductor, and keep the other end tied to a heavy object. However, when throwing the short-circuit wire, attention must be paid to preventing arc injury or broken wire from endangering personnel safety.

(4) If the victim touches the live high voltage wire broken on the ground, the rescue workers should wear insulating shoes or keep feet together and jump close to the victim; otherwise they should not get closer than 8 meters from the breakwire point to prevent injury by step voltage.

3. Precautions for disconnecting from power supply

(1) Do not use metal and other wet articles as first-aid tools.

(2) Do not touch the skin and damp clothing of the victim directly without any insulation measures.

(3) Use one hand and stand on a dry wooden bench, wooden board or insulating mat to prevent electric shock in the process of removing the victim from the power supply.

(4) Take measures to prevent the victim from falling after disconnecting from the power supply when she/he is standing or high up.

(5) Consider the temporary lighting problem for first aid after cutting off the power supply in case of an electric shock at night.

(Ⅱ) Treatment of victims after disconnecting from power supply

After the victim is disengaged from the power supply, the first responders should quickly judge the victim's injury and carry out symptomatic rescue. Meantime, try to contact the doctor of the medical emergency center (medical department) to the scene for treatment. The method of judging the victim's injury is as follows:

1. Determine if the victim is conscious and call for help

(1) Pat the victim gently on the shoulder and shout, "Hello! What's happened to you? Wake up," as shown in Fig. 8-3.

(2) Call for help: Once the victim is known to be unconscious, immediately shout: "Help! Someone got electrocuted. Call 120!" and call those around to help.

Note: The above actions should be completed within 10s; do not pat on the shoulder too hard to prevent the aggravation of possible fractures and other injuries; be sure to call someone else to help, since it is impossible to perform cardiopulmonary resuscitation (CPR) alone for long, and CPR tends to go out of shape after fatigue. Besides assistance with CPR, those called should call the ambulance station or call someone with ambulance training immediately for help.

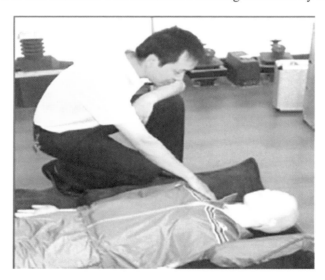

Fig. 8-3 Pat on the shoulder and shout loudly

2. Lie the body position flat

The victims may be in a variety of positions. The correct rescue position should be supine: Keep the victim lying flat on his/her head, neck, and trunk, free from twisting, with hands placed on both sides of the trunk, as shown in Fig. 8-4.

If the victim falls face down, s/he should be carefully turned while calling for help, so that the victim is in a supine position. Take care to protect the neck while turning, hold the neck in one hand

and support the shoulder with the other hand, so that the victim's head, neck and chest can be turned smoothly and straight to the supine position. Lay the limbs flat on a solid surface.

Once in the supine position, untie the victim's coat and expose his/her chest (or underwear left only) to facilitate the observation of the chest movement and external compression. Keep the victim warm in cold days.

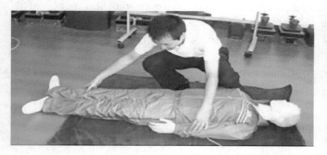 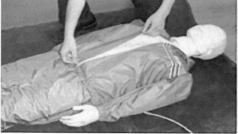

Fig. 8-4 Lie the victim flat, untie his/her coat and expose his/her chest

3. Determine if the victim is breathing

If the victim loses consciousness, check him/her for breathing by observing, listening, and sensing within 10 seconds. The responder kneels on one side of the victim, puts his/her ear close to the victim's mouth and nose, and keeps his/her head on the victim's chest, as shown in Fig. 8-5.

Observing: Observe the victim's chest and abdominal wall for breathing movement.

Listening: Listen for exhalation at the nose and mouth of the victim.

Sensing: Use your face to feel any expiratory air at the mouth and nose of the victim.

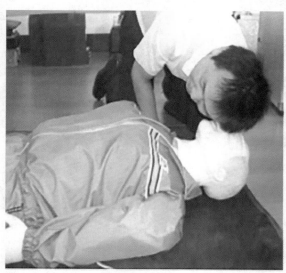

Fig. 8-5 Determine if the victim is breathing by observing, listening, and sensing

4. Determine if the victim has a pulse

After checking the victim's consciousness and breathing, his/her pulse should be checked to determine how well his/her heart is beating. The specific methods are as follows: Place one hand on the victim's forehead to keep his/her head tilted back, and keep the tips of the index and middle fingers of the other hand first palpating the median part of trachea (or the Adam's apple for males), then slide 2 ~ 3 cm to both sides to the depression, and gently palpate for carotid pulsation, as shown in Fig. 8-6.

This procedure is for medical workers only. Those other than specialists can start external chest compressions directly while unconscious and breathing.

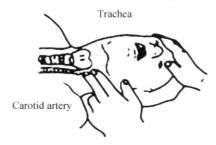

Fig. 8-6 Palpate carotid artery

Notes:

① Do not palpate the carotid artery too hard, so as not to push the carotid artery and hinder the palpating;

② Do not palpate both carotid arteries at the same time, which, otherwise, may result in the interruption of blood supply to the head;

③ Do not compress the trachea to avoid respiratory obstruction; do not examine for more than 10 s;

④ Carotid pulsation not palpated: the heartbeat has stopped, or the wrong position is palpated;

⑤ Palpate for pulsation: pulses and heartbeat palpated, or the wrong sensation of palpation (mistaking the pulsation of your fingers as the victim's pulse);

⑥ Comprehensive assessment: In case of unconsciousness, no breathing, mydriasis, cyanosis or pallor, coupled with no pulse, the heart can be judged to have stopped;

⑦ Check the brachial artery of infants and young children, since it is not easy to palpate their carotid arteries as a result of cervical obesity. The brachial artery is located at the midpoint between the armpit and the elbow joint on the inside of the upper arm. You can feel the pulse by pressing lightly on the inside with your index and middle fingers.

Based on the above assessment of consciousness, breathing, and heartbeat, the next first aid plan is determined, as shown in Tab. 8-4.

Tab. 8-4 First aid measures for victims in different states

Consciousness	Heartbeat	Breathing	Symptomatic treatment measures
Conscious	Yes	Yes	Lie still, keep warm, and observe closely
Unconscious	Cardiac arrest	Yes	External chest compression
Unconscious	Yes	Cardiac arrest	Mouth-to-mouth (nose) resuscitation
Unconscious	Cardiac arrest	Cardiac arrest	Administer CPR

(Ⅲ) Cardiopulmonary resuscitation (CPR)

If the victim is unconscious, without any breathing and heartbeat, perform immediate CPR to help restore the basic characteristics of life. CPR includes external chest compression (artificial circulation) and mouth-to-mouth (nose) resuscitation.

1. External chest compression (artificial circulation)

(1) Compression position: The correct compression position is an important premise to ensure the effect of chest compressions. The index and middle fingers of the right hand go up along the victim's lower margin of the right costal arch to find the midpoint where the ribs meet the sternum; keep two fingers together and place your middle finger at the midpoint where the ribs meet the sternum, as shown in Fig. 8-7 (a); the palm base of the other hand is placed on the sternum close to the upper edge of the index finger, which is the correct position for compression, as shown in Fig. 8-7 (b).

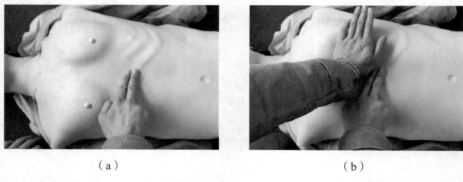

(a)　　　　　　　　　　　(b)

Fig. 8-7　Pinpoint the compression position

(2) Compression posture: The correct compression posture is the basic guarantee to achieve the effect of chest compression. During the compression, the palm bases overlap, and the fingers of both hands are crossed and cocked; compress and relax for the same time; keep your arms straight and press down vertically by your own weight. The compression should be smooth and rhythmical, without interruption. Do not lean on the victim's chest during the compression interval, so that the chest can fully bounce back after each compression, as shown in Fig. 8-8.

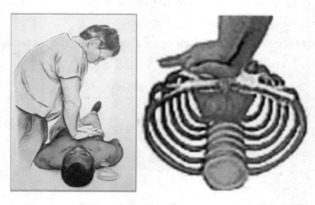

Fig. 8-8　Compression gesture

(3) Position: The victim should lie on his/her back on a hard bed or on the ground. In case of a spring bed, a hard plate should be placed on his/her back. The hard plate should be long and wide enough to keep the victim's body not moved in the process of sternum compression. However, do not delay the compression time as a result of finding the pad.

(4) Compression depth: usually 5-6 cm for adult victims; 5-6 cm for teenagers, just as the adults; one-third of the anterior-posterior diameter of the chest for children and infants, approximately 5cm for children and 4cm for infants.

(5) Compression frequency: 100-120 times per min. Compression to breathing ratio: 30∶2 for adults (2 ventilations for every 30 compressions); 15∶2 (2 ventilations for every 15 compressions) for infants and children with two rescuers, and 30∶2 (2 ventilations for every 30 compressions) for one rescuer.

(6) Compression interval: In order to ensure full recovery of the chest after each compression, the rescuer should have his/her hands removed from the patient's chest wall during the compression interval; otherwise, it will prevent the patient's chest wall from bouncing back.

(7) Common mistakes:

① Besides the palm base, the fingers are pressed against the chest wall, which is prone to fracture;

② If the compression position is not correct, downward pressure is easy to break the xiphoid process and even liver rupture; Lateral pressure is prone to the fracture of rib or costal cartilage, giving rise to pneumothorax and hemothorax;

③ The pressure is not vertical, resulting in ineffective compression or fracture of costal cartilage, especially rocking compression is more prone to serious complications;

④ Bend your elbows in the compression, making the pressure and depth not enough;

⑤ Impact compression with strong pressure is of poor effect, and prone to fracture;

⑥ Palms are crossed, not overlapped;

⑦ During relaxation, the chest fails to relax sufficiently and still under pressure, making it difficult for blood to return to the heart;

⑧ The compression speed increases or decreases involuntarily, affecting the compression effect;

⑨ Both hands do not leave the victim's chest wall during the compression interval.

2. Clear the airway

Some victims may have foreign bodies in their mouths and nasal passages, such as blood, vomit, or dentures, which may block their airways. In this case, we must turn them sideways, remove the foreign matter quickly with our fingers or gauze.

When injured, the victim has a soft tongue and stacked on the airway, making it difficult to breathe. Therefore, the airway should be cleared before artificial respiration. The airway is mainly cleared by raising the head and jaw: place one hand on the victim's forehead, and the index and middle fingers of the other hand near the angle of mandible, push his/her head back with your hands and raise the base of his/her tongue, as shown in Fig. 8-9.

Notes:

① Do not use pillows or other items under the victim's head, otherwise it will aggravate the airway obstruction, reduce or even disappear the blood flow to the brain during chest compression;

② When raising the head and jaw, adults should be raised for 90°, 60° for children and 30° for infants;

③ When picking out foreign bodies of the mouth or nasal cavity, do not push foreign bodies deep into the throat.

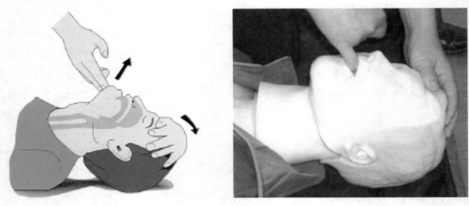

Fig. 8-9   Clear the airway

3. Mouth-to-mouth (nose) resuscitation

(1) Keep the airway unobstructed, pinch the victim's nose with the thumb and index finger of the hand on the forehead to prevent the air from escaping from the mouth through the nostrils, take a deep breath and hold it, wrap the victim's mouth with your lips, ventilate hard, fast and deep into the victim's mouth, and observe the victim's chest carefully for movement.

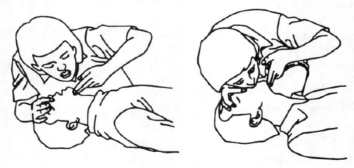

Fig. 8-10   Mouth-to-mouth resuscitation

(2) After one ventilation, disengage from the victim's mouth, and relax your hand pinching the nose, and see the victim's chest bounce back.

(3) Every ventilation cycle requires two consecutive ventilations, each 1-1.5 s, and the cycle should be completed in 5 s.

(4) In the ventilation, artificial respiration is effective for those with thoracic bulge; no

thoracic bulge may be ascribed to a lack of airway clearance, or air leakage in the nostrils, or insufficient ventilation, or airway obstruction, which should be corrected in time, as shown in Fig. 8-10.

Notes: ① Don't ventilate too much at a time, about 600 mL (6–7 mL/kg), and more than 1,200 mL will cause gastric dilation;

② The ventilation rate may vary with age for children, about 500 ml, subject to the thoracic bulge;

③ For those with pulse and no breathing, ventilate every five seconds, and 12 ventilations per minute;

④ Mouth-to-nose resuscitation is suitable for victims with severe jaw and lip trauma, trismus, lower frontal bone fracture, etc. The mouth-to-mouth insufflation is not applicable;

⑤ Please note in the first aid operation for infants and young children that due to the ligament laxity and muscular flaccidity in infants and young children, their head should not be excessively backward to avoid pressure on the trachea, affecting the airway patency. Use one hand to support their neck to keep the airway straight; on the other hand, their mouth and nose openings are small and close together. The responder can put his mouth against the opening of their mouth and nose, and perform mouth-to-mouth (nose) resuscitation.

(Ⅳ) Redetermination in the process of first aid

(1) Keep an eye out for changes when performing CPR to an electric shock victim. After 2 min of compression and ventilation (equivalent to five cycles of 30∶2 during the single rescue), re-determine whether the victim's breathing and heartbeat have recovered within 5–10 s by observing, listening and sensing.

(2) If the carotid artery is pulsating but not breathing, stop the external chest compression and ventilate twice in a row (each time 1–1.5 s), then ventilate once every 4–5 s (i.e., about 12–16 times/min). If neither pulse nor breathing is restored, keep performing CPR.

(3) The successful manifestations of CPR include the pupil decreasing from large to small, flushing complexion, visual eye movement, carotid artery still pulsating after stopping compression, spontaneous breathing, coma gradually becoming shallow or struggling.

(4) In the entire rescue process, make a rejudgment every few minutes, and the judgment must not exceed 5 ~ 10s. The on-site rescue must continue until medical workers take over the rescue.

Ⅴ. Overview of first aid for electric shock

(Ⅰ) First aid procedures

In summary, the flow chart of first aid procedures for unconscious victims without breathing and heartbeat is as shown in Fig. 8-11.

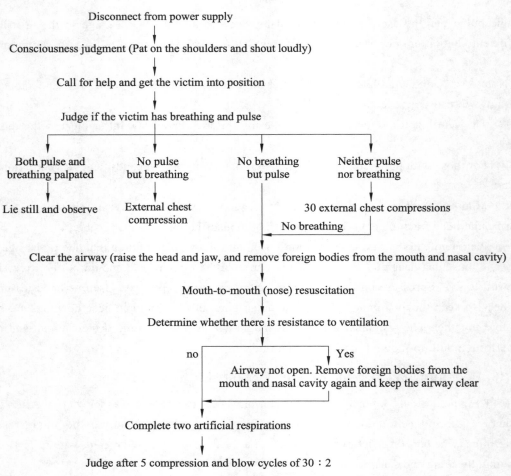

Fig. 8-11 First aid procedures

(Ⅱ) Requirements for two-person resuscitation

(1) The two should coordinate, ventilation should be done within the relaxation time of chest compression.

(2) The compression frequency is 100–120 times per min.

(3) For adults, the compression-ventilation ratio is 30 : 2, i.e., 30 heart compressions followed by 2 breathings.

(4) In order to achieve tacit cooperation, please count "1, 2, 3, 4... 29, ventilate", when the ventilator hears "29", get ready and ventilate into the victim's mouth after hearing "ventilate", and so the cycle continues.

(5) In addition to artificial respiration, responders should sense the victim's facial temperature and observe his/her pupil, so as to keep abreast of the victim.

(Ⅲ) Movement and transfer of victims during rescue

(1) CPR must be done on the spot, and never move the victims for the sake of convenience. If you really need to move, the rescue should not be interrupted for more than 30 s.

(2) When the victims are moved or taken to hospital, keep them lying flat on a stretcher and pad their back with a flat, hard plank. Continue first aid during the movement or transfer to hospital, and continue with CPR in the case of cardiac and respiratory arrest. Do not stop until medical workers take over the treatment.

(3) Conditions should be created by filling a plastic bag with the crushed ice and wrapping around the victim's head, exposing his/her eyes, so as to cool his/her brain, and strive for complete recovery of heart, lung and brain.

(Ⅳ) Treatment of victims after recovering

(1) If the victim's heartbeat and breathing have been recovered after rescue, CPR may be suspended. However, early cardiac and respiratory recovery may result in another sudden arrest, in need of close monitoring, and ready to rescue again.

(2) After initial recovery, the victims may be delirious, confused or agitated. Try to make them quiet.

任务三　紧急救护常识及应用

在电力生产、使用、维护中，除人体触电造成的伤害外，还会发生高空坠落、机械卷轧、交通挤轧、摔跌等意外造成的局部外伤，因此在现场中，还应会做适当的外伤处理，以防止细菌侵入引起严重感染或摔断的骨尖刺破皮肤或周围组织，任何迟疑、拖延或不正确的救护都会给伤员带来危害。因此，电力工人应该了解现场外伤救护的基本常识，学会急救的简单方法，以减少伤员的痛苦，避免可能发生的伤残，从而达到现场自救、互救的目的。

一、紧急救护的基本要求

创伤急救原则上是先抢救，后固定，再搬运，并注意采取措施，防止伤情加重或污染。需要送医院救治的，应立即做好保护伤员措施后送医院救治。

抢救前先使伤员躺平，判断全身情况和受伤程度，如有无出血、骨折和休克等。

外部出血立即采取止血措施，防止失血过多而休克。外观无伤，但呈休克状态，神志不清或昏迷者，要考虑胸腹部内脏或脑部受伤的可能性。

为防止伤口感染，应用清洁布片覆盖。救护人员不得用手直接接触伤口，更不得在伤口内填塞任何东西或随便用药。

搬运时应使伤员平躺在担架上，腰部束在担架上，防止跌下。平地搬运时，伤员头部在后，上楼、下楼、下坡时头部在上，搬运中应严密观察伤员，防止伤情突变。

二、止血急救

（一）止血的意义

血液是流动于心脏和血管里的液体，是维持人体生命活动不可或缺的重要组成部分。正常情况下，成人的血液占其体重的 8% 左右，即 4 500～5 000 mL。个体在电力生产、基建和日常生活中，很难避免创伤出血，但只要是小伤口、出血量少，对人体健康并无多大影响。但如果是较大的动脉血管受到损伤，则会大出血。如果抢救或处理不当，伤员就可能因出血过多而有生命危险。一般急性失血达到人体总血量的 10%（相当于 450～500 mL）时，伤员除了心跳略快以外，并无其他特殊症状；当失血量达总血量的 20% 时，即会出现头晕、头昏、脉搏增快、血压下降、出冷汗、肤色变白、神志不清、脉搏细弱无力等状况，甚至危及生命。在现场工作中，若发生创伤，并伴有大出血情况，则必须抓紧时间，迅速、准确、有效地给以止血，这对于抢救伤员生命具有极为重要的意义。

（二）损伤性出血的分类

所谓损伤性出血是指由于人体受到损伤，血液从损伤部位的血管外流。

1. 以血管分类

（1）动脉出血。其特点是出血呈鲜红色，出血速度快且量多，不易凝固，血流从断裂动脉血管内呈喷射状流出。

（2）静脉出血。出血呈暗红色，速度较慢或点滴出血，容易控制。

（3）毛细血管出血。出血血流很慢，呈渗血状，在 6~8 min 均能自行凝固停止。

2. 以损伤类型分类

（1）外出血，指受外伤时血液从损伤的血管流向体外。

（2）内出血，指血管破裂后血液积滞在体腔内、体外看不到的出血，如腹部发生外伤后肝脾破裂、骨盆骨折引起腹膜出血等。

（三）止血的方法

现场发生的创伤大部分是外出血，有时也是内出血。在现场进行急救主要是针对外出血的，故这里只讲述外出血的止血法。外出血的常用止血方法主要有以下几种：

（1）抬高患肢位置法，适用于肢体出血量较小，其方法是将患肢抬高，使其超过心脏位置，目的是增加静脉回流和减少出血量。

（2）加压包扎止血法。加压包扎是一种常用的有效止血法，大多数创伤性出血经加压包扎均能止住或减少出血。其方法是：① 先用数块面积大于伤口面积的灭菌纱布覆盖在伤口上，然后用手指或手掌用力加压，假如出血量不多，经直接加压止血后大多能够奏效。现场无消毒纱布时可用清洁的手帕或布片代替，也可从衣服上剪下最清洁的部分，用以代替纱布加压包扎，然后将出血肢体抬高。② 加压 10~30 min 后，一般都能止血。出血停止后不必调换原来的纱布（或其他包垫物），让血染的纱布留在原处不动，以防更换时引起再出血。如怀疑尚有少量渗血，则可在原纱布上再重叠放置纱布数块，略加压力包扎，然后送医院再进行处理，这个方法用于四肢止血是合适和安全的。

（3）指压止血法，即用手指压迫"止血点"止血。"止血点"就是身体的主要动脉经过而又靠近骨骼的"搏动"部位。这是最方便而又及时的临时止血法，适用于现场止血急救。具体做法是在伤口的靠近心脏端找到出血肢、体部位的止血点，用手指用力向骨头压迫，这样就会阻断血流来源而达到急救止血的目的。此法适用于面部、颈部和四肢动脉的出血。

① 面部、颈部出血。面部出血：供应侧面部血液的血管是颜面动脉，当此处出血时，应用手指压住下颌角（下巴颏）前一横指处的血管，如图 8-12（a）所示。颈部出血：四个手指并拢，在颈部凹陷处可以触及颈动脉的搏动，手指放在搏动处，拇指放在伤员颈后部，前后手指共同用力，将颈动脉向颈椎方向加压（手指要固定于搏动点上，不能揉搓），如图 8-12（b）所示。

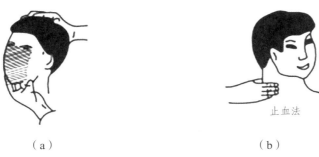

(a)   (b)

图 8-12 面部、颈部出血指压止血法

② 上肢出血。上肢止血时，首先要找到肱动脉的止血点位置，上肢止血点如图 8-13（a）所示。若上臂出血，其止血法为，一只手抬高患肢，另一只手四个手指将肱动脉压向肱骨上，如图 8-13（b）所示。若前臂出血，则将患肢抬高，用四个手指压在肘窝处肱二头肌内侧的肱动脉，如图 8-13（c）所示。

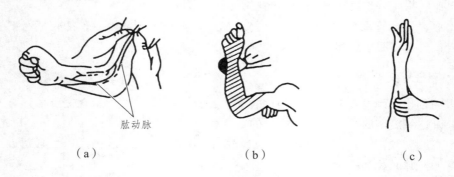

图 8-13　上肢出血指压止血法

③ 下肢出血。下肢出血时，首先要找到股动脉止血点位置。止血点在腹股沟的中点稍下方（用手指可试出股动脉的搏动），下肢止血点部位如图 8-14（a）所示；若大腿出血，则可用双手拇指向后用力压迫大腿出血止血点部位股动脉，如图 8-14（b）所示。

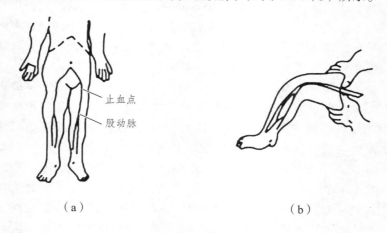

图 8-14　下肢出血指压止血法

④ 肩、腋部、手指、脚出血。肩、腋部出血：用拇指压迫同侧锁骨上窝，将锁骨下动脉压向第一肋骨，如图 8-15（a）所示；手指出血：将患肢抬高，用食指、拇指分别压迫手指两侧的指动脉，如图 8-15（b）所示；脚出血：用两手拇指分别压迫足背动脉和内踝与跟腱之间的胫后动脉，如图 8-15（c）所示。

⑤ 高处坠落、撞击、挤压可能有胸腹内脏破裂出血。受伤者外观无出血但常表现出面色苍白、脉搏细弱、气促、冷汗淋漓、四肢厥冷、烦躁不安，甚至神志不清等休克状态，应迅速躺平，抬高下肢，保持温暖，速送医院救治。若送院途中时间较长，可给伤员喝少量糖盐水。

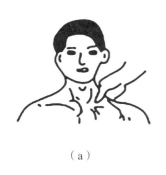
（a）

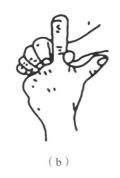
（b）

（c）

图 8-15 肩、腋部、手指、脚出血指压止血法

4. 现场伤口的简单包扎

（1）包扎的目的。伤口是细菌入侵人体的门户，哪怕只是破一个小口，病菌也会乘机侵入人体生长、繁殖、放出毒素，使伤口感染，如果不及时包扎，轻者伤口化脓，重者全身感染，甚至危及人的生命安全。因此，当现场有人受伤后，在送往医院之前施行一些简单的包扎是很有必要的。

（2）对包扎的要求。首先动作要轻，不要碰撞伤口，以免增加伤员的疼痛和出血；其次包扎要迅速，松紧合适、方法得当；还要注意不得用水冲洗伤口、去掉血迹，也不准用手和脏物触摸伤口。

（3）包扎步骤。首先要弄清伤口位置和受伤情况，然后依不同伤情进行对症救护和包扎。使伤口暴露，并检查伤情，再进行包扎。在伤口暴露过程中，如需脱衣服时，应先脱未受伤的一侧，然后再脱负伤的一侧。若伤情严重不能脱衣时，亦可沿衣缝将衣服剪、撕开。若衣服已粘在伤口上，则不能用力拉，也不要用水浸湿揭下，在紧急情况下，可在衣服外面包扎。

（4）包扎材料。包扎伤口的常用材料是绷带、三角巾、四头带。如果现场无这些材料，则可临时用干净的手绢、毛巾、衣物代替。包扎时要用干净的一面接触伤口，然后尽快去医院（卫生所）更换消毒敷料进行重新包扎。

（5）不同部位的简单包扎。

① 头面部伤包扎：可用三角巾或四头带，打结时尽可能打在以下部位，即下颌下、后脑勺下或前额的眉弓处，以免包扎松落，如图 8-16 所示。

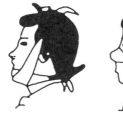

图 8-16 头面部伤包扎法

② 膝关节伤包扎：用三角巾折成适合于伤部宽度的条带，斜放在伤口，用条带两端分别压住上、下两边，缠绕肢体一周，然后在肢体内侧或外侧打结，此法同样也适用于上肢包扎，如图 8-17 所示。

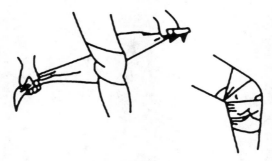

图 8-17　膝关节伤包扎法

三、骨折急救

骨骼是人体中最坚硬的组织，它除作为身体的支架外，还起着保护人体脏器的作用。骨折时，不但骨骼本身受到破坏，骨骼附近的其他软组织，如纤维、韧带、肌肉、神经、血管等也会受到不同程度的损伤。

（一）骨折的分类

人体全身有 206 块骨，均可能发生各类骨折。所谓骨折，就是骨质或骨小梁发生完全或不完全的断裂。现介绍骨折的主要分类。

1. 按骨折端与皮肤、肌肉的关系分类

（1）闭合性骨折：骨折端未刺出皮肤，与外界空气不相通，如图 8-18（a）所示。

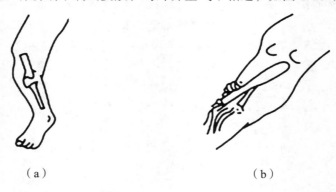

（a）　　　　　　　　　　　（b）

图 8-18　闭合性骨折和开放性骨折

（2）开放性骨折：骨折端刺出皮肤、肌肉，与外界空气相通，如图 8-18（b）所示。

2. 从骨折断裂的程度分类

（1）完全性骨折。

（2）不完全性骨折。

（二）骨折的症状与判断

如果发现有人因摔伤、挤伤而出现以下症状时，就可初步确定是发生了骨折。

1. 局部症状

（1）有局部痛感。如果有局部压痛或间接叩击振动痛感，可能是骨折端刺激骨膜及其周围软组织的神经末梢所致，一般疼痛的部位可能就是骨折的部位。

（2）局部畸形。若受伤处出现缩短、旋转或成角畸形，这可能是由于外力作用、肌肉收缩、肢体重量作用等，骨骼完全折断和骨折端发生不同程度的移位。

（3）局部软组织肿胀且呈现青紫色。这是由于骨折出血和渗出液所致。此外，骨折错位和重叠，在外表上也形成局部肿胀。

（4）骨擦音或骨擦感。伤员自己动作时，骨折端互相摩擦，可听到骨擦音或有骨擦感。

（5）功能受限。若下肢骨折，则不能站立；若肋骨骨折，则呼吸困难、剧痛；若关节附近骨折，则不能伸屈；脊椎骨折，不能坐立。

2. 全身症状

（1）休克。当脊椎骨折、骨盆骨折、大的管形骨发生骨折后，伤员常由于失血量大而发生休克。

（2）体温升高。经常在骨折两三天出现体温升高，一般体温不应超过 39 ℃，如超过时，应检查是否有其他并发症，如伤口感染、其他器官受损等。

（3）肢体瘫痪。主要是神经组织被骨折端压迫损伤所致。在现场发现伤员出现上述症状时，要想到可能是发生了骨折，应做好骨折急救工作，然后送医院进行救治。

（三）骨折的现场急救

1. 骨折急救的基本原则

（1）现场急救的目的是防止伤情恶化，为此，千万不要让已经骨折的肢体活动，不能随便移动骨折端，以防锐利的骨折端刺破皮肤、周围组织、神经、大血管等。首先，应将受伤的肢体进行包扎和固定。

（2）对于开放性骨折的伤口，最重要的是防止伤口污染。为此，现场抢救者不要在伤口上涂任何药物，不要冲洗或触及伤口，更不能将外露骨端推回皮内。

（3）抢救者应保持镇静，正确地进行急救操作，应取得伤员的配合。现场严禁将骨折处盲目复位。

（4）待全身情况稳定后再考虑固定、搬运。骨折固定材料常采用木制、塑料和金属夹板。如果现场没有现成的夹板，则可就地取材，采用木板、竹竿、手杖、伞柄、木棒、树枝等物代替。骨折固定时，应注意要先止血，后包扎，再固定。选择的夹板长度应与肢体长度相对称。夹板不要直接接触皮肤，应采用毛巾、布片垫在夹板上，以免神经受压损伤。

（5）现场骨折急救仅是将骨折处作临时固定处理，在处理后应尽快送往医院救治。

2. 几个部位骨折的急救

（1）上臂部肱骨发生骨折、前臂部尺骨、桡骨骨折。

上臂部肱骨发生骨折：使受伤上臂紧贴胸廓，并在上臂与胸廓之间用折叠好的围巾或干毛巾衬垫好；将肘关节屈曲 90°，使前臂依托在躯干部，用一条三角巾将前臂悬挂于颈项部；取一与上臂长度相当的一条木板置于上臂外侧，在木板与上臂之间用毛巾等物衬垫；最后用两条绷带（或其他布条）将上臂与胸廓上下环行缚住。

前臂部尺骨、桡骨骨折：取与前臂长度相当的两块木板，用毛巾等柔软衣物衬垫好后，一条置于前臂掌侧，一条置于前臂的背侧；用三条绷带（或其他布条）将两块木板扎缚好，大拇指须暴露于外；夹板固定后，使肘关节屈曲90°，再用一块三角巾将前臂悬挂在颈项部，如图8-19所示。

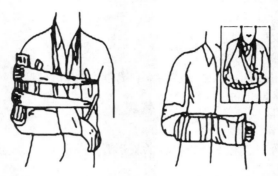

图8-19 上臂部肱骨发生骨折，前臂部尺骨、桡骨骨折急救

（2）大腿部肱骨骨折。

一人使骨折的上下部肢体保持稳定不动，另一人在断骨远端沿骨的长轴方向向下方轻轻牵引，不得旋转；用折好的被单放在两腿之间，将两下肢靠拢；用与下肢等长的一块短夹板放在伤肢内侧；用自腋窝起直达足跟的一块长夹板放在伤肢的外侧；用宽布带将两侧夹板包括躯干多处进行固定；最后将固定好的伤员再固定于木板上，同时用枕头将下肢稍微垫高，如图8-20所示。

图8-20 大腿部肱骨骨折急救

（3）小腿部胫、腓骨骨折。

小腿部胫、腓骨骨折：胫骨及腓骨在膝关节以下，再下部分即为踝关节及庶、趾骨。固定方法为：用两块夹板分别置于小腿内、外侧，骨折突出部位要加垫；自膝关节以上至踝关节以下进行固定；最后用绷带卷或布卷、毛巾等物，如图8-21所示。

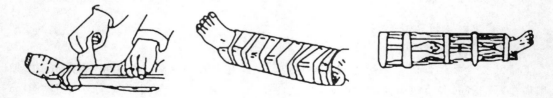

图8-21 小腿部胫、腓骨骨折急救

（4）颈椎骨折。

颈椎骨折：让伤者躺平，不要抬头、摇头、转动、搀扶活动、行走或翻身脱衣，否则，转动头部可能立刻导致伤员瘫痪，甚至突然死亡；救护者可位于伤员头部，两手稳定垂直地将头部向上牵引，并将可脱卸的环形颈圈或小枕置于伤员的颈部，以维持牵引不动；用较厚的（或多册）书籍或沙袋等堆置头部两侧，使其头部不能左右摇动；用绷带将伤员额部连同书籍等再次固定于木板担架上，如图8-22所示。

图8-22　颈椎骨折急救

3. 骨折伤员的搬运

在现场进行止血、包扎或骨折固定之后，要搬运伤员去医院救治，搬运的方法正确与否对伤员的伤情及以后的救治效果好坏都有直接关系。搬运伤员的原则是，让伤员舒适、平稳，而且力争将有害影响减低到最小程度。

（1）将一般伤员搬上担架的做法。

将一般伤员搬上担架的做法：两担架员跪下右腿，一人用一只手托住伤员头部和肩部，另一只手托住腰部；另一人一只手托住骨盆，另一只手托住膝下；二人同时起立，把伤员轻放于担架上，如图8-23所示。

图8-23　一般伤员搬运

（2）颈椎骨折伤员的搬运。

颈椎骨折伤员的搬运：对这种病人的搬运更需注意，一不小心可能造成病人立即死亡。搬运方法是由3~4人一起搬动，其中1人专管头部牵引固定，使头部保持与躯干成直线位置，以维持颈部不动；其余3人蹲在伤员的同侧，其中2人托住躯干，1人托住下肢，一齐起立，将伤员轻放在担架上，如图8-24所示。

图 8-24　颈椎骨折伤员搬运

（3）颈椎骨折伤员的运送。

颈椎骨折伤员的运送：使伤员平躺在担架上，并将其腰部束在担架上，防止跌下，如图 8-25（a）所示。平地运送时，伤员头部在后；上楼、下楼、下坡时，让伤员头部在上，如图 8-25（b）所示；没有采用任何工具和保护措施的情况下运送，伤员易加重伤情甚至死亡。

（a）

（b）

图 8-25　颈椎骨折伤员运送

四、烧伤急救

电灼伤、火焰烧伤或高温气、水烫伤均应保持伤口清洁。伤员的衣服鞋袜用剪刀剪开后除去。伤口全部用清洁布片覆盖，防止污染。四肢烧伤时，先用清洁冷水冲洗，然后用清洁布片或消毒纱布覆盖送医院。强酸或碱灼伤应立即用大量清水彻底冲洗，迅速将被侵蚀的衣物剪去。为防止酸、碱残留在伤口内，冲洗时间一般不少于 10 min。

未经医务人员同意，灼伤部位不宜敷涂任何东西和药物。

送医院途中，可给伤员多次少量口服糖盐水。

五、冻伤急救

冻伤使肌肉僵直，严重者深及骨骼，在救护搬运过程中动作要轻柔，不要强使其肢体弯曲，以免加重损伤，应使用担架，将伤员平卧并抬至温暖的室内救治。

将伤员身上潮湿的衣服剪去后用干燥柔软的衣服覆盖，不得烤火或搓雪。

全身冻伤者呼吸和心跳有时十分微弱，不应误认为死亡，应努力抢救。

六、动物咬伤急救

（一）毒蛇咬伤

（1）毒蛇咬伤后，不要惊慌、奔跑、饮酒，以免加速蛇毒在人体内扩散。

（2）咬伤大多在四肢，应迅速从伤口上端向下方反复挤出毒液，然后在伤口上方（近心端）用布带扎紧，将伤肢固定，避免活动，以减少毒液的吸收。

（3）有蛇药时可先服用，再送往医院救治。

（二）犬咬伤

（1）犬咬伤后应立即用浓肥皂水冲洗伤口，同时用挤压法自上而下将残留伤口内唾液挤出，然后再用碘酒涂搽伤口。

（2）少量出血时，不要急于止血，也不要包扎或缝合伤口。

（3）尽量设法查明该犬是否为"疯狗"，对医院制订治疗计划有较大帮助。

七、溺水急救

发现有人溺水应设法迅速将其从水中救出，呼吸心跳停止者用心肺复苏法坚持抢救。

口对口人工呼吸因异物阻塞发生困难，而又无法用手指除去时，可用两手相叠，置于脐部稍上正中线上（远离剑突）迅速向上猛压数次，使异物退出，但也不可用力过大。

溺水死亡的主要原因是窒息缺氧。由于淡水在人体内能很快经循环吸收，而气管能容纳的水量很少，因此在抢救溺水者时不应因"倒水"而延误抢救时间，更不应仅"倒水"而不用心肺复苏法进行抢救。

八、高温中暑急救

烈日直射头部，环境温度过高，饮水过少或出汗过多等可以引起中暑现象，其症状一般为恶心、呕吐、胸闷、眩晕、嗜睡、虚脱，严重时抽搐、惊厥甚至昏迷。

应立即将病员从高温或日晒环境转移到阴凉通风处休息。用冷水擦浴，湿毛巾覆盖身体、电扇吹风，或在头部置冰袋等方法降温，并及时给病人口服盐水。严重者送医院治疗。

九、有害气体中毒急救

气体中毒开始时有流泪、眼痛、呛咳、咽部干燥等症状，应引起警惕。稍重时头痛、气促、胸闷、眩晕，严重时会引起惊厥昏迷。

怀疑可能存在有害气体时，应立即将人员撤离现场，转移到通风良好处休息。抢救人员进入险区必须戴防毒面具。

已昏迷病员应保持气道通畅，有条件时给予氧气吸入。呼吸心跳停止者，按心肺复苏法抢救，并联系医院救治。

迅速查明有害气体的名称，供医院及早对症治疗。

Task Ⅲ First aid knowledge and application

In the production, use and maintenance of electricity, there are even local injuries caused by accidents such as high-altitude falling, mechanical rolling, traffic crowding, and falling, in addition to injuries caused by electric shock. Therefore, appropriate trauma treatment should be done at the scene to prevent bacterial immersion and serious infection or broken bone tips from piercing the skin and surrounding tissues. Any hesitation, delay or incorrect medical care may be detrimental to the victims. Therefore, electric workers should know the basic knowledge of on-site trauma rescue, and master the simple first aid methods, in order to reduce the suffering of the injured, and avoid possible disability for the purpose of self-rescue and mutual rescue on the spot.

Ⅰ. Basic requirements for first aid

In principle, trauma must be rescued first, then fixed, and handled; moreover, measures are taken to prevent exacerbation or contamination. Those in need of medical attention should be sent to the hospitals immediately with proper measures.

Lay the victims down before rescue, and judge the general condition and extent of injury, such as bleeding, fracture and shock.

In case of external bleeding, take immediate hemostatic measures to prevent excessive blood loss and even shock. Those of no apparent injury, but in a state of shock, delirium or coma may suffer from chest, abdominal or visceral or brain injury.

Cover the wound with a piece of clean cloth to prevent it from getting infected. Rescue workers are not allowed to touch the wound directly with their hands, nor put anything in the wound or administer drugs casually.

When handling, the injured should lie flat on the stretcher, with their waist braced on the stretcher to keep them from falling. The victim's head should be in the back during the flat handling, and on the top when going upstairs, downstairs and downhill. The injured person should be closely observed during the handling to prevent sudden change of injury.

Ⅱ. Emergent hemostasis

(Ⅰ) Significance of hemostasis

Blood is the fluid that flows through the heart and blood vessels and is an indispensable part of maintaining human life activities. Under normal circumstances, the blood in adults accounts for about 8% of their body weight, or rather, 4,500−5,000 mL. In power production, infrastructure and daily life, it is difficult for individuals to avoid traumatic hemorrhage. However, small wounds and little blood loss have little impact on human health. In the event that larger arteries are damaged, there will be massive hemorrhage. If rescue or treatment is not done properly, the victim may be in danger from excessive bleeding. Generally, when acute blood loss is 10% of the total blood volume of the human body (equivalent to 450–500 mL), the victim has no other special symptoms other than a slightly rapid heartbeat; when the blood loss reaches more than 20% of the total blood volume, the victim may suffer from dizziness, vertigo, rapid pulse, drop of blood pressure, cold sweat, pallor, unconsciousness, thin weak pulses, and even life-threatening symptoms. In case of any trauma (accompanied by massive haemorrhage) in the field work, we must stop the bleeding quickly, accurately and effectively in the first time. This is of great significance to save the lives of victims.

(II) Classification of traumatic hemorrhage

The so-called traumatic hemorrhage refers to the outflow of blood from the blood vessels at the injury site for the human body is damaged.

1. Classification by blood vessel

(1) Arterial hemorrhage is characterized by bright red, rapid and heavy bleeding, not easy to coagulate, and blood flows out of the severed artery in a jet pattern.

(2) Venous hemorrhage is dark red, slow or spotting, easy to control.

(3) Capillary bleeding features very slow blood flow and oozing blood, able to clot in 6–8 min.

2. Classification by injury type

(1) External hemorrhage refers to the flow of blood outside the body from the damaged vessel in the event of trauma.

(2) Internal hemorrhage refers to blood accumulation in the body cavity after the rupture of blood vessels, which is invisible in vitro. For example, liver and spleen rupture after abdominal trauma, and peritoneal hemorrhage due to pelvic fracture.

(III) Methods of hemostasis

Most of the trauma at the scene was external hemorrhage, and sometimes internal hemorrhage. First aid at the scene is mainly for external hemorrhage, and only the hemostasis methods of external hemorrhage are described here. The commonly used hemostasis methods for external hemorrhage mainly include the following:

(1) Elevation of the injured limb is applicable to small amount of limb bleeding. Elevate the injured limb above the heart position to increase the venous return and reduce blood loss.

(2) Pressure bandaging. Pressure bandaging is a commonly used and effective way to stop bleeding. Most traumatic bleeding can be stopped or reduced by pressure bandaging, specifically:

① First cover the wound with several pieces of sterile gauze larger than the area of the wound, and then apply pressure with your fingers or palms. The small amount of bleeding can be mostly stopped by this method. In the absence of sterilized gauze on site, a clean handkerchief or pieces of cloth is OK. Alternatively, you can cut the cleanest part of your clothes (instead of gauze) to apply pressure bandaging, and then elevate the bleeding limb.

② In general, 10 to 30 min of pressure can stop bleeding. It is not necessary to replace the original gauze (or other packing material) after the bleeding stops, and leave the bloody gauze in place to prevent bleeding again during replacement. If a small amount of bleeding is suspected, you can overlap several pieces of gauze on the existing gauze, apply some pressure to bandage, and then send the victim to hospital for treatment. This method is suitable and safe for hemostasis of extremities.

(3) Hemostasis by finger pressing, that is, apply pressure to the "hemostatic point" with your finger to stop the bleeding. The "hemostatic point" is the "pulsating" part of our body where the main arteries pass and near the bones. This is the most convenient and timely temporary hemostatic method, applicable to on-site hemostatic first aid. The specific practice is to find hemostatic points on the bleeding limb and body near the heart of the wound, and press your fingers firmly against the bone. This will block the source of blood flow for the purpose of first-aid hemostasis. This method is applicable to bleeding from the arteries of the face, neck and limbs.

① Bleeding in the face and neck. Facial bleeding: The vessel that supplies blood to the side is the facial artery. In case of bleeding here, apply finger pressure to the blood vessel in the anterior transverse finger of the mandibular angle (chin), as shown in Fig. 8-12 (a). Cervical bleeding: Keep four fingers together and touch the pulsation of carotid artery in the depression of the victim's neck. Put your fingers on the pulsation and your thumb on the back of the victim's neck. Use your front and rear fingers together to press the carotid artery towards the cervical spine (fingers should be fixed on the pulsation point, not rubbed), as shown in Fig. 8-12 (b).

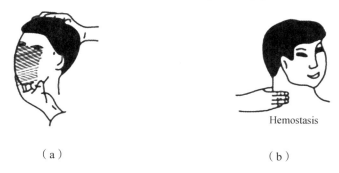

Fig. 8-12 Hemostasis by finger pressing for facial and cervical bleeding

② Bleeding in the upper limbs. To stop bleeding in the upper limb, first locate the hemostatic point of the brachial artery. Fig. 8-13 (a) shows the hemostatic point of the upper limb. If the upper arm is bleeding, elevate the injured limb with one hand; four fingers of the other hand press the brachial artery against the humerus, as shown in Fig. 8-13 (b). If the forearm is bleeding, elevate the injured limb and use four fingers to press against the brachial artery inside the biceps at the elbow fossa, as shown in Fig. 8-13 (c).

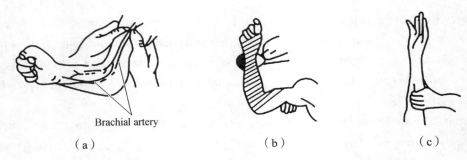

Fig. 8-13　Hemostasis by finger pressing for bleeding in the upper limbs

③ Bleeding in the lower limbs. When the lower limb is bleeding, first locate the hemostatic point of the femoral artery. The hemostatic point is slightly below the midpoint of the groin (the pulsation of the femoral artery can be sensed with fingers), and the hemostatic points in the lower limbs are as shown in Fig. 8-14 (a). If the thigh is bleeding, compress the femoral artery backward with both thumbs, as shown in Fig. 8-14 (b).

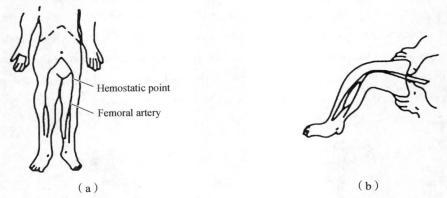

Fig. 8-14　Hemostasis by finger pressing for bleeding in the lower limbs

④ Bleeding in the shoulders, axillae, fingers, and feet. Shoulder and axillary bleeding: Apply thumb pressure to ipsilateral supraclavicular fossa and press the subclavian artery toward the first rib, as shown in Fig. 8-15 (a). Finger bleeding: Elevate the injured limb and press the digital arteries on both sides of the finger with the index finger and thumb, respectively, as shown in Fig. 8-15 (b). Foot bleeding: use both thumbs to compress the dorsal foot artery and the posterior tibial artery between the medial ankle and Achilles tendon, as shown in Fig. 8-15 (c).

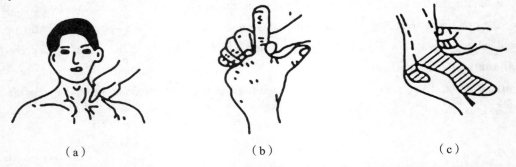

Fig. 8-15　Hemostasis by finger pressing for bleeding in the shoulder, axilla, fingers and feet

⑤ Falling from height, impact, crushing may cause rupture and bleeding of internal organs in the chest and abdomen. The injured appears to have no bleeding, but often shows shock states such as pale complexion, weak pulse, shortness of breath, dripping cold sweat, cold limbs, dysphoria, and even unconsciousness. Keep them quickly lying down, raise their lower limbs, keep them warm, and send them to hospital for treatment. If there is a long way to the hospital, the injured may be given a small amount of sodium chloride and dextrose injection.

(Ⅳ) Simple dressing of wounds at the scene

(1) The purpose of bandaging. The wound is the gateway for bacteria to invade the human body. Even if it is just a small cut, the bacteria will invade our human body to grow, reproduce, release toxins, and make the wound infected. If not bandaged in time, the injured will suffer wound suppuration, general infection, and even life-threatening situation. Therefore, it is necessary to apply simple bandaging to the injured at the scene before taking them to hospital.

(2) Requirements for bandaging. First of all, be gentle. Don't touch the wound, so as not to increase the pain and bleeding of the injured; the bandaging should be done quickly and properly, with appropriate tightness; do not rinse the wound with water to remove blood, nor touch the wound with your hands or dirt.

(3) Dressing procedures. First of all, we need to know where the wound is and what the injury is, and then carry out symptomatic rescue and bandaging depending on the injuries. Expose, examine, and bandage the wound. During the wound exposure, undress the uninjured side first and then the injured side. If the badly injured cannot be undressed, cut and tear their clothes along the seam. If the clothes are stuck to the wound, do not pull hard nor wet with water. In case of emergency, bandage the outside of clothes.

(4) Dressing materials. The commonly used materials for wound dressing are bandage, triangular binder, and four head belt. If such materials are not available on site, just use clean handkerchiefs, towels, and clothes instead. When dressing, contact the wound with the clean side, and then go to the hospital (health clinic) as soon as possible to replace the sterile dressing and re-bandage.

(5) Simple dressing of different parts.

① Head and face dressing: Use a triangular binder or four head belt, and tie it at the submandible, below the back of the head, or the arch of the forehead to prevent loose bandaging, as shown in Fig. 8-16.

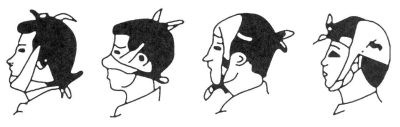

Fig. 8-16　Head and face dressing

② Knee injury dressing: Fold a triangular binder to fit the width of the wound, place it diagonally over the wound, press the upper and lower sides of the strip on each end, wrap it around the limb, and then tie a knot on the inside or outside of the limb. This method also applies to the upper limb dressing, as shown in Fig. 8-17.

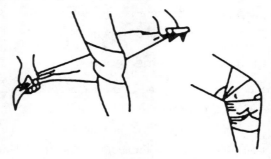

Fig. 8-17 Knee injury dressing

III. First aid for fracture

Bones are the hardest tissues in the human body. In addition to being a support for the body, they play a role in protecting human organs. Besides the damage to bones, other soft tissues near the bones, such as fibers, ligament, muscles, nerves, and blood vessels, will be damaged to varying degrees, in the event of fracture.

(I) Classification of fracture

Our human body has 206 bones, which are prone to fractures. The so-called fracture is a complete or incomplete fracture of bone or bone trabecula. The following gives the classification of fracture.

1. **Classification by the relationship between fracture end and skin and muscle**

(1) Closed fracture: The fracture end does not pierce the skin, and is not communicated with the outside air, as shown in Fig. 8-18 (a).

(a)　　　　　　　　　　　　　(b)

Fig. 8-18 Closed fracture and open fracture

(2) Open fracture: The fracture end pierces the skin and muscle, and communicates with the

outside air, as shown in Fig. 8-18 (b).

2. Classification by degree of fracture

(1) Complete fracture

(2) Incomplete fracture

(Ⅱ) Symptoms and diagnosis of fracture

If someone is found to have the following symptoms due to a fall or crowded injury, she can be preliminarily confirmed to have a fracture.

1. Local symptoms

(1) Local pain. The local tenderness or indirect percussion vibration pain may be caused by the stimulation of nerve endings of the periosteum and surrounding soft tissues at the fracture end. The most common area of pain is probably the site of fracture.

(2) Local deformity. Any shortening, rotation or angular deformity at the injury may be ascribed to complete fracture of bones and varying degrees of displacement at the fracture end arising from external force, muscle contraction, limb weight, etc.

(3) Swelling and purplish local soft tissues, as a result of bleeding and exudation from the fracture. In addition, the fracture dislocation and overlapping form local swelling on the surface.

(4) Bony crepitus or bone scraping sensation. When the injured moves on their own, the fracture ends rub against each other, and they can hear bony crepitus or have the bone scraping sensation.

(5) Functional limitation. If the lower extremities are fractured, the injured cannot stand; if the ribs are fractured, they will suffer from dyspnea and sharp pain; if the fracture is near the joint, they cannot bend or stretch; if the spine is fractured, they cannot sit up.

2. Systemic symptoms

(1) Shock. In case of spinal fracture, pelvic fracture, or large tubular bone fracture, the victims often go into shock due to high blood loss.

(2) Elevated temperature. The victims will often suffer fervescence within two or three days of fracture, generally not higher than 39 °C. Otherwise, examine for other complications, such as wound infection and damage to other organs.

(3) Paralysis of limbs. It is mainly ascribed to nerve tissue injury caused by compression of the fracture end. In case of the above symptoms at the scene, consider the possibility of fracture, and do first aid for fracture, and then take the victims to hospital for treatment.

(Ⅲ) First aid for fracture on the scene

1. Basic principles

(1) The purpose of first aid on the scene is to prevent further injury. To this end, never move the broken limbs and the fracture end, to prevent the sharp fracture end from piercing the skin, surrounding tissues, nerves, great vessels, etc. First, the injured limb should be bandaged and

immobilized.

(2) For wounds with open fractures, the most important thing is to prevent contamination of the wound. The first responders must not apply any medicine to the wound, rinse or touch the wound, nor push the exposed end back into the skin.

(3) The responders must keep calm, perform first aid correctly, and get the cooperation of the injured. Blind restoration of the fracture is strictly prohibited on site.

(4) After the whole body condition is stable, consider fixing and handling. Wooden, plastic and metal splints are often used for fracture fixation. In the absence of ready-made splints at the scene, use the locally available planks, bamboo poles, walking sticks, umbrella handles, wooden sticks, branches, etc. Stop the bleeding, bandage and then fix the fracture. The length of the splint used should be proportional to that of the limb. Do not contact the splint directly with the skin; instead, pad a towel or cloth on the splint to avoid nerve damage.

(5) The first aid on the scene is only temporary fixation of the fracture, and the victims should be sent to the hospital as soon as possible.

2. First aid for fractures in several parts

(1) Humerus fracture in the upper arm, radius and ulna fracture in forearm.

Humerus fracture in the upper arm: Hold the injured upper arm against the thorax and line the upper arm with a folded scarf or dry towel between the upper arm and the thorax; bend the elbow joint by 90°, keep his/her forearm resting on the trunk, and hang the forearms on his/her neck with a triangular binder; take a board equal to the length of the upper arm and place it on the outside of the upper arm, and line it with towels and other things between the plank and the upper arm; finally, use two bandages (or other strips of cloth) to tie the upper arm up and down to the thorax.

Radius and ulna fracture in forearm: Take two planks the length of forearms and pad them with soft clothes such as towels. Place one on the volar side of the forearm and the other on the back side of the forearm. Tie the two planks together with three bandages (or other strips of cloth), leaving the thumbs exposed; after the splint is fixed, bend the elbow by 90°, and then hang his/her forearm over the neck with a triangular binder, as shown in Fig. 8-19.

Fig. 8-19　First aid to humerus fracture in the upper arm, radius and ulna fracture in forearm

(2) Humerus fracture in the thigh.

One person stabilizes the broken upper and lower limbs, and the other person gently pulls down the distal end of the broken bone along the long axis of the bone, and cannot rotate; place a folded sheet between his/her legs to bring his/her legs together; place a short splint equal to the length of the lower limb on the inside of the injured limb; place a long splint from the armpit to the heel on the outside of the injured limb; fix the splints on both sides, including the trunk, with a wide cloth belt; finally, reattach the fixed victim to the plank, and place his/her legs slightly higher with a pillow, as shown in Fig. 8-20.

Fig. 8-20　First aid for humerus fracture in the thigh

(3) Fracture of tibia and fibula in the lower leg.

Fracture of tibia and fibula in the lower leg: The tibia and fibula are below the knee joint, and then the lower part is the ankle joint, metatarsus and phalanx. The fixation method is to place two splints on the inside and outside of the lower leg, respectively, and pad the protruding part of fracture; make fixation from above knee joint to below the ankle joint; finally, apply the gauze roller bandage, yardage roll, or towel, as shown in Fig. 8-21.

Fig. 8-21　First aid for fracture of tibia and fibula in the lower leg

(4) Fracture of cervical vertebrae.

Cervical fracture: Lay the victims down, avoid looking up, shaking head, turning, supporting, walking or rolling over for undressing; otherwise, turning their head may cause immediate paralysis or even sudden death. The responder can be located on the victim's head, pull their head up steadily and vertically with both hands, and place a detachable collar or pillow on the victim's neck to maintain the traction; pile thick (or a few) books or sandbags on both sides of the head, so that his/her head cannot shake from side to side; bandage the victim's frontal part and books to the stretcher again, as shown in Fig. 8-22.

Fig. 8-22　First aid for cervical fracture

3. Handling of fractured victims

After hemostasis, bandaging or fracture fixation at the scene, the victim must be handled to the hospital for treatment. The correct handling has a direct bearing on the victim's injury and the effect of future treatment. The victims are handled in such a way to keep them comfortable and immobilized, and strive to minimize harmful effects.

(1) The practice of loading general victims onto a stretcher.

The practice of loading general victims onto a stretcher: Two stretcher-bearers kneel on their right legs, one supports the victim's head and shoulders with one hand and his/her waist with the other; the other supports the victim's pelvis with one hand and his/her knee with the other. The two stretcher-bearers stand up at the same time and load the victim onto the stretcher, as shown in Fig. 8-23.

Fig. 8-23 Handling of general victims

(2) Handling of victims with cervical fracture.

Handling of victims with cervical fracture: Pay more attention to the handling of such patients, since a careless move can result in immediate death. The victim is handled by 3 to 4 stretcher-bearers, one of whom specializes in head traction and fixation, so that the head is kept in a straight line with his/her trunk to keep the neck still. The other three squat on the same side of the victim, with two supporting the victim's trunk and the last man supporting his/her lower limbs. They stand up at the same time and place the victim gently on the stretcher, as shown in Fig. 8-24.

Fig. 8-24 Handling of victims with cervical fracture

(3) Transport of victims with cervical fracture.

Transport of victims with cervical fracture: Lay the victim flat on a stretcher, and strap his/her waist to the stretcher to prevent falling, as shown in Fig. 8-25 (a). The victim's head should be in the back during the flat handling, and on the top when going upstairs, downstairs and downhill, as shown in Fig. 8-25 (b). If transported without any tools and protection measures, the victim is prone to aggravated injury or even death.

（a）

（b）

Fig. 8-25　Transport of victims with cervical fracture

Ⅳ. First aid for burns

Keep the wound clean for electric injury, flame burns or hot gas or water scalds. The victim's clothes, shoes and socks are cut open with scissors and removed. All wounds are covered with clean cloth to prevent contamination. Rinse limbs with clean cold water when burned, then cover them with clean cloth or sterile gauze and send the victims to hospital.

Strong acid or alkali burns should be immediately washed thoroughly with plenty of water, and the eroded clothing should be quickly cut away. In order to prevent acid and alkali residue in the wound, the flushing time is generally not less than 10 min.

Without the consent of medical workers, anything and drugs should not be applied to the burned area.

On the way to the hospital, the victims may be orally given a small amount of sodium chloride and dextrose injection multiple times.

Ⅴ. First aid for frostbite

Frostbite stiffens the muscles, sometimes to the bones. Be gentle during the rescue, and do not force their limbs to bend, so as to avoid further damage. Use a stretcher lay the victims down and carry them to a warm room for treatment.

Cut the wet clothes off the victims and cover them with dry, soft clothes. Do not warm them or rub snow.

Those with frostbite are sometimes very weak in breathing and heartbeat. They should not be mistaken for death; try to save them.

VI. First aid for animal bites

(I) Venomous snake bites

(1) When bitten by a venomous snake, do not panic, run, or drink to prevent the venom from spreading faster through the human body.

(2) Most of the bites are on the extremities. Squeeze the venom quickly and repeatedly from the top of the wound down, and then tie a cloth strap over the wound (proximal to the heart) to immobilize the injured limb, thereby reducing the absorption of venom.

(3) Take the drug for snakebite (if any) and then send the victim to hospital for treatment.

(II) Dog bite

(1) After the dog bites, rinse the wound immediately with strong soapy water, squeeze saliva from the remaining wound from top to bottom, and then smear the wound with iodine.

(2) When there is a small amount of bleeding, do not rush to stop it, nor bandage or stitch the wound.

(3) Try to find out if the dog is a "rabid dog", which is of great help for the hospital to give a treatment plan.

VII. First aid for drowning

If a person is found drowning, try to pull him out of the water quickly. Apply CPR to those with cardiac arrest.

When mouth-to-mouth resuscitation is unfeasible due to foreign body obstruction and cannot be removed with fingers, place your hands on top of each other on the midline slightly above the umbilical cord (away from the xiphoid process) and push up fast and hard several times to make the foreign body withdraw, but do not push too hard.

Drowning is mainly ascribed to asphyxia. Fresh water circulates quickly through the body, and the trachea can hold very little water. Therefore, do not delay the rescue time because of "pouring water" in the rescue of drowning victims, nor fail to perform CPR as a result of "pouring water".

VIII. First aid for heat stroke

The sun beating down on our head, too high ambient temperature, drinking too little water or excessive sweating can cause heat stroke. It is generally manifested as nausea, vomiting, chest distress, dizziness, drowsiness, collapse, and in severe cases convulsions, seizures and even coma.

Move the victims immediately from high temperature or sun exposure to a cool and ventilated place. Cool down with a cold bath, a wet towel covering the body, a fan, or an ice pack over the head, and give the victims oral saline promptly. In severe cases, take them to hospital for treatment.

IX. First aid for harmful gas poisoning

Gas poisoning starts with tears, eye pain, bucking, dry throat and other symptoms. Be on alert. Then it is progressed to headache, shortness of breath, chest distress, dizziness, and even convulsion or coma.

When there is a suspicion that harmful gases may be present, personnel should be immediately evacuated from the scene and transferred to a well-ventilated place. Responders must wear gas masks when entering danger zones.

For unconscious victims, keep their airways open and give oxygen inhale when possible. In case of respiratory and cardiac arrest, apply CPR, and contact the hospital for treatment.

Identify the names of harmful gases quickly for early symptomatic treatment in the hospital.

任务四　排除作业现场安全风险实训

一、作业任务

以"110 kV 检修大厅"为工作环境，按照《国家电网公司电力安全工作规程（变电部分）》的要求，单人进行，分析并写出"10 kV 配电变压器绝缘电阻测试作业现场存在的安全风险及排除措施"。

二、引用标准及文件

（1）《国家电网公司电力安全工作规程（变电部分）》。
（2）《国家电网公司十八项电网重大反事故措施》。

三、作业条件

有能胜任本次作业任务的工作负责人；作业人员精神状态良好，熟悉《国家电网公司电力安全工作规程（变电部分）》中电气试验部分的相关规定，熟悉保证安全的组织措施和技术措施，熟悉风险点分析及预控方法。

四、作业前准备

1. 工器具及材料选择

10 kV 配电变压器工位 1 个（已装设好围栏、悬挂好标示牌）、进行 10 kV 配电变压器绝缘电阻测试所需的工器具（放电棒、手摇式兆欧表、绝缘手套、验电器、安全帽、铜线若干）、中性笔 1 支、笔记本 1 个、工作负责人红背心 1 件、安全技术交底卡 1 份。

2. 危险点及预防措施

（1）低压触电。
危险点：大厅内其他设备二次部分带电，有低压触电危险。
预防措施：作业前对作业人员进行危险点告知，并签署安全技术交底卡，提醒作业人员作业时应与带电部位保持足够的安全距离，禁止进入未经许可的间隔，严禁乱摸乱碰未经许可的设备。
（2）机械伤人。
危险点：大厅内设备较多，可能同时有多个班组作业，有机械伤人危险。
预防措施：作业前对作业人员进行危险点告知，并签署安全技术交底卡，提醒作业人员禁止在作业现场嬉戏打闹，禁止未经许可触碰其他设备。

3. 作业人员分工

现场工作负责人（监护人）：×××
现场作业人员：×××

五、作业规范及要求

（1）现场作业时，由工作负责人交代清楚本次作业的作业任务、危险点及防控措施，作业人员清楚后在安全技术交底卡上签字确认。

（2）作业人员在现场工作负责人的监护下单人进行，写出 10 kV 配电变压器绝缘电阻测试存在的风险因素及排除措施。

六、作业流程及标准（表 8-5）

表 8-5　排除作业现场安全风险流程及评分标准

班级		姓名		学号		考评员		成绩	
序号	作业名称	质量标准			分值/分	扣分标准		扣分	得分
1	工作准备								
1.1	着装穿戴	穿工作服、戴安全帽			2	未穿工作服、未戴安全帽各扣 1 分；着装穿戴不规范，每处扣 0.5 分，扣完为止			
2	环境风险分析预控								
2.1	环境风险分析	正确分析作业现场是否存在有害粉尘、有毒气体、噪声、辐射、易燃易爆、异常天气等风险因素；正确分析作业现场温度、湿度是否适宜作业			7	未分析到作业环境风险源每处扣 2 分，分析不准确扣 1 分，扣完为止			
2.2	环境风险排除措施	正确写出环境风险因素排除措施			7	未写出环境风险源排除措施或措施不正确每项扣 2 分，措施不准确每项扣 1 分，扣完为止			
3	场地风险分析预控								
3.1	场地风险分析	正确分析作业场地是否存在高空坠物、机械伤人、设备故障、触电等风险因素			6	未分析到作业场地风险源每处扣 2 分，分析不准确扣 1 分，扣完为止			
3.2	场地风险排除措施	正确写出场地风险因素排除措施			6	未写出场地风险源排除措施或措施不正确每项扣 2 分，措施不准确每项扣 1 分，扣完为止			

续表

序号	作业名称	质量标准	分值/分	扣分标准	扣分	得分
4		工器具风险分析预控				
4.1	工器具风险分析	正确分析着装规范方面存在的风险因素；正确分析安全工器具存在的风险因素；正确分析所用摇表、配电变压器、短接铜线等是否存在危险因素	7	未分析到作业工器具存在的风险源每处扣2分，分析不准确扣1分，扣完为止		
4.2	工器具风险排除措施	正确写出工器具风险因素排除措施	7	未写出作业工器具存在的风险源排除措施或措施不正确每项扣2分，措施不准确每项扣1分，扣完为止		
5		安全措施风险分析预控				
5.1	安全措施风险分析	正确分析设备突然来电、验电、接地方面存在的风险因素；正确分析装设围栏方面存在的风险因素；正确分析悬挂标示牌方面存在的风险因素	9	未分析到作业工器具存在的风险源每处扣2分，分析不准确扣1分，扣完为止		
5.2	安全措施风险排除措施	正确写出工器具风险因素排除措施	9	未写出作业工器具存在的风险源排除措施或措施不正确，每项扣2分；措施不准确，每项扣1分，扣完为止		
6		作业人员风险分析预控				
6.1	人员风险分析	正确分析人员精神状态、安全意识、工作态度方面是否存在风险因素；正确分析人员技能水平、知识水平方面是否存在风险因素；正确分析人员分工协作方面是否存在风险因素。	7	未分析到作业人员存在的风险源每处扣2分，分析不准确扣1分，扣完为止		
6.2	人员风险排除措施	正确写出作业人员风险因素排除措施	7	未写出作业人员风险源排除措施或措施不正确每项扣2分，措施不准确每项扣1分，扣完为止		

续表

序号	作业名称	质量标准	分值/分	扣分标准	扣分	得分
7	作业过程风险分析预控					
7.1	作业过程风险分析	正确分析作业方法方面是否存在风险因素；正确分析作业过程是否存在突发危险因素；正确分析作业过程是否存在突然来电危险	13	未分析到作业过程存在的风险源每处扣2分，分析不准确扣1分，扣完为止		
7.2	作业过程风险排除措施	正确写出作业过程风险因素排除措施	13	未写出作业过程风险源排除措施或措施不正确每项扣2分，措施不准确每项扣1分，扣完为止		
合计			100			

七、排除作业现场安全风险记录卡（表8-6）

表8-6　排除作业现场安全风险记录卡

序号	分类	存在的风险因素	风险排除措施	备注
1	作业环境			
2	作业场地			
3	作业工器具			
4	安全措施			
5	作业人员			
6	作业过程			
7	其他			

Task Ⅳ Training in eliminating safety risks on the job site

Ⅰ. Operating tasks

Taking the "110 kV maintenance hall" as the working environment, students are expected to analyze and write "Safety risks and elimination measures of insulation resistance test for 10 kV distribution transformers" alone in accordance with the *Electric Power Safety Working Regulations (Power Transformation) of State Grid Corporation of China*

Ⅱ. Referenced standards and documents

(1) *Electric Power Safety Working Regulations (Power Transformation) of State Grid Corporation of China*;

(2) *18 Major Anti-accident Measures for Power Grid of State Grid Corporation of China*.

Ⅲ. Operating conditions

A person in charge of work is competent for this task; the operators are in good spirits, acquainted with the electrical test part in the *Electric Power Safety Working Regulations (Power Transformation) of State Grid Corporation of China*, the organizational and technical measures to ensure safety, as well as risk point analysis and pre-control methods.

Ⅳ. Preparation before operation

1. Selection of tools and instruments and materials

A 10 kV distribution transformer station (the fence and sign board already available), tools and instruments for insulation resistance test of 10 kV distribution transformer (discharging rod, hand-operated megohmmeter, insulating gloves, electroscope, safety helmet, copper wires), a gel pen, a notebook, a red vest for the person in charge of work, and a safety technical disclosure card.

2. Dangerous points and preventive measures

(1) Low voltage electric shock.

Dangerous points: The secondary parts of other equipment in the hall are electrically charged, posing a risk of low voltage electric shock.

Prevention and control measures: Before the operation, inform the operators of dangerous points, have them sign the safety technical disclosure card, and remind them to maintain a sufficient safe distance from the live parts during the operation. Do not enter unauthorized intervals, nor touch unauthorized equipment.

(2) Mechanical injury.

Dangerous points: There are many devices in the hall, and more than one team may carry out simultaneous operation, which may cause mechanical injury.

Prevention and control measures: Before the operation, inform the operators of dangerous points, have them sign the safety technical disclosure card, and remind them not to play on the job site. Do not touch other devices without permission.

3. Division of labor among operators

Person in charge of on-site work (supervisor): ×××

On-site operator: ×××

Ⅴ. Operating specifications and requirements

(1) During the field operation, the person in charge of work shall explain clearly the operation tasks, dangerous points and prevention and control measures of this operation, and the operators must sign the safety technical disclosure card for acknowledgement.

(2) The operators shall write the risk factors and elimination measures of insulation resistance test for 10 kV distribution transformers alone under the supervision of the person in charge of on-site work.

Ⅵ. Operation procedures and standards (see Tab. 8-5)

Tab. 8-5 Procedures and scoring criteria for eliminating safety risks on the job site

Class		Name		Student ID		Examiner		Score		
S/N	Operation name	Metrics			Points	Deduction criteria		Deduction		Score
1	Work preparation									
1.1	Wearing	Wear work clothes and a safety helmet			2	Deduct 1 point if failing to wear work clothes and safety helmet; Deduct 0.5 points if the wearing does not meet the requirements until 0				
2	Environmental risk analysis, prevention and control									
2.1	Environmental risk analysis	Analyze any risk factors on the worksite, such as harmful dust, toxic gas, noise, radiation, inflammable and explosive articles, and unusual weather; Analyze whether the temperature and humidity of the worksite are suitable for operation			7	Deduct 2 points if failing to analyze the risk source of the working environment; deduct 1 point if the analysis is not accurate until 0				
2.2	Environmental risk elimination measures	Write the correct elimination measures for environmental risk factors			7	Deduct 2 points if failing to write environmental risk source elimination measures or giving the wrong measures; deduct 1 point if failing to write proper measures until 0				

Continued

S/N	Operation name	Metrics	Points	Deduction criteria	Deduction	Score
3		On-site risk analysis, prevention and control				
3.1	On-site risk analysis	Correctly analyze the presence of risk factors, such as falling objects from high altitudes, mechanical injury, equipment failure, and electric shock at the job site	6	Deduct 2 points if failing to analyze the risk source at the job site; deduct 1 point if the analysis is not accurate until 0		
3.2	On-site risk elimination measures	Write the proper measures to eliminate on-site risk factors	6	Deduct 2 points if failing to write on-site risk source elimination measures or giving the wrong measures; deduct 1 point if failing to write proper measures until 0		
4		Risk analysis, prevention and control for tools and instruments				
4.1	Risk analysis for tools and instruments	Correctly analyze the risk factors in the proper dressing; Correctly analyze the risk factors in the safety tools and instruments; Correctly analyze the presence of risk factors in the megameter, distribution transformer, and short-circuit copper wires used	7	Deduct 2 points if failing to analyze the risk source of working tools and instruments; deduct 1 point if the analysis is not accurate until 0		
4.2	Risk elimination measures for tools and instruments	Correctly write the measures to eliminate the risk factors for tools and instruments	7	Deduct 2 points if failing to write measures for eliminating risk sources in tools and instruments or giving the wrong measures; deduct 1 point if failing to write proper measures until 0		
5		Risk analysis, prevention and control for safety measures				
5.1	Risk analysis for safety measures	Correctly analyze the risk factors in sudden power supply, verification of live parts and grounding of equipment; Correctly analyze the risk factors in the erection of fences; Correctly analyze the risk factors in the hanging of sign boards	9	Deduct 2 points if failing to analyze the risk source of working tools and instruments; deduct 1 point if the analysis is not accurate until 0		
5.2	Risk elimination measures for safety measures	Correctly write the measures to eliminate the risk factors for tools and instruments	9	Deduct 2 points if failing to write measures for eliminating risk sources in tools and instruments or giving the wrong measures; deduct 1 point if failing to write proper measures until 0		
6		Operator risk analysis, prevention and control				

Continued

S/N	Operation name	Metrics	Points	Deduction criteria	Deduction	Score
6.1	Operator risk analysis	Correctly analyze the presence of risk factors in the operator's mental state, safety consciousness and working attitude; Correctly analyze the presence of risk factors in the operator's skills and knowledge level; Correctly analyze any risk factors in the division of labor and cooperation	7	Deduct 2 points if failing to analyze the risk source of operators; deduct 1 point if the analysis is not accurate until 0		
6.2	Operator risk elimination measures	Write the proper elimination measures for operator risk factors	7	Deduct 2 points if failing to write operator risk source elimination measures or giving the wrong measures; deduct 1 point if failing to write proper measures until 0		
7		Operational process risk analysis, prevention and control				
7.1	Operational process risk analysis	Correctly analyze any risk factors in the working methods; Correctly analyze any sudden risk factors in the operational process; Correctly analyze any sudden power supply in the operational process	13	Deduct 2 points if failing to analyze the risk source in the operational process; deduct 1 point if the analysis is not accurate until 0		
7.2	Risk elimination measures in the operational process	Write the proper elimination measures for risk factors in the operational process	13	Deduct 2 points if failing to write the measures for elimination of risk sources in the operational process or giving the wrong measures; deduct 1 point if failing to write proper measures until 0		
Total			100			

VII. Log sheet for eliminating safety risks on the job site (see Tab. 8-6)

Tab. 8-6 Log sheet for eliminating safety risks on the job site

S/N	Category	Risk factors	Risk elimination measures	Remarks
1	Working environment			
2	Job site			
3	Working tools and instruments			
4	Safety measures			
5	Operators			
6	Operation process			
7	Miscellaneous			

任务五 触电急救实训

一、作业任务

李某家插线板电线绝缘层破损，李某由于缺乏安全用电常识，有险不知险，继续使用绝缘层破损的插线板。某日，李某使用该插线板从屋里引电到院落里，不小心碰触到裸露出来的导线，致使李某触电倒下，将插线板电线压在了身下。

用模拟人代替李某，模拟以上触电场景，单人操作，对李某实施触电急救。

二、引用标准及文件

（1）《国家电网公司电力安全工作规程（变电部分）》。
（2）《电力行业紧急救护技术规范》。
（3）《国际心肺复苏（CPR）与心血管急救（ECC）指南 2015》。

三、作业条件

有能胜任本次作业任务的工作负责人；作业人员精神状态良好，熟悉《国家电网公司电力安全工作规程（变电部分）》，熟悉《电力行业紧急救护技术规范》，熟悉《国际心肺复苏（CPR）与心血管急救（ECC）指南 2015》中关于心肺复苏的最新规定。

四、作业前准备

1. 工器具及材料选择

电脑心肺复苏模拟人，酒精卫生球（签），一次性 CPR 屏障消毒面膜，干燥木棒、金属杆各 1 根，2 m 及以上无卷曲电线 1 根，中性笔 1 支，笔记本 1 个。

2. 危险点及预防措施

危险点：操作过程中模拟人及室内照明设备带电，有低压触电危险。
预防措施：作业前对作业人员进行危险点告知，提醒作业人员作业时防止低压触电。

3. 作业人员分工

现场工作负责人（监护人）：×××
现场作业人员：×××

五、作业规范及要求

（1）迅速将触电者脱离低压电源。解救触电者时救护者必须首先懂得自我保护，若出现有可能导致救护者触电的情况，立即终止任务，考核成绩记为 0 分。
（2）正确进行脱离电源后的处理。要求正确判断触电者神志、呼吸及脉搏状况。
（3）设触电者无神志、呼吸及心跳，要求结束判断后立即运用现场心肺复苏法进行紧急救护。

（4）在完成 5 个 CPR 压吹循环后，要求对触电者的情况进行判定，并口述瞳孔、脉搏和呼吸情况。

（5）要求操作程序正确、动作规范并成功将触电者救活。

（6）"人工循环（体外按压）"项目中，按压错误次数超过 30 次，考核成绩记为不合格。

（7）本模块操作时间为 3 min，时间到立即终止任务。

六、作业流程及标准（表 8-7）

表 8-7 触电急救流程及评分标准

班级		姓名		学号		考评员		成绩	
序号	作业名称	质量标准			分值/分	扣分标准		扣分	得分
1	开场白（15 s）	着装整洁、规范，表情端庄；要有上、下、左、右观察动作，口述"周围环境安全"；报告考官：××考生参加考试；考官指示"开始"后进入比赛操作流程			1	着装不规范扣 0.5 分；不报告考官扣 0.5 分；不按要求口述扣 0.5 分			
2	迅速脱离电源	立即拉开电源开关或拔除电源插头，或用有绝缘柄的电工钳或有干燥木柄的斧头切断电线，断开电源；用带有绝缘胶柄的钢丝钳、绝缘物体或干燥不导电物体等工具将触电者迅速脱离电源（可任选一种操作）			4	任何使救护者或触电者处于不安全状况的行为不得分，操作时间超过 10 s 不得分			
3	判断伤员意识及呼叫（10 s）	意识判断：轻拍伤员双肩，分别对病人双耳呼叫"同志，醒醒"，确认意识丧失；6 s 内完成			4	未操作一项扣 2 分；其中如果某一项有操作动作，但操作不规范扣 1 分，两项操作都不规范扣 2 分；操作时间超过 5 s 扣 2 分			
		呼救：转身招手摆臂，并大声呼叫"快来人，救命啊，帮忙打 120 啊"；4 s 内完成			2	操作不规范扣 1 分；操作时间超过 4 s 扣 1 分			
4	摆好伤员体位（5 s）	患者取仰卧位，置于地面或硬板上；靠近患者跪地，操作者左腿平病人肩部，双膝略开与肩同宽			2	未操作一项或操作不规范扣 1 分；操作时间超过 5 s 扣 1 分			

续表

序号	作业名称	质量标准	分值/分	扣分标准	扣分	得分
5	判断伤员呼吸(12 s)	用仰头抬颌法开放气道,在观察过程中要求气道始终保持开放位置	2	气道未开放扣2分; 整个观察过程始终保持气道在开放位置,中途有间断扣1分		
		看:看伤员的胸部、腹部有无起伏动作,3~5 s完成	2	操作不规范扣1分; 操作时间超过5 s或少于3 s扣1分		
		听:用左耳贴近伤员的口鼻处,听有无呼气声音,与"看"同时进行	2	操作不规范扣1分		
		试:用左侧脸颊贴近患者口鼻,感觉病人有无气流呼出;3~5 s完成	2	操作不规范扣1分; 操作时间超过5 s或少于3 s扣1分		
		口述"病人无自主呼吸"	1	不操作扣1分		
6	现场心肺复苏CPR					
6.1	CPR操作频率任务:首次按压30次	按压部位:胸骨中下1/3或两乳头连线与胸骨交叉点	9	按压位置不正确扣3分		
		按压手法:一手掌根部放在胸骨按压部位,另一手掌根部放在此手背上,双手掌根重叠,十指相扣,手指不接触胸壁,手臂与胸骨水平垂直;按压间隙,双手应离开胸壁		不符合要求每项扣1分		
		按压深度:利用上身重量垂直下压,使胸骨下陷至少5 cm,不得超过6 cm并高声报数		未报数扣1分; 深度不够或超过6 cm(即按压过轻或过重,每次扣0.5分,扣完为止; 按压每多1次或少1次扣0.5分; 累计扣分最高不超过本项满分		
		按压频率:按压频率100~120次/min,按压与放松时间比1∶1		频率低于100或高于120次/min(与6.4合并扣分),按压与放松时间比不正确超过2次扣2分		

续表

序号	作业名称	质量标准	分值/分	扣分标准	扣分	得分
6.1	CPR操作频率任务：首次按压30次	按压与呼吸比例：按压与呼吸比例单双人均30∶2；方法：按压用力适当、节奏均匀、持续进行，迅速放松时胸廓复原，放松时手掌根部不离开胸壁		压吹比例错误扣1分；不符合要求每项扣1分		
6.2	开放气道（5 s）	清理气道：检查口鼻腔有无分泌物、活动假牙、异物，并进行清理开放气道	2	不规范扣1分；操作时间超过2 s扣1分		
		仰头抬颏法：抢救者站在病人右侧，左手置于病人前额上用力向后压，同时右手食指和中指放在病人下颌骨下缘，将颏部向上向前抬起（头颈部损伤者禁用）；双手托颌法：操作者双手将下颌角抬起，并使病人头后仰，下颌骨前移使气道打开（用于颈椎损伤或疑有颈椎损伤者）	4	未采用仰头抬颏法通畅气道扣2分，气道未打开或打开无效扣2分；未保持气道在整个吹气过程中始终通畅，扣1分；其他操作不规范一项扣1分；操作时间超过3 s扣1分		
6.3	口对口人工呼吸2次（6 s）	吹气：操作者一手捏住病人鼻孔，平静吸气，双唇紧紧包绕住病人口唇用力吹气，连续2次，每次持续1~1.5 s，每次吹气量500~600 mL/次（吹气时，病人胸部隆起即可，避免过度通气）；吹气毕放开鼻孔，让气体自然由口鼻逸出，吹气和放气时间比1∶1	4	少吹一次气或未吹进气一次扣2分，多吹一次扣1分；吹气量不足或过大一次扣1分；操作不规范一项扣1分；操作时间超过6 s扣1分		
6.4	CPR操作频率任务：完成4个30∶2的按压：吹气循环	继续完成4个30∶2的压吹循环；按压质量标准同6.1中按压质量标准；吹气质量标准同6.3中按压质量标准	49	按压扣分标准同6.1中按压扣分标准；吹气扣分标准同6.3中吹气扣分标准		

续表

序号	作业名称	质量标准	分值/分	扣分标准	扣分	得分
7	抢救过程中的再判定（15 s）	判定是否恢复：用看、听、试方法对伤员呼吸进行判断，摸颈动脉搏动判断心跳，查看瞳孔来判断意识等；口述"病人自主呼吸恢复、触及大动脉搏动，心肺复苏成功"	5	有动作，但未到位或方法不对，每一项扣1分，最多不超过2分；有一项未操作扣2分；操作时间超过15 s扣1分		
8	退场	整理模拟人后退场	1	未操作扣1分		
9	熟练程度	操作熟练，各项之间无停顿，无不必要动作	4	各项之间有一次停顿扣1分；有一项不必要动作扣1分；一项操作程序顺序错误扣2分		
合计			100			

Task V　Training in first aid for electric shock

Ⅰ. Operating tasks

The plug board wire insulation in farmer Li's home was damaged. Li continued to use the plug board due to lack of knowledge about safe use of electricity. One day, Li used the plug board to feed electricity from the house to the courtyard, but accidentally touched the exposed conductor. Li fell down by electric shock, and put the plug board wire under him.

Replace Li with a dummy to simulate the above electric shock scene. Please implement first aid to him alone.

Ⅱ. Referenced standards and documents

(1) *Electric Power Safety Working Regulations (Power Transformation) of State Grid Corporation of China*;

(2) *Electric Power Trade Urgently Saving Technique Rules*;

(3) *International Guidelines for Cardiopulmonary Resuscitation (CPR) and Emergency Cardiovascular Care (ECC) 2015*.

Ⅲ. Operating conditions

A person in charge of work is competent for this task; the operators are in good spirits, acquainted with the *Electric Power Safety Working Regulations (Power Transformation) of State Grid Corporation of China*, the *Electric Power Trade Urgently Saving Technique Rules*, and the latest regulations on CPR in the *International Guidelines for Cardiopulmonary Resuscitation (CPR) and Emergency Cardiovascular Care (ECC) 2015*.

Ⅳ. Preparation before operation

1. Selection of tools and instruments and materials

A computerized CPR simulator, alcohol cotton balls (swabs), a disposable CPR barrier disinfection mask; a dry stick, a metal bar, a 2 m or longer non-crimped wire, a gel pen, and a notebook.

2. Dangerous points and preventive measures

Dangerous points: The simulator and indoor lighting equipment are electrically charged in the operation process, posing a risk of low voltage electric shock.

Prevention and control measures: Before the operation, inform the operators of dangerous points, and remind them to prevent low voltage electric shock during operation.

3. Division of labor among operators

Person in charge of on-site work (supervisor):　×××

On-site operator:　×××

V. Operating specifications and requirements

(1) Quickly disengage the victim from the low-voltage power supply. The responder must first know how to protect himself. In case of any potential electric shock to the responder, terminate the task immediately, and the appraisal result will be 0 point.

(2) Properly handle the victim after disconnecting from power supply. The responder is required to correctly judge the victim's consciousness, breathing and pulse.

(3) If the victim has no consciousness, breathing and heartbeat, the responder must apply on-site CPR immediately after judgment.

(4) After five CPR cycles, the responder must determine the victim's condition, and dictate his pupil, pulse, and respiration.

(5) The responder must follow the correct operation procedures, with standard operation, and save the victim successfully.

(6) In the "artificial circulation (external chest compression)", more than 30 compression errors will end up with "Fail" in the appraisal result.

(7) The operation duration of this module is 3 min, and the task is terminated immediately when time is up.

VI. Operation procedures and standards (see Tabl. 8-7)

Tab. 8-7 Procedures and scoring criteria of first aid for electric shock

Class		Name		Student ID		Examiner		Score	
S/N	Operation name	Quality standard		Points	Deduction criteria		Deduction		Score
1	Opening remarks (15 s)	Dress neatly and properly, with dignified expression; Have up, down, left and right observation movements, and dictate "The ambient environment is safe"; Report to the examiner: Candidate ×× take the exam; Enter the competition operation process after the examiner indicates "Start"		1	Deduct 0.5 points if not dressing properly; Deduct 0.5 points if failing to report to the examiner; Deduct 0.5 points if not dictating as required				
2	Get away from power supply quickly	Immediately disconnect the power switch or remove the power plug, or use electrician's pliers with an insulated handle or an axe with a dry wooden handle to cut the wires and disconnect the power supply; use the tools such as the cutting pliers with an insulating handle, an insulator, or a dry non-conductive object to quickly disengage the victim from the power supply (either of the two operations)		4	Any act that places the responder or victim in an unsafe situation will be scored 0, and the operation lasting more than 10 s will be scored 0				

Continued

S/N	Operation name	Quality standard	Points	Deduction criteria	Deduction	Score
3	Determine the victim's consciousness and shout (10 s)	Consciousness judgment: Pat on the shoulders and say "Comrade, wake up" to the victim's ears to confirm the loss of consciousness. Done in 6 s	4	Deduct 2 points for failing to do any operation; Deduct 1 point if any operation is not standard, and deduct 2 points if two operations are not standard; Deduct 2 points if the operation exceeds 5 s		
		Call for help: Turn around and swing arms, and call loudly "Help! Call 120!" Done in 4 s	2	Deduct 1 point if the operation is not standard; Deduct 1 point if the operation exceeds 4 s		
4	Get the victim into position (5 s)	Place the victim in a supine position on the ground or a hard board; Kneel close to the victim, the operator's left leg is level with the victim's shoulder, and keep both knees slightly shoulder-width apart	2	Deduct 1 point if any operation is missing or the operation is not standard; Deduct 1 point if the operation exceeds 5 s		
5	Determining the victim's breathing (12 s)	Open the airway by raising the victim's head and jaw, and keep it always open during the observation	2	Deduct 2 points if failing to open the airway; Keep the airway always open during the observation, deduct 1 point if there is any discontinuity		
		Observing: Observe the victim's chest and abdomen for any breathing movement, done in 3-5 s	2	Deduct 1 point if the operation is not standard; Deduct 1 point if the operation exceeds 5 s or is less than 3 s		
		Listening: Put your left ear close to the victim's nose and mouth and listen for exhalation, done simultaneously with "observing"	2	Deduct 1 point if the operation is not standard		
		Sensing: Put your left cheek close to the victim's nose and mouth and feel if the victim exhales. Done in 3-5 s	2	Deduct 1 point if the operation is not standard; Deduct 1 point if the operation exceeds 5 s or is less than 3 s		
		Dictate "The victim is not breathing on his own"	1	Deduct 1 point if no operation is done		
6	On-site CPR					
6.1	CPR frequency task: First do 30 compressions	Compression site: Mid-lower third of the sternum or the intersection of the two-nipple connection with the sternum	9	Deduct 3 points if the compression site is incorrect		

Continued

S/N	Operation name	Quality standard	Points	Deduction criteria	Deduction	Score
6.1	CPR frequency task: First do 30 compressions	Compression technique: Place the base of one palm on the sternum compression site, and the base of the other palm on the back of the hand, the bases of two palms are overlapping and the fingers interlocked but not in contact with the chest wall, and the arm is horizontal and perpendicular to the sternum. Both hands leave the victim's chest wall during the compression interval		Deduct 1 point for any non-conforming item		
		Compression depth: Use the upper body weight to compress down vertically, make the sternum sink at least 5 cm, not more than 6cm and count off loudly		Deduct 1 point if failing to count off; Deduct 0.5 points if any compression depth is not enough or more than 6cm until 0; Deduct 0.5 points for every extra or less compression; The cumulative deduction shall not exceed the full mark of this task		
		Compression frequency: 100–120 times/min, compression and relaxation time ratio 1∶1		Deduct 2 points if the frequency is less than 100 or more than 120 times/min (combined with 6.4) and the compression and relaxation time ratio is incorrect for more than 2 times		
		Compression-ventilation ratio: 30∶2		Deduct 1 point if the compression-ventilation ratio is wrong		
		Method: Apply compression with appropriate force, uniform rhythm and continuous approach, see the chest recovery at rapid release, keep your base of palms not leaving the chest wall when relaxed		Deduct 1 point for any non-conforming item		
6.2	Open the airway (5 s)	Clear the airway: Check the mouth and nose for secretions, movable dentures, foreign bodies, and clear them to keep the airway open	2	Deduct 1 point if the operation is not standard; Deduct 1 point if the operation exceeds 2 s		

Continued

S/N	Operation name	Quality standard	Points	Deduction criteria	Deduction	Score
6.2	Open the airway (5 s)	Raising head and jaw: Stand on the right side of the victim, put your left hand on the victim's forehead and press back hard, and place your right index and middle fingers on the lower margin of the victim's mandible to lift chin up and forward (not for the head and neck injury); Jaw-thrust maneuver: The operator lifts the victim's angle of mandible with both hands and tilts the victim's head back, moving the mandible forward to open the airway (for cervical injury or suspected cervical injury)	4	Deduct 2 points if failing to open the airway by raising head and jaw, deduct 2 points if the airway is not kept open or does not open effectively; deduct 1 point if failing to keep the airway open throughout the ventilation process; Deduct 1 point if any of other operations is not standard; Deduct 1 point if the operation exceeds 3 s		
6.3	Mouth-to-mouth resuscitation 2 times (6 s)	Ventilation: The operator holds the victim's nostril with one hand, inhales calmly, and put your lips firmly around the victim's mouth and ventilate hard for two consecutive times; each time lasts 1–1.5 s, 500–600 mL per time (the ventilation should keep the victim's chest bulged, not excessive ventilation). After ventilation, release the nostrils and let the gas escape naturally from the mouth and nose. The ventilation and air bleeding ratio is 1∶1	4	Deduct 2 points for a short or no ventilation, and deduct 1 point for an extra ventilation; Deduct 1 point for insufficient or excessive ventilations; Deduct 1 point if any operation is not standard; Deduct 1 point if the operation exceeds 6 s		
6.4	CPR frequency task: Complete four 30∶2 cycles	Continue to complete four cycles of 30∶2; Compression quality standards are the same as those in 6.1; Ventilation quality standards are the same as those in 6.3	49	Compression deduction standards are the same as those in 6.1; Ventilation deduction standards are the same as those in 6.3		
7	Redetermination in the process of first aid (15 s)	Determine the recovery: Judge the victim's breathing by observing, listening and sensing, judge the heartbeat by the carotid artery pulse, and judge the victim's consciousness by looking at his pupil, etc. narrate "The victim recovered on his own breathing, his main artery is pulsation, and CPR succeeded"	5	Deduct 1 point if any action is not in place or the method is wrong, a maximum of 2 points to be deducted; Deduct 2 points if failing to do any operation; Deduct 1 point if the operation exceeds 15 s		

Continued

S/N	Operation name	Quality standard	Points	Deduction criteria	Deduction	Score
8	Demobilization	Tidy up the simulator and demobilize	1	Deduct 1 point if no operation is done		
9	Proficiency level	Skillful operation, no pause between operations, free from any necessary action	4	Deduct 1 point if there is a pause between operations; Deduct 1 point for any unnecessary action; Deduct 2 points if an operating procedure is out of order		
Total			100			

任务六 伤口包扎实训

一、作业任务

王先生，25岁，右前臂被机械击伤，可见一长约 4 cm 的伤口，有活动性出血，局部畸形，反常活动。

请用橡皮止血带、夹板等为患者（医学模拟人）进行止血、固定处理。两人一组。

考试时间：11 min。

二、引用标准及文件

（1）《国家电网公司电力安全工作规程（变电部分）》。
（2）《电力行业紧急救护技术规范》。

三、作业条件

有能胜任本次作业任务的工作负责人；作业人员精神状态良好，熟悉《国家电网公司电力安全工作规程（变电部分）》，熟悉《电力行业紧急救护技术规范》。

四、作业前准备

1. 工器具及材料选择

橡皮止血带、急救医药箱、模拟人、夹板、三角巾、绷带等。

2. 危险点及预防措施

危险点：处理不当造成伤员二次伤害。

预防措施：作业人员经过系统培训，熟练掌握伤口包扎相关知识，熟练操作包扎过程，防止造成伤员二次伤害。

3. 作业人员分工

现场工作负责人（监护人）：×××
现场作业人员：×××

五、作业规范及要求

（1）包扎过程若对伤员造成二次伤害，立马终止考试，考试成绩记为 0 分。
（2）包扎过程关心、体恤伤员，注意缓解伤员的紧张情绪。
（3）作业人员熟练操作，在规定时间内完成操作。

六、作业流程及标准（表8-8）

表8-8 伤口包扎流程及评分标准

班级		姓名		学号		考评员		成绩		
序号	作业名称	质量标准			分值/分	扣分标准			扣分	得分
1	操作前准备	准备橡皮止血带、急救医药箱、模拟人、夹板、三角巾、绷带等			3	少准备一件工器具扣1分，扣完为止				
2	职业素质	仪表端庄，举止大方			2	仪表、举止不符合要求，视情况扣1~2分				
		着装（工作服）整洁规范			2	着装不规范视情况扣1~2分				
		报告考官：××考生准备完毕，请求开始考试			2	未报告考官，扣2分				
3	判断伤员情况	快速检测患者的主要生命体征			3	未检查患者生命体征，扣3分；检查不规范扣1~2分				
		检查患肢：暴露右臂了解伤口及畸形情况			3	未检查患肢，扣3分；检查不规范扣1~2分				
		告知患者操作的目的，并取得患者的配合			2	未告知患者操作目的，扣2分				
		关注患者的疼痛程度并给予适当的处理			2	未关注患者的疼痛程度、未给予适当的处理各扣1分				
		缓解焦虑紧张情绪			3	未缓解患者焦虑紧张情绪扣3分				
4	止血、固定操作过程									
4.1	抬高患肢	使用止血带前抬高患肢2~3 min；提问：为什么要抬高患肢2~3 min？			10	未抬高患肢扣3分；回答不正确扣7分，回答不准确扣3分				
4.2	止血带位置选择	止血带位置选择右上臂中上1/3处			5	止血带位置选择错误扣5分				
4.3	绕扎止血带	在扎止血带处置衬垫物			5	未放置衬垫物扣5分				
		绕扎松紧程度以控制出血、右侧桡动脉摸不到搏动为宜			5	绕扎松紧程度不符合要求扣5分				
		记录使用止血带的开始时间			5	未记录使用止血带的开始时间扣5分				

续表

序号	作业名称	质量标准	分值/分	扣分标准	扣分	得分
4.4	捆扎、固定夹板	充分暴露右前臂，伤口创面用无菌纱布或棉垫覆盖并固定	6	未操作该步骤扣6分，操作不规范扣3分		
		夹板长度超过肘关节和腕关节，置于前臂四侧	6	未操作该步骤扣6分，操作不规范扣3分		
		固定前用毛巾等软物铺垫在夹板与肢体间	6	未操作该步骤扣6分，操作不规范扣3分		
		用绷带捆扎固定夹板，上端固定至肘部，下端固定至手掌	8	未操作该步骤扣8分，操作不规范扣4分		
		先捆扎骨折中部，然后捆扎下部，再捆扎上部，松紧度以绷带上下可移动1cm为宜	8	未操作该步骤扣8分，操作不规范扣4分		
4.5	悬吊绷带或三角巾	用绷带或三角巾悬吊于胸前	5	未操作该步骤扣5分，操作不规范扣2分		
5	告知注意事项	操作结束后告知患者相关注意事项	5	未操作该步骤扣5分，注意事项告知不全面扣2分		
6	操作结束	整理工器具、清理现场	3	未操作该步骤扣3分，操作不规范扣1分		
		汇报考官	1	未汇报考官扣1分		
合计			100			

Task Ⅵ Training in wound dressing

Ⅰ. Operating tasks

Wang, 25 years old, suffered a mechanical injury to his right forearm, showing a 4cm long wound with active bleeding, local deformity, and anomalous activity.

Please stop bleeding and fix the patient (medical simulator) with rubber compression cords, splints, etc. Work in pairs.

Exam time: 11min.

Ⅱ. Referenced standards and documents

(1) *Electric Power Safety Working Regulations (Power Transformation) of State Grid Corporation of China*;

(2) *Electric Power Trade Urgently Saving Technique Rules*.

Ⅲ. Operating conditions

A person in charge of work is competent for this task; the operators are in good spirits, acquainted with the *Electric Power Safety Working Regulations (Power Transformation) of State Grid Corporation of China*, and the *Electric Power Trade Urgently Saving Technique Rules*.

Ⅳ. Preparation before operation

1. Selection of tools and instruments and materials

Rubber compression cord, medical first aid kit, simulator, splint, triangular binder, bandage, etc.

2. Dangerous points and preventive measures

Dangerous points: Any improper handling will pose secondary injury.

Prevention and control measures: After systematic training, operators have mastered the knowledge of wound dressing, and become skilled in the dressing process to prevent secondary injury.

3. Division of labor among operators

Person in charge of on-site work (supervisor): ×××

On-site operator: ×××

Ⅴ. Operating specifications and requirements

(1) If the victim suffers secondary injury during the dressing process, the exam will be terminated immediately and the exam result will be 0.

(2) Care and compassion for the victim during the dressing process, try to ease his tension.

(3) The operator is skilled in operation and completes the operation within the specified time.

Ⅵ. Operation procedures and standards (see Tab. 8-8)

Tab. 8-8 Wound dressing procedures and scoring criteria

Class		Name		Student ID		Examiner		Score		
S/N	Operation name	Quality standard			Points	Deduction criteria		Deduction		Score
1	Pre-operation preparation	Prepare rubber compression cord, medical first aid kit, simulator, splint, triangular binder, bandage, etc.			3	Deduct 1 point if missing any tools and instruments until 0				
2	Professional quality	Dignified appearance, well-behaved			2	Deduct 1–2 point(s) if the appearance and behavior do not meet the requirements				
		Dress (work clothes) neatly and properly			2	Deduct 1–2 point(s) if not dressing properly				
		Report to the examiner: Candidate ×× is ready and requests to begin examination			2	Deduct 2 points if failing to report to the examiner				
3	Assess the victim's condition	Check the victim's key vital signs rapidly			3	Deduct 3 points if failing to check the victim's key vital signs; Deduct 1–2 point(s) if the checking is not standard				
		Examine the injured limb: Expose the right arm to know injuries and deformities			3	Deduct 3 points if failing to examine the injured limb; Deduct 1–2 point(s) if the checking is not standard				
		Inform the victim of the purpose of the procedure and obtain his cooperation			2	Deduct 2 points if failing to inform the victim of the purpose of the procedure				
		Notice the victim's level of pain and give proper treatment			2	Deduct 1 point if failing to notice the victim's level of pain or give proper treatment				
		Relieve anxiety and tension			3	Deduct 3 points if failing to relieve the victim's anxiety and tension				
4		Hemostasis and fixation procedure								
4.1	Elevate the injured limb	Elevate the injured limb for 2–3 minutes before applying the tourniquet; Ask: Why do we elevate the injured limb for 2–3 min?			10	Deduct 3 points if failing to elevate the injured limb; Deduct 7 points for incorrect answers, and deduct 3 points for inaccurate answers				
4.2	Tourniquet location	The tourniquet should be placed on the mid-upper third of the right upper arm			5	Deduct 5 points for wrong tourniquet location				
4.3	Wrap the tourniquet	Place padding where the tourniquet is applied			5	Deduct 5 points if no padding is placed				
		The wound should be wrapped in such a way to control bleeding and feel no pulsation in the right radial artery			5	Deduct 5 points if the wrapping tightness does not meet the requirements				

Continued

S/N	Operation name	Quality standard	Points	Deduction criteria	Deduction	Score
4.3	Wrap the tourniquet	Record the start time of applying the tourniquet	5	Deduct 5 points if failing to record the start time of applying the tourniquet		
4.4	Strapping and fixing splints	Fully expose the right forearm, cover the wound with sterile gauze or cotton pad and then fix the wound	6	Deduct 6 points if failing to perform this step, and deduct 3 points if the operation is not standard		
		The splint extends beyond the elbow and wrist joints and is placed on four sides of the forearm	6	Deduct 6 points if failing to perform this step, and deduct 3 points if the operation is not standard		
		Before fixing, use a towel or other soft materials to lay between the splint and the limbs	6	Deduct 6 points if failing to perform this step, and deduct 3 points if the operation is not standard		
		Secure the splint with a bandage, the upper end to the elbow and the lower end to the palm	8	Deduct 8 points if failing to perform this step, and deduct 4 points if the operation is not standard		
		Start by strapping the middle of the fracture, then the lower part, and the upper part. The tightness should be able to move 1cm up and down the bandage	8	Deduct 8 points if failing to perform this step, and deduct 4 points if the operation is not standard		
4.5	Suspensory bandage or triangular binder	Dangle over the chest with a bandage or triangular binder	5	Deduct 5 points if failing to perform this step, and deduct 2 points if the operation is not standard		
5	Inform precautions	Inform the victim of relevant precautions after the operation	5	Deduct 5 points if failing to perform this step, and deduct 2 points if the precautions are not fully informed		
6	End of operation	Tidy up the tools and instruments and clean up the site	3	Deduct 3 points if failing to perform this step, and deduct 1 point if the operation is not standard		
		Report to the examiner	1	Deduct 1 points if failing to report to the examiner		
Total			100			

模块九　防火与防爆

　　火是物质燃烧过程中所进行的强烈氧化反应，而且其能量会以光和热的形式释放，此外还会产生大量的生成物。火的可见部分称作焰，可以随着粒子的振动而有不同的形状，在温度足够高时能以等离子体的形式出现。火必须有可燃物、点火源、氧化剂三项并存才能维持，缺一不可，根据质量守恒定律，火不会使被燃烧物的原子消失，只是通过化学反应转变了被燃烧物的分子形态。当火焰失控时，则被称为火灾。

　　学习目标：

（1）了解火灾、爆炸相关事故机理。

（2）知道各种消防设施与消防器材的特点。

（3）掌握消防器材的选用、检查、使用方法。

Module IX Fire and explosion protection

Fire is a strong oxidation reaction that occurs during the combustion of substances. Its energy is released in the form of light and heat. In addition, a large number of products are produced. Flame, the visible part of the fire, can take on different shapes as the particles vibrate, and even plasma when the temperature is high enough. Fire cannot survive without any of combustible, ignition source, and oxidizing agent. According to the law of conservation of mass, fire does not make the atoms of the object being burned disappear, but changes the molecular form of what's being burned through chemical reactions. If out of control, the flame will give rise to a fire.

Learning objectives:

(1) Understand fire and explosion related accident mechanism.

(2) Know the characteristics of fire-fighting equipment and facilities.

(3) Master the selection, inspection and use methods of fire-fighting facilities.

任务一 认识火灾和爆炸

火灾和爆炸事故往往会带来人身伤亡或者设备损坏，通常都是重大事故。电气火灾在火灾、爆炸中占很大比例。而电气火灾发生的原因是多种多样的，有物的因素，比如接触不良、过载、短路、漏电、静电火花等，也有人的各种因素。电气设备安装使用不当、保养不良，雷击和静电是造成电气火灾的主要原因。

一、燃烧与火灾

（一）燃烧、火灾的定义

燃烧：物质与氧化剂之间的放热反应，它通常同时释放出火焰和可见光。

火灾：在时间和空间上失去控制的燃烧所造成的灾害。[引用自《消防词汇 通用术语》（GB/T 5907.1—2014）]

燃烧和火灾发生的必要条件是同时具备氧化剂（助燃物）、可燃物、点火源，即火的三要素。这三个要素中，缺少任何一个，燃烧都不可能发生持续，即火的三要素是燃烧的必要条件。

常见的氧化剂除了空气外，还有液氧（O_2）、氯（Cl_2）、溴（Br_2）、高锰酸钾（$KMnO_4$）、氯酸钾（$KClO_3$）、硝酸钾（KNO_3）、过氧化氢（H_2O_2）等。

（二）燃烧的过程

可燃物在燃烧过程中，包括许多吸热、放热的化学过程和物理过程。在燃烧的过程中，热量通过热传导、热辐射、热对流的方式进行传递，温度变化是非常复杂的。

1. 燃烧过程

最初一段时间，大部分热量用于对燃烧物质的熔化、蒸发、分解，温度上升很缓慢。当温度达到一定高度，可燃物与氧化剂开始发生氧化反应。如果氧化反应不够猛烈，所产生的热量不足以抵消系统向外散失的热量，则系统温度会逐渐降低，不会发生燃烧。当外界继续对系统进行加热，温度继续上升，系统吸收和发出的热量与系统向外界散失的热量相等，此时处于一个平衡状态。当温度继续升高一点，则会打破平衡，可燃物会发生剧烈氧化，产生大量的热量，这时候即使停止加热，可燃物也会发生剧烈氧化反应持续提高温度，当达到某个温度，就会出现明火燃烧起来。

2. 燃烧形式

根据可燃物的聚集状态不同，燃烧可以分为以下四种形式：

（1）扩散燃烧：可燃气体从管道、容器等缝隙流向空气时，可燃气体分子与空气中的氧气分子相互扩散、混合，当浓度达到爆炸极限范围内的可燃气体遇到点火源，便能形成稳定火焰的燃烧，称为扩散燃烧。例如家庭天然气炉灶的火焰，就是属于扩散燃烧。

（2）混合燃烧：可燃气体和助燃气体一起在管道、容器和空间中扩散、混合，混合气体的浓度在爆炸范围内，遇到火源即可发生燃烧，称为混合燃烧。混合燃烧相比扩散燃烧，实际上是一种爆燃，煤气、天然气等可燃气体泄漏后遇到明火发生的爆燃就是属于混合燃烧，失去控制的混合燃烧往往会造成不可控的冲击波，造成较大的人员伤亡和经济损失。

（3）蒸发燃烧：可燃液体、固体在火源或者热源的作用下，蒸发出可燃蒸气发生氧化反应而进行的燃烧，称为蒸发燃烧。例如煤油、蜡烛等物质的燃烧，均为蒸发燃烧。

（4）分解燃烧：可燃物质在燃烧前需要先遇热分解出可燃气体，再进行的燃烧，称为分解燃烧。

（三）燃烧的分类

（1）依据《火灾分类》（GB/T 4968—2008），按物质的燃烧特性，火灾可分为6类：

A 类火灾：固体物质火灾，通常具有有机物质（非金属），一般在燃烧中能够产生灼热灰烬，如木材、棉花、纸张等固体燃烧造成的火灾。

B 类火灾：液体火灾和可熔化固体火灾，如汽油、柴油、煤油、沥青、石蜡、乙醇等液体燃烧造成的火灾。

C 类火灾：气体火灾，如煤气、天然气、甲烷、乙烷、乙炔、氢气等气体可燃物造成的火灾。

D 类火灾：金属火灾，如钾、钙、钠、镁等金属的燃烧造成的火灾。

E 类火灾：带电火灾，是指带电物体燃烧造成的火灾，如工作中的发电机、电缆、家用电器等燃烧造成的火灾。

F 类火灾：烹饪器具内烹饪物，如动物油脂、植物油等在烹饪过程中发生的火灾。

（2）火灾按照其造成的人员伤亡、受灾户数和经济损失，可分为三类：

①具有以下情况之一的，称为特大火灾：死亡人数10人及以上，死亡加重伤20人及以上，受灾户数50户及以上，财物损失100万元及以上。

②具有以下情况之一的非特大火灾，称为重大火灾：死亡3人及以上，死亡加重伤10人及以上，受灾户数30户及以上，财物损失30万元及以上。

③不具有前两项情形的火灾事故，称为一般火灾。

（四）火灾基本概念及参数

（1）闪燃：可燃物表面火灾可燃液体上方，在很短的时间内重复出现火焰一闪即灭的现象。闪燃往往是持续燃烧的先兆。

（2）阴燃：没有火焰和可见光的燃烧。阴燃长时间得不到控制则有可能会转化为明火，但是它不能被视觉察觉，很多火灾发生之前，可燃物都处于长时间阴燃的状态。红外测温可以发现阴燃现象，鼻嗅也能发现阴燃产生的异味。

（3）爆燃：伴随爆炸的燃烧波，以亚音速传播。混合燃烧往往是以爆燃的形式出现。

（4）自燃：可燃物在空气中没有外来火源的作用下，靠自热火灾外热发生燃烧的现象，根据热源的不同，可以分为自热自燃与受热自燃两种。白磷放在空气中的自动燃烧就是属于自燃。

（5）闪点：材料加热到释放出气体瞬间着火，并出现火焰的最低温度。闪点是衡量物质火灾危险的重要参数。一般情况下闪点越低，火灾危险性越大。

（6）燃点：应用外部热源使物质表面起火并持续燃烧一定时间所需的最低温度。燃点对可燃固体和闪点较高的液体具有重要意义，在控制燃烧时，需将可燃物的温度降至燃点以下。

一般情况下燃点越低，火灾危险性越大。

（7）自燃点：物质在没有外来火源、与空气接触的情况下发生燃烧的最低温度。由于不需要外来火源，所以自燃点一般比燃点更高。液体和固体可燃物受热分解挥发物越多，其自燃点就越低。固体可燃物粉碎得越细，其自燃点就越低。一般情况下，可燃物密度越大，闪点就越高，自燃点就越低，比如下列油的密度：汽油＜煤油＜轻柴油＜重柴油＜蜡油＜渣油，它们的闪点依次升高，自燃点却依次降低。（燃点与自燃点的区别在于是否需要明火）

（8）引燃能、最小点火能：释放能够触发初始燃烧化学反应的能量。

（9）着火延滞期（诱导期）：可燃物和氧化剂的混合物在高温下从开始暴露到起火的时间，单位用 ms 表示。

（五）典型火灾发生规律

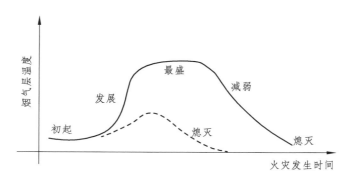

图 9-1　火灾发展的过程

典型火灾事故分为初起期、发展期、最盛期、减弱期和熄灭期，如图 9-1 所示。最初期是指火灾发生的最初阶段，主要特征是冒烟、引燃，这一时期往往难以被人发现；发展期是指火势由小到大发展的阶段，这一阶段火灾释放出的热量与时间平方成正比，轰燃就发生在这一阶段，很多火灾被发现时，就已经处于这一阶段了；最盛期的火灾燃烧方式是通风控制火灾，火势大小由建筑物的通风情况决定；熄灭期是火灾从最盛期开始逐渐消减直至熄灭的阶段，原因可能是燃料不足、灭火系统作用等。有一部分火灾可能由于建筑物通风不足、灭火系统作用、可燃物不足等，达不到最盛期，而是在发展期之后就熄灭了。

二、爆炸

（一）爆炸及其分类

爆炸：物质的一种极其迅速的物理、化学的能量释放或者转化过程，是系统蕴藏的或者瞬间形成的大量能量，并且在有限的体积和极短的时间内，骤然释放或者转化的现象。在这种释放或转化过程中，系统的能量将转化为机械能、光能、热辐射能等形式。

1. 爆炸的过程

爆炸过程表现为两个阶段：第一阶段中，物质的潜在能量以一定方式转化为强烈的压缩势能；第二阶段，被压缩的物质急剧膨胀，对外做功，从而引起周围介质的变化和破坏。

2. 爆炸的特征

一般来说，爆炸具有以下特征：
（1）爆炸过程高速进行。
（2）爆炸点附近压力急剧升高（最主要的特征），多数爆炸伴有温度升高。
（3）发出或大或小的响声。
（4）周围介质发生震动或邻近的物体遭到破坏。

3. 爆炸的分类

爆炸可以由不同的原因引起，按照能量来源，爆炸可以分为 3 类：物理爆炸、化学爆炸和核爆炸。

按照爆炸反应相的不同，爆炸可以分为以下 3 类。

（1）气相爆炸：包括可燃气体和助燃气体混合的爆炸，气体的分解爆炸，液体被喷成雾状物在剧烈燃烧时引起的爆炸（也称为喷雾爆炸），飞扬悬浮于空气中的可燃粉尘引起的爆炸，等。气相爆炸的分类见表 9-1。

表 9-1 气相爆炸类别

类别	爆炸机理	举例
混合气体爆炸	可燃气体和助燃气体以适当的浓度混合，由于燃烧或爆炸的传播而引起的爆炸	空气与氢气、甲烷、煤气、天然气等混合产生的爆炸
气体的分解爆炸	单一气体由于分解反应产生大量热量引起的爆炸	乙炔、乙烯、氯乙烯在分解时产生爆炸
粉尘爆炸	空气中飞散的易燃性粉尘，由于剧烈燃烧引起的爆炸	空气中悬浮的铝粉、煤粉、淀粉等引起的爆炸
喷雾爆炸	空气中易燃液体被喷成雾状，在剧烈燃烧时引起的爆炸	油压机喷出的油雾、喷漆作业时引起的爆炸

（2）液相爆炸：包括聚合爆炸、蒸发爆炸以及由不同的液体混合所引起的爆炸。液相爆炸分类举例见表 9-2。

（3）固相爆炸：包括爆炸性固体化合物及其他爆炸性固体物质的爆炸。固相爆炸分类举例见表 9-2。

表 9-2 液相、固相爆炸类别

类别	爆炸机理	举例
混合危险物质爆炸	氧化性物质与还原性物质或其他物质混合造成的爆炸	硝酸和油脂、液氧和煤粉、高锰酸钾和浓酸等
易爆化合物	有机过氧化物、硝基化合物、硝酸酯等燃烧引起的爆炸	丁酮过氧化物、三硝基甲苯、乙炔酮等
导线爆炸	过载、短路电流过导线，使导线过热，金属、绝缘迅速气化导致的爆炸	导线因电流过载而引起的爆炸

续表

类别	爆炸机理	举例
蒸汽爆炸	过热、液体快速蒸发导致的爆炸	熔融的矿渣与水接触，钢水与水接触等
固相转化时造成的爆炸	固相相互转化时放出热量造成空气急速膨胀	无定形锑转化成结晶锑时放热产生的爆炸

（二）爆炸的破坏作用

（1）冲击波：爆炸形成的高温、高压、高能量密度的气体产物，以极高的速度向周围膨胀，强烈压缩周围的静止空气，使其压力、密度和温度突然升高，产生波状气压向周围扩散冲击。这种冲击波能够对附近建筑物造成破坏，其破坏程度与冲击波能量大小、冲击波传播速度、建筑物的坚固度、与爆炸中心的距离等因素有关。

（2）碎片冲击：爆炸的机械破坏作用会使得各种容器、设备、建筑等材料的碎片在相当大的范围内飞散而造成伤害。

（3）震荡作用：爆炸发生时，特别是猛烈的爆炸往往会引起短暂的地震波。

（4）次生事故：发生爆炸时，如果存放有可燃物，则会造成火灾；高空作业人员会因为冲击波、振荡作用而发生高处坠落事故；粉尘作业场所会由于冲击波出现扬尘，造成范围更大的二次爆炸等事故。

三、特大火灾、爆炸事故案例

（1）洛阳东都商厦特大火灾事故：2000年12月25日21时35分，河南省洛阳市老城区东都商厦发生特大火灾事故，26日0时45分大火最终被扑灭。此次火灾造成309人死亡、7人受伤，直接经济损失275万元。火灾由现场施工人员烧焊作业时电焊火花导致，火灾发生后，现场工作人员没有及时报警和疏散现场人员，火灾从二层迅速蔓延到四层东都娱乐城，导致娱乐城人员失去逃生的机会，致其吸入过多浓烟窒息死亡，见图9-2。

图9-2 洛阳东都商厦特大火灾事故

（2）新疆克拉玛依特大火灾事故：1994年12月8日，新疆克拉玛依市发生恶性火灾事故，造成 325 人死亡、132 人受伤。死者中小学生 288 人，家长、教师及工作人员 37 人。事故发生当时，克拉玛依市友谊馆正在举行文艺演出活动，共 796 人参加该活动，其中大部分为学生。在演出过程中，18 时 20 分左右，舞台上方的 7 号光柱灯突然烤燃了附近的纱幕，接着引燃了大幕。随后火势迅速蔓延，导致各种易燃材料燃烧后产生大量有害气体，由于现场人员众多，而且很多安全门紧锁，无法迅速逃生，导致混乱不堪，从而酿成惨剧，见图 9-3。

图 9-3　新疆克拉玛依特大火灾事故

（3）天津港滨海新区特别重大火灾爆炸事故：2015 年 8 月 12 日，位于天津市滨海新区天津港的瑞海公司危险品仓库发生火灾爆炸事故，本次事故造成 165 人遇难，8 人失踪，798 人受伤，304 幢建筑物、12 428 辆商品汽车、7 533 个集装箱受损。事故已核定的直接经济损失 68.66 亿元。经国务院调查组认定，"8·12 天津滨海新区爆炸事故"是一起特别重大生产安全责任事故，见图 9-4、图 9-5。

图 9-4　天津港滨海新区特别重大火灾爆炸事故（1）　　图 9-5　天津港滨海新区特别重大火灾爆炸事故（2）

Task Ⅰ Learn about fire and explosion

Fire and explosion often result in casualties or equipment damage, and are usually major accidents. Electrical fire accounts for a large proportion of fire and explosion. The electrical fire is ascribed to a variety of reasons, including physical factors (such as poor contact, overload, short circuit, electric leakage, and electrostatic spark) and human factors. Improper installation and use and poor maintenance of electrical equipment, lightning strikes and static electricity are the main contributors to electrical fire.

Ⅰ. Combustion and fire

(Ⅰ) Definition

Combustion is an exothermic reaction between a substance and an oxidizing agent, usually emitting both flame and visible light.

Fire is a disaster caused by burning out of control both in time and space. [Quoted from *Fundamental Terminology of Fire Protection* (GB/T 5907.1—2014)]

The necessary conditions for combustion and fire are the presence of oxidizing agent (comburent), combustible and ignition source, i.e., the three elements of fire. Without any one of these three elements, combustion cannot be sustained, that is, the three elements are the necessary conditions for combustion.

In addition to air, common oxidizing agents include liquid oxygen (O_2), chlorine (Cl_2), bromine (Br_2), potassium permanganate ($KMnO_4$), potassium chlorate ($KClO_3$), potassium nitrate (KNO_3), hydrogen peroxide (H_2O_2), etc.

(Ⅱ) Process of combustion

During the process of combustion, the combustible involves many endothermic and exothermic chemical and physical processes. In the process of combustion, heat is transferred by means of heat conduction, radiation and convection, and the temperature change is very complicated.

1. Combustion process

At first, most of the heat is used to melt, evaporate, and decompose the burning materials, and the temperature rises very slowly. At a certain temperature, the combustible and the oxidizing agent begin to oxidize. If the oxidation reaction is not strong enough and the heat generated is not enough to offset the heat lost from the system, the system temperature will gradually decrease and combustion will not occur. When the outside world continues to heat the system and the temperature keeps rising, the heat absorbed and emitted by the system is equal to the heat loss to the outside world, and it is in an equilibrium state. A little higher temperature will tip the balance, and the combustible oxidizes violently, producing a lot of heat. At this time, even if you stop heating, the combustible will undergo violent oxidation reactions to keep raising the temperature. At a certain temperature, there will be an open flame.

2. Combustion forms

By the state of the combustible accumulation, combustion can be divided into the following four forms:

(1) Diffusion combustion: When the combustible gas flows to the air from the gap of the pipeline, container, the combustible gas molecules and the oxygen molecules in the air diffuse and mix with each other; when meeting the ignition source, the combustible gas with the concentration of the explosive limit can form a stable flame of combustion. For example, the flame of a gas range for family purposes.

(2) Mixed combustion: Combustible gas and combustion-supporting gas are diffused and mixed together in pipelines, containers and spaces; when the concentration of gas mixture is within the explosion range, and combustion can occur when the gas mixture encounters an ignition source. Compared with diffusion combustion, mixed combustion is actually a deflagration. The deflagration of coal gas, natural gas and other combustible gases after they leak into an open flame is a mixed combustion. If out of control, mixed combustion often causes uncontrollable shock waves, resulting in great casualties and economic losses.

(3) Evaporation combustion: combustion by evaporation of combustible vapor and oxidation reaction from combustible liquid and solid under the action of fire or heat source. For example, the burning of kerosene, candles, etc.

(4) Decomposition combustion: Combustible materials need to be decomposed into combustible gases by heat before combustion.

(Ⅲ) Classification of combustion

(1) According to the *Classification of Fires* (GB/T 4968—2008), fire can be divided into 6 categories by combustion characteristics of the substance:

Class A fire: solid fire, usually with organic materials (non-metallic), normally producing hot ash in the combustion, e.g., the fire caused by the burning of wood, cotton, paper and other solids.

Class B fire: liquid fire and meltable solid fire, the fire caused by the combustion of liquids, such as gasoline, diesel, kerosene, asphalt, paraffin, and ethanol.

Class C fire: gas fire, the fire caused by gas combustibles, such as gas, natural gas, methane, ethane, acetylene, and hydrogen.

Class D fire: metal fire, such as the fire caused by the burning of potassium, calcium, sodium, magnesium and other metals.

Class E fire: live fire, the fire caused by the burning of charged objects, such as generators, cables, and household appliances at work.

Class F fire: the fire caused by those in the cooking utensils, such as animal fat and vegetable oil during the cooking process.

(2) Fire can be fallen into three categories by the casualties, the number of households affected

and economic losses:

① Extraordinarily serious fire accidents feature any of the following circumstances: 10 or more deaths, 20 or more deaths and severe wounds, 50 or more households affected, and RMB 1 million or more property loss.

② Major fire accidents have any of the following circumstances: 3 or more deaths, 10 or more deaths and severe wounds, 30 or more households affected, and RMB 300,000 or more property loss.

③ Fire accidents without the first two conditions are called general fires.

(Ⅳ) Basic concepts and parameters of fire

(1) Flash: The phenomenon that the flame flares and goes out repeatedly in a short period of time above the combustible liquid on the surface of the combustible. Flash is often a precursor to sustained combustion.

(2) Smouldering: the combustion without flame or visible light. If not controlled for a long time, smouldering could turn into an open flame; yet it cannot be visually detected. Many fires are preceded by long periods of smouldering combustibles. The smouldering can be detected by infrared temperature measurement, and smelled by noses.

(3) Deflagration: a wave of combustion accompanying an explosion that spreads at subsonic speeds. Mixed combustion is often in the form of deflagration.

(4) Spontaneous combustion: In the absence of external ignition source in the air, the combustible burns by external heat of self-heating fire. Depending on heat sources, it can be divided into self-heating and heated spontaneous combustion. For example, spontaneous combustion of white phosphorus in the air.

(5) Flash point: The lowest temperature at which a material is heated to the point at which the gas released catches fire instantly and a flame appears. It is an important parameter to measure the material fire risk. In general, the lower the flash point, the greater the fire risk.

(6) Fire point: The minimum temperature at which an external heat source is applied to cause the surface of a substance to ignite and continue burning for a certain period of time. The fire point is of great significance to combustible solids and liquids with high flash point. When controlling combustion, the temperature of the combustible should be lowered below the fire point. In general, the lower the fire point, the greater the fire risk.

(7) Spontaneous ignition point: the lowest temperature at which a substance burns in the absence of an external ignition source and in contact with air. Since no external ignition source is required, the spontaneous ignition point is generally higher than the fire point. The more liquid and solid combustibles are heated to decompose the volatiles, the lower their spontaneous ignition points. The finer the solid combustibles are crushed, the lower their spontaneous ignition points. In general, the higher the combustible density, the higher the flash point, the lower the spontaneous ignition point. For example, the density of the following oils: gasoline < kerosene < light diesel < heavy diesel < wax oil < residual oil, their flash points rise in turn, but the spontaneous ignition

points decrease in turn. (The difference between the fire point and the spontaneous ignition point is whether an open flame is required)

(8) Ignition energy, minimum ignition energy: the release of energy that can trigger the initial combustion chemical reaction.

(9) Retardation period of ignition (induction period): the time from initial exposure of a mixture of combustibles and oxidizing agents to fire at high temperatures, expressed in ms.

(V) Typical fire occurrence rules

Typical fire accidents are divided into incipience, development, climax, weakening and extinction periods, as shown in Fig. 9-1. Incipience refers to the initial stage of fire, mainly characterized by smoking and ignition. This period is often difficult to be detected. Development is the stage when a fire grows from small to large. The heat released by a fire at this stage is proportional to the square of time. Flashover occurs at this stage. Many fires are already in this stage when they are discovered. The combustion mode in the climax stage is ventilation-controlled, and the size of fire is determined by the ventilation of structures. Extinction is the stage when the fire tapers off from the climax until extinguished as a result of fuel shortage or the fire extinguishing system. Some of the fires may not reach the climax period due to poorly ventilated structures, fire extinguishing systems, insufficient combustibles and other reasons, but extinguished after the development stage.

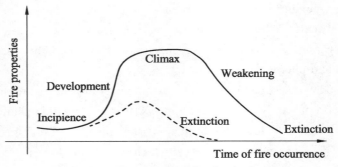

Fig. 9-1 The process of fire development

II. Explosion

(I) Explosion and its classification

Explosion refers to an extremely rapid physical and chemical energy release or conversion process of matter. It is the sudden release or conversion of a large amount of energy contained in the system or formed instantaneously in a limited volume and very short time. In this release or conversion process, the energy of the system will be converted into mechanical energy, light energy, thermal radiation energy and other forms.

1. Explosion process

The explosion process is divided into two stages: In the first stage, the potential energy of matter is transformed into strong compressive potential energy in a certain way; in the second stage, the compressed material expands rapidly and does work externally, thereby causing changes and destruction of the surrounding media.

2. Characteristics of explosion

In general, explosion features the following:

(1) The explosion proceeds at high speed.

(2) Pressure builds up dramatically near the blast site (the most important feature), and most explosions are accompanied by an increase in temperature.

(3) Loud or small noises are made.

(4) The surrounding media vibrate or adjacent objects are damaged.

3. Classification of explosion

Explosion may be ascribed to different causes, and can be fallen into physical, chemical and nuclear explosion by the source of energy.

Explosion can be fallen into the following three categories by the reaction phase:

(1) Gas phase explosion: including explosion involving the mixture of combustible gas and combustion-supporting gas; decomposition explosion of gas; the explosion when the liquid (sprayed into a mist) burns violently, also known as spray explosion; and the explosion caused by combustible dust suspended in the air. The classification of gas phase explosions is as shown in Tab. 9-1.

Tab. 9-1　Category of gas phase explosion

Category	Explosion mechanism	Example
Gas mixture explosion	An explosion caused by the mixture of combustible gas and combustion- supporting gas in appropriate concentrations, as a result of the spread of combustion or explosion	Explosion caused by mixing air with hydrogen, methane, coal gas, natural gas, etc.
Decomposition explosion of gas	Explosion of a single gas resulting from a decomposition reaction that produces a lot of heat	Explosion of acetylene, ethylene, and vinyl chloride during decomposition
Dust explosion	Explosion caused by airborne combustible dust in violent combustion	Explosion caused by aluminum powder, coal powder, and starch suspended in the air
Spray explosion	Flammable liquid in the air is sprayed into a mist, causing an explosion when it burns violently	Explosion caused by oil mist from oil press and painting operation

(2) Liquid phase explosion: including polymerization explosion, evaporative explosion, and explosion caused by the mixing of different liquids. See Tab. 9-2 for examples of liquid phase explosion.

(3) Solid phase explosion: The explosion of explosive solid compounds and other explosive solid substances. See Tab. 9-2 for examples of solid phase explosion.

Tab. 9-2   Categories of liquid and solid phase explosion

Category	Explosion mechanism	Example
Mixed dangerous substance explosion	Explosion caused by mixing oxidizing substances with reducing substances or other substances	Nitric acid and grease, liquid oxygen and pulverized coal, potassium permanganate and concentrated acid, etc.
Explosive compound	Explosion caused by combustion of organic peroxides, nitro compounds, nitric esters, etc.	Butanone peroxide, trinitrotoluene, ethisterone, etc.
Conductor explosion	Explosion arising from the rapid gasification of metal and insulation when the conductor is overheated since the overload, short circuit current flows through the conductor	Explosion of a conductor caused by current overload
Steam explosion	Explosion caused by the rapid evaporation of liquid when overheated	Molten slag in contact with water, molten steel in contact with water, etc.
Explosion caused by solid transformation	Rapid expansion of air caused by the heat released in solid transformation	Explosion resulting from exothermic conversion of amorphous antimony into crystalline antimony

(II) Destructive effects of explosion

(1) Shock wave: The gas product of high temperature, pressure and energy density formed by the explosion expands around at a very high rate, strongly compressing the surrounding still air; thus its pressure, density and temperature are suddenly rising, resulting in a wave of air pressure that spreads around. This shock wave can cause damage to nearby structures, and the damage degree is related to the energy of shock wave, the velocity of shock wave, the robustness of structures, and the distance from the explosion center.

(2) Debris impact: The mechanical damage of explosion will cause the debris of containers, equipment, buildings and other materials to fly in a considerable range, thus giving rise to injury.

(3) Shock action: An explosion, especially a violent explosion, often causes a brief seismic wave.

(4) Secondary accidents: An explosion will cause a fire, if the combustible is available; those working at heights will fall from height because of shock wave and shock; fugitive dust will be seen in the dust site due to the shock wave, resulting in a wider range of secondary explosions and other accidents.

III. Cases of extraordinarily serious fire and explosion accidents

(1) Extraordinarily serious fire accident in Dongdu Commercial Building of Luoyang City: At 21:35 pm on December 25, 2000, an extraordinarily serious fire occurred in the Dongdu Commercial Building in the old town of Luoyang City, Henan Province. The fire was finally extinguished at 00:45 am on December 26. The fire killed 309 people, injured 7 others and caused a direct economic loss of RMB 2.75 million. The fire was ascribed to welding sparks during welding operation. After the fire, the scene staff did not alarm and evacuate those at the scene in time. The first spread quickly from the second floor to Dongdu Entertainment Center on the fourth floor,

where the staff and guests had no chance to escape. In the end, they died from smoke inhalation and asphyxia, as shown in Fig. 9-2.

Fig. 9-2　Extraordinarily serious fire accident in Dongdu Commercial Building of Luoyang City

(2) Extraordinarily serious fire accident in Karamay, Xinjiang: On December 8, 1994, a vicious fire accident occurred in Karamay, Xinjiang, killing 325 people and injuring 132 others. Among the dead were 288 primary and secondary school students and 37 parents, teachers and staff. At the time of accident, 796 people, most of them were students, were attending an artistic performance at the Friendship Hall in Karamay. At about 18:20, the No. 7 light pillar above the stage suddenly burned the nearby veil screen and then set the curtain on fire. Subsequently, the fire spread rapidly, resulting in a large number of harmful gases after the combustion of flammable materials. There were so many spectators at the scene, and many emergency exits were locked, making them unable to escape quickly. The scene was chaotic, leading to tragedy, as shown in Fig. 9-3.

Fig. 9-3　Extraordinarily serious fire accident in Karamay, Xinjiang

(3) Catastrophic fire and explosion accident in Tianjin Binhai New Area: On August 12, 2015, a fire and explosion occurred in the dangerous goods store of Ruihai Company in Tianjin Port, Binhai New Area, Tianjin. The explosion, with a total energy of about 450 tons of TNT equivalent, killed 165 people, left eight missing, injured 798, and damaged 304 structures, 12,428 commercial vehicles, and 7,533 containers. The accident has confirmed direct economic losses of RMB 6.866 billion. According to the investigation team of the State Council, the "explosion accident in Tianjin Binhai New Area" was a particularly serious production safety responsibility accident, as shown in Fig. 9-4 and Fig. 9-5.

Fig. 9-4 Catastrophic fire and explosion accident in Tianjin Binhai New Area (1)

Fig. 9-5 Catastrophic fire and explosion accident in Tianjin Binhai New Area (2)

任务二　消防设施与消防器材的选择与使用

《中华人民共和国消防法》规定：消防设施是指火灾自动报警系统、自动灭火系统、消防栓系统、防烟排烟系统以及应急广播和应急照明、安全疏散设施等。

消防器材，指用于灭火、防火及火灾事故的器材。

一、消防设施概述

（一）火灾自动报警系统

自动消防系统包括探测、报警、联动、灭火、减灾等功能。而火灾自动报警系统主要完成探测和报警功能。

火灾自动报警系统一般分为三类：区域火灾报警系统、集中报警系统、控制中心报警系统。

区域报警系统包括火灾探测器、手动报警按钮、区域火灾报警控制器、火灾报警装置和电源部分。该系统比较简单，使用广泛，一般用于二级保护对象，例如行政事业单位、工矿企业的要害部门和娱乐场所等。

集中报警系统由一台集中报警控制器、两台以上区域报警控制器、火灾报警装置和电源部分组成。一般适用于一、二级保护对象，例如宾馆、饭店、大型建筑群等。

控制中心报警系统除了上述两种系统中应有的装置外，还增加了消防联动控制设备，被联动的设备包括火灾警报装置、火警电话、应急照明、应急广播排烟通风、消防电梯、固定灭火装置等，一般适用于一级、特级保护对象，例如大型宾馆、饭店、办公大楼、综合商场等。

（二）自动灭火系统

（1）水灭火系统：包括室内外消防栓系统、自动喷淋灭火系统、水幕、水喷雾灭火系统等，一般设计为喷头受热即可喷水，是目前使用最为广泛的一种灭火系统，见图9-6。

（2）气体自动灭火系统：以气体作为灭火介质的灭火系统。对灭火剂的要求有：化学稳定性好、耐储存、腐蚀性小、不导电、毒性低、蒸发后不留痕迹等。

图9-6　自动消防喷淋头

（3）泡沫灭火系统：空气机械泡沫灭火系统，按发泡倍数可分为低倍泡沫灭火系统（发泡倍数在20倍以下）、中倍数泡沫灭火系统（发泡倍数在21~200倍）、高倍数泡沫灭火系统（发泡倍数在201~1 000倍）。

（三）消防栓

消防栓，也叫消火栓，是一种固定式消防设施，主要作用是控制可燃物、隔绝氧化剂、降低温度、消除着火源，分室内消防栓和室外消防栓。消防栓套装一般由"消防箱＋消防水带＋水枪＋接扣＋栓＋卡子"等组合而成，消防栓主要供消防车从市政给水管网或室外消防给水管网取水实施灭火，也可以直接连接水带、水枪出水灭火。该设施不仅仅适用于初起火灾，只要满足安全要求，火灾发展期、最盛期、减弱期就都能够使用。所以，室内外消防栓系统是扑救火灾的重要消防设施之一，如图 9-7 所示。

图 9-7　室内（左）和室外（右）消防栓（消火栓）

（四）防烟排烟系统

火灾中除了火，还有烟气也是十分有害的，烟气包括烟雾、有毒气体和热气，不但影响消防员的扑救，还会对人体呼吸系统造成非常大的伤害。火灾时，把烟气排出建筑物外，就需要设置防排烟系统，见图 9-8。

图 9-8　防烟排烟管道

（五）火灾应急广播和警报装置

火灾警报装置是指发生火灾时，自动或手动向人们发出警告信号的装置。火灾应急广播是指火灾时，指挥现场人员进行疏散的广播设备。

二、消防器材

（一）灭火剂

灭火剂是能够有效地破坏燃烧条件（燃烧三要素），终止燃烧的物质。灭火剂被喷射到燃烧物或燃烧区域时，通过一系列物理、化学作用，停止燃烧反应，从而起到灭火作用。

1. 水系灭火剂

水（H_2O）是最常见的一种灭火剂，它既能够单独用来灭火，也可以在其中添加化学物质配置成混合液使用。这种在水中加入化学物质制成的灭火剂统称为水系灭火剂。

水灭火的原理一方面是水的比热容非常大，导热性能较强，所以在接触燃烧物后，能够迅速吸收热量，使得燃烧物温度迅速下降至燃点以下，从而使燃烧终止；另一方面，水在受热时会沸腾汽化，形成大量水蒸气笼罩在燃烧物周围，可以阻隔空气进入，减少氧含量，使得燃烧缺氧而窒息熄灭。

在使用水灭火时，加压水枪能够把水喷射到非常远的地方，所以对消防员而言这种方法较为安全。并且高水压具有冲击作用，能够冲过燃烧表面进入内部，从而使得未着火的与已经着火的部位隔离开来，增强灭火效果，见图9-9。

图9-9 消防员使用水枪灭火

但是由于水的一些特性，部分火灾是不能使用水系灭火剂的，包括：

（1）密度小于水和不溶于水的易燃液体造成的火灾，比如汽油、煤油、柴油、苯类、醇类、酮类等不完全溶于水，并且密度比水小的液体燃烧，如果使用水扑救：第一，水会沉淀至下层，无法起到隔绝空气的作用；第二，由于水的流动性强，会带动燃烧液体流动，扩大火势；第三，下层的水被加热后会引起爆沸，造成可燃液体飞溅，扩大火势。

（2）遇水即燃的化学物质造成的火灾，例如钾、钙、钠、碳化钙等物质，会与水发生剧烈化学反应，甚至发生爆炸，所以不能用水系灭火剂灭火，最好使用沙土或金属专用灭火剂。

金属钠与水发生反应，如图9-10所示。

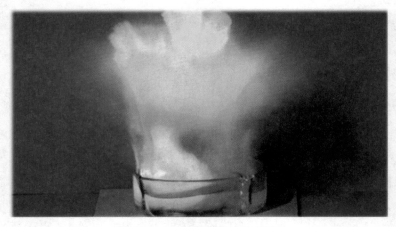

图9-10　金属钠与水发生反应

（3）硫酸、盐酸、硝酸引发的火灾，不能用水流冲击，否则会使酸液飞溅，流出后遇到可燃物质有引发二次火灾或爆炸的危险，飞溅在人身上还会灼伤人体。

（4）电气火灾在未切断电源前，不能使用水系灭火剂灭火，因为水是良导体，容易造成触电伤害。

（5）高温状态下的化工设备发生火灾不能使用水系灭火及扑救，以防高温设备遇水后骤冷，引起形变爆裂。

2. 气体灭火剂

气体灭火剂在19世纪末就开始使用。气体灭火剂有释放后对设备无损害的优点，其防护对象从20世纪开始逐步向各种不同领域扩展。由于二氧化碳（CO_2）气体来源较广，利用隔绝空气后的窒息作用可以成功抑制火灾，因此早期的气体灭火剂主要是二氧化碳气体。

二氧化碳气体不含水、不导电、无腐蚀性、无污染，对绝大多数物质没有破坏作用，所以可以用来扑灭精密仪器和一般电气火灾。它还适用于扑救可燃液体或固体物质的火灾。但是有一些火灾是不能使用二氧化碳气体作为灭火剂扑救的：

（1）活泼金属，如钾、钙、钠、铝等引发的火灾。

（2）金属过氧化物，如过氧化钾、过氧化钠等引发的火灾。

（3）有机过氧化物，如过氧化苯甲酰、过氧化苯甲酸等化工材料引发的火灾。

（4）氯酸盐、硝酸盐、高锰酸盐等氧化剂造成的火灾。

因为二氧化碳喷出时，周围温度会降低，空气中的水分会凝结成液滴，而上述物质与水相遇会立刻发生剧烈反应，释放大量的热量，同时还会释放出氧气，使二氧化碳的窒息作用受到影响，因此，对于上述物质引发的火灾，使用二氧化碳灭火剂扑灭的效果较差。

在研究二氧化碳灭火剂时，西方发达国家也在开发新型的气体灭火剂，卤代烷1211、1301灭火剂具有优秀的灭火性能，因此，在一段时间内卤代烷灭火器统治了整个灭火领域。后来人们发现卤代烷会与臭氧发生反应，导致臭氧层出现空洞，破坏地球环境，使人类生存环境恶化。所以，国家环保局于1994年发出关于非必要场所停止配置卤代烷灭火器的通知。

在淘汰卤代烷灭火剂后，人们寻找了多种环保气体灭火剂来作为代替。其中七氟丙烷

（CF_3CHFCF_3）最具推广价值，该灭火剂属于含氟灭火剂，简称 FM-200。这种气体有灭火剂无色、无味、低毒、不导电、无污染、不会对财物和精密设施造成损坏、灭火效率高等优点，现在在大量使用。

除此之外，还有一些气体灭火剂，如 IG-541 混合气体灭火剂也在使用，它是由氮气（N_2）、氩气（Ar）、二氧化碳（CO_2）自然组合成的一种混合物，以气态形式存在，喷射的时候不会造成视野不清，而且对人体完全无毒。

3. 泡沫灭火剂

泡沫灭火剂有化学泡沫和空气泡沫灭火剂两大类。化学泡沫灭火剂是通过硫酸铝[$Al_2(SO_4)_3$]和碳酸氢钠（$NaHCO_3$）的水溶液发生化学反应，产生二氧化碳而形成泡沫。空气泡沫灭火剂是由含有表面活性剂的水溶液在泡沫发生器中通过机械作用产生的，泡沫中所含气体为空气，也称为机械泡沫灭火剂。

空气泡沫灭火剂种类繁多，根据发泡倍数不同可分为低倍泡沫、中倍泡沫和高倍泡沫灭火剂。高倍泡沫灭火系统代替低倍泡沫灭火系统是当今发展的趋势，它发泡倍数非常高（201～10 000倍），能在短时间内用泡沫填充整个着火空间，特别适用于大空间火灾，如仓库、厂房、油轮、冷库等，并且具有灭火速度快的优点，如图 9-11 所示。而低倍泡沫与此不同，主要靠泡沫覆盖可燃物表面，将空气隔绝而灭火，并且伴有水渍，所以它对液化烃流淌、船舶、仪器仪表、高精设备、电气火灾等无能为力。

图 9-11　高倍数消防泡沫

4. 干粉灭火剂

干粉灭火剂由一种或者多种具有灭火能力的细微无机粉末组成，主要包括活性灭火组分、疏水组分、惰性填料。粉末颗粒大小以及分布对灭火效果有很大的影响。干粉灭火剂对火焰和可燃物有窒息、冷却、化学抑制等作用，其中化学抑制作用是灭火的基本原理，起主要灭火作用。干粉灭火剂中的灭火组分是燃烧反应的非活性物质，当进入燃烧区域内，捕捉并终止燃烧反应的自由基，降低燃烧反应的速率，当干粉浓度足够高、与火焰接触面积足够大时，链式燃烧反应将被终止，从而使火焰熄灭。

干粉灭火剂与水系灭火剂、泡沫灭火剂、二氧化碳灭火剂相比，在灭火速率、灭火面积、灭火成本上，都有一定优越性。由于干粉灭火剂有灭火速率快，制作工艺过程简单，使用温度范围宽泛，对环境无特殊要求，使用方便，不需要外界动力、水源，无毒无污染，安全等特点，目前在手提式灭火器和固定灭系统上得到广泛的应用。

5. 其他特殊灭火剂

7150灭火剂：它是一种专门用于扑救D类初起火灾时使用的灭火剂，即扑救可燃金属火灾。其主要成分是三甲氧基硼氧六环，为无色透明的可燃液体，热稳定性较差，在火焰温度作用下能分解或燃烧。它的燃烧耗氧量极高，能够迅速消耗金属周围空间的氧气，并且分解产生的硼酐在高温下融化成玻璃状液体，覆盖在金属表面，阻隔金属与氧气的接触，从而窒息灭火。

四氯化碳灭火剂：四氯化碳为无色透明液体，不助燃、不自燃、不导电、沸点低（80 ℃）。当四氯化碳落到火区中时，迅速蒸发，由于其蒸汽密度大（约为空气的5倍），能密集在火源四处包围着正在燃烧的物质，起到了隔绝空气的作用。四氯化碳灭火剂是一种阻燃能力很强的灭火剂，特别适用于带电设备的E类初起火灾的灭火。四氯化碳灭火剂禁止用来扑救电石和钾、钠、铝、镁等火灾。

原位膨胀石墨灭火剂：石墨经处理后的变体，是一种新型的金属灭火剂。石墨是一种碳异构体，无毒、无腐蚀性。当温度低于150 ℃时，其密度基本稳定；温度超过150 ℃时，其密度变小，开始膨胀；当温度达到800 ℃时，体积膨胀达50倍。它是专门用于扑救D类初起火灾的灭火剂，当碱金属或轻金属发生火灾后，在燃烧材料表面喷涂可膨胀石墨灭火剂。在高温作用下，石墨体积迅速膨胀，在燃烧材料表面形成海绵泡沫。同时，与燃烧金属接触的部分金属被液态金属润湿，形成金属碳化物或部分石墨层间化合物，形成空气绝缘隔膜，阻止燃烧。

消防沙和灰铸铁末（屑）：这是两种非专门制造的灭火剂，它们单独应用于规模很小的磷、镁、钠等火灾，起隔绝空气或从火焰中吸热的作用，可以灭火或控制火灾的发展，消防沙还能够用于组织B类火灾蔓延。

（二）灭火器

灭火器由筒体、器头、喷嘴、把手等部件组成，借助驱动压力可将所填充的灭火剂由喷嘴喷出，达到灭火目的。灭火器由于结构简单，价格低廉，操作方便，轻便灵活，是扑救初起火灾的重要消防器材。

灭火器种类很多，按移动方式分，可以分为手提式、推车式和悬挂式；按驱动动力源来分，可以分为储气瓶式、储压式、化学反应式；按充装的灭火剂不同可主要分为以下6类。

1. 水系灭火器

水系灭火器也称为清水灭火器，以清洁的水加入一定的添加剂作为灭火剂，采用储气瓶的方式，利用二氧化碳钢瓶中的气体作为动力，将灭火剂喷射到可燃物上，达到窒息灭火的作用。清水灭火剂适用于扑灭可燃固体物质引发的火灾，即A类火灾。需要注意的是，由于水的导电性，一定不能用于扑灭电气火灾及E类火灾。由于水系灭火器使用范围较窄，用途不广，所以使用较少，清水灭火剂通常用于消防设施，而非消防器材。

2. 泡沫灭火器

泡沫灭火器分为化学泡沫灭火器和机械泡沫灭火器。泡沫灭火器适合扑救油脂类、石油产品等 B 类初期火灾以及木材等 A 类初期火灾，不能扑救 B 类水溶性火灾，例如乙醇造成的火灾，也不能扑灭 C 类、D 类、E 类火灾。

化学泡沫灭火器内充装有酸性和碱性两种化学药剂。使用时，两种溶液混合引起化学反应产生泡沫，喷射出去扑灭火焰。按移动方式可分为手提式（图 9-12）、舟车式、手推式（图 9-13）。

空气泡沫灭火器也称为机械泡沫灭火器，具有良好的热稳定性，抗烧时间长，灭火能力比化学泡沫高 3~4 倍，性能优良，保存期长，使用方便，是取代化学泡沫灭火器的换代产品。它根据不同需要分别充装蛋白泡沫、氟氮泡沫等，用来扑灭各种油类及极性溶剂引起的初期火灾。目前消防公司推出了一种空气泡沫灭火器，俗称"水基灭火器"，是可以使用于 36 kV 以下电压的 E 类电气火灾，也可用于 C 类气体火灾，这种灭火器抗复燃性非常好，但是相比其他灭火器，其价格较为昂贵。

图 9-12　手提式水基灭火器　　　　图 9-13　手推式泡沫灭火器

3. 酸碱灭火器

酸碱灭火器是一种内部装有 65%的工业硫酸和碳酸氢钠的水溶液作为灭火剂的灭火器。使用时，使两种药物混合发生反应，产生二氧化碳，在其压力下喷出灭火。该类灭火器适用于扑救 A 类物质的初期火灾，不能用于 B 类、C 类、D 类、E 类火灾。

4. 二氧化碳灭火器

二氧化碳灭火器是在灭火器内部充装业态二氧化碳，使用时打开阀门，二氧化碳在自身压力作用下喷出，填充可燃物周围空间，隔绝空气，造成窒息灭火。

一般当氧气含量低于 12%或二氧化碳浓度达到 30%~35%时，燃烧即可终止。1 kg 二氧化碳在常温下能生成 500 L 二氧化碳气体，足以使得 1 m^3 空间范围内的火焰熄灭。二氧化碳灭火器有无毒、无污染、不留痕迹、绝缘等特点，所以通常用于扑救 600 V 以下的带电电器、精密仪器、贵重设备的 E 类电气初期火灾，也可以用于扑灭小范围、封闭空间引发的 A 类、B 类初期火灾。

由于二氧化碳灭火器在喷出时，会使周围空间温度骤降，所以其灭火器把手通常使用隔热材料制成的大型喷嘴。二氧化碳灭火器与干粉、泡沫等灭火器相比，内部压力较高，所以对灭火器筒体密封工艺要求也较高，灭火器本体价格较为昂贵。手提式二氧化碳灭火器如图9-14所示。

5. 卤代烷灭火器

凡是内部充满卤代烷灭火剂的灭火器统称为卤代烷灭火器。其主要通过抑制燃烧的化学反应过程，使燃烧中断达到灭火目的。卤代烷灭火器种类较多，有1211灭火器、1301灭火器、2402灭火器、1202灭火器等。其中我国曾经主要使用1211灭火器，主要用于扑救部分A类、B类、C类初起火灾，尤其适用于扑救精密仪器、计算机、文物等产生的初起火灾，其灭火效率大概为二氧化碳灭火器的5倍。卤化烷不能扑救金属火灾和在惰性介质中自身供氧燃烧硝化纤维、火药等的火灾；也不

图9-14 手提式二氧化碳灭火器

能扑灭金属氢化物如氢化钾、氢化钠火灾及自行分解的化学物质，如过氧化物、联氨等。由于卤代烷灭火剂对大气污染较为严重，并且其蒸汽有一定毒性，所以逐渐已被淘汰，我国目前生产、使用得很少。

6. 干粉灭火器

干粉灭火器是以液态二氧化碳或液态氮气作为动力，将灭火器内干粉灭火剂喷出进行灭火。该类灭火器主要通过对火焰的抑制作用灭火，按使用范围可分为普通干粉灭火器和多用干粉灭火器两类。

普通干粉灭火器也称BC干粉，指碳酸氢钠干粉、改性钠盐、氨基干粉等，主要用于扑灭B类、C类、E类初期火灾。多用干粉也称为ABC干粉，指磷酸铵盐干粉、聚磷酸铵干粉等，是目前使用最为广泛的一种灭火器。它不仅适用于扑救B类、C类、E类初起火灾，还适用于扑灭一般A类火灾，但是都不能扑灭金属D类火灾。手提式ABC干粉灭火器如图9-15所示。

图9-15 手提式ABC干粉灭火器

（三）火灾探测器

物质在燃烧过程中，通常会产生烟雾，释放出气溶胶类燃烧气体，并且形成大量的红外线和紫外线，导致周围温度升高。这些烟雾、温度、火焰、燃烧气体统称为火灾参量。

火灾探测器的基本功能就是对烟雾、温度、火焰、燃烧气体等火灾参量做出有效的反应，通过敏感元件，将火灾参量转化为数字电信号传递给报警控制器。

根据对不同火灾参量的响应，可以分为感光式火灾探测器、感烟式火灾探测器（图9-16）、感温式火灾探测器、复合式火灾探测器、可燃气体火灾探测器等。

（四）消防水带

消防水带是提供消防用水或输送泡沫灭火剂的必备器材，广泛应用于消防车、消防泵、消防栓等设备上，见图9-17。

图 9-16　感烟式火灾探测器　　　图 9-17　消防水带

（五）消防水枪

消防水枪是灭火时用来射水的工具，其作用是加快水的流速，改变水流形状。有不同的口径，应用于消防车、消防泵、消防栓，见图9-18。

图 9-18　消防水枪

（六）消防车

目前我国使用的消防车有水罐消防车、泡沫消防车、干粉消防车、CO_2 消防车、干粉泡沫水罐联用消防车、火灾照明车、曲臂登高消防车等。

（七）消防沙箱（桶）

消防专用黄沙箱（桶）是用纯铁皮制作的箱子（桶），结构简单，里面所盛的是专用的消防沙，主要是利用阻止 B 类火灾蔓延，或者用于扑灭规模很小的磷、镁、钠等初期火灾，使用简单，见图9-19。

图 9-19　消防沙箱

Task II Selection and use of fire-fighting equipment and facilities

The *Fire Control Law of the People's Republic of China* stipulates that fire-fighting equipment refer to automatic fire alarm system, automatic fire extinguishing system, fire hydrant system, smoke control and exhaust system, emergency broadcast and lighting, safety evacuation facilities, etc.

Fire-fighting facilities refers to those used for fire fighting, fire protection and fire accidents.

I. Overview of fire-fighting equipment

(I) Automatic fire alarm system

The automatic fire control system includes detection, alarm, linkage, fire fighting, disaster reduction and other functions. The automatic fire alarm system mainly functions as detection and alarm.

The automatic fire alarm system is generally fallen into three categories: local alarm system, remote alarm system, and control center alarm system.

The local alarm system consists of a fire detector, a manual alarm button, a local alarm controller, a fire alarm device, and a power supply. The system is simple and widely used for second-class protected objects, such as administrative institutions, key departments of industrial and mining enterprises, and entertainment venues.

The remote alarm system is composed of a remote alarm controller, two or more local alarm controllers, a fire alarm device, and a power supply. It is generally applicable to first- and second-class protected objects, such as hotels, restaurants, and large complexes.

In addition to the devices in the above two systems, the control center alarm system is further equipped with fire linkage control equipment. Those linked include the fire alarm device, fire telephone, emergency lighting, emergency broadcast, smoke exhaust and ventilation, fire elevator, and fixed fire-extinguishing unit. It is generally suitable for first-class and special-class protected objects, such as large hotels, restaurants, office buildings, and shopping complexes.

(II) Automatic fire extinguishing system

(1) The water fire extinguishing system includes indoor and outdoor fire hydrant system, automatic sprinkler system, water curtain, and water spray system. It is generally designed to spray water when the nozzle is heated. It is now the most widely used fire extinguishing system, as shown in Fig. 9-6.

(2) The automatic gas fire extinguishing system refers to the fire extinguishing system with gas as the fire extinguishing media. The extinguishing agents are expected to feature good chemical stability, storage stability, little corrosion, non-conductivity, low toxicity, and no trace after evaporation.

Fig. 9-6　Automatic fire sprinkler head

(3) The foam fire extinguishing system refers to air mechanical foam fire extinguishing system. By the foam expansion ratio, it can be divided into the low-expansion foam fire extinguishing system (below 20 times of expansion), medium-expansion foam fire extinguishing system (21-200 times of expansion), and high-expansion foam fire extinguishing system (201-1,000 times of expansion).

(Ⅲ) Fire hydrant

Fire hydrant, a fixed fire-fighting facility, is mainly to control the combustible, isolate the oxidizing agent, reduce temperature, eliminate the ignition source. There are indoor and outdoor fire hydrants. The fire hydrant kit is generally composed of a fire box, fire hoses, fire hose nozzles, buckles, hydrants, and clips. Fire hydrants are mainly used by fire trucks to take water from the municipal water supply network or outdoor fire water supply network, or can be directly connected to the fire hoses and fire hose nozzles to put out a fire. The facility is not just for incipient fire; as long as the safety requirements are met, it is applicable to the development, climax, and weakening periods. Therefore, the indoor and outdoor fire hydrant systems are one of the important fire-fighting equipment, as shown in Fig. 9-7.

Fig. 9-7　Indoor (left) and outdoor (right) fire hydrants

(Ⅳ) Smoke control and exhaust system

Besides fire, smoke is very harmful. The smoke includes smog, toxic gas, and heat, which affects the firefighter's rescue, and even causes great damage to the human respiratory system. In case of fire, a smoke control and exhaust system is required to exhaust the smoke outside the structures, as shown in Fig. 9-8.

Fig. 9-8　Smoke control and exhaust pipe

(Ⅴ) Fire emergency broadcast and alarm device

The fire alarm device refers to a device that automatically or manually sends a warning signal in case of fire. Fire emergency broadcast refers to the broadcast equipment that directs the evacuation of personnel on the scene in case of fire.

Ⅱ. Fire-fighting facilities

(Ⅰ) Extinguishing agents

Extinguishing agents are substances that can effectively destroy the combustion conditions (the three elements for combustion) and terminate the combustion. When sprayed into the combustion materials or the combustion area, the extinguishing agents can stop the combustion reaction through a series of physical and chemical processes, thus extinguishing the fire.

1. Water extinguishing agent

Water (H_2O) is one of the most common extinguishing agents. It can be used alone to extinguish the fire, or be prepared into a mixture of chemicals. This extinguishing agent made of adding chemicals to the water is collectively referred to as water extinguishing agent.

The principle of water fire extinguishing lies in, on the one hand, the very large specific heat capacity of water and strong heat-conducting property. After contact with the comburent, water can quickly absorb heat, so that the temperature of the comburent rapidly drops below the fire point, and the combustion is terminated. On the other hand, water will boil and vaporize when heated, forming a large amount of water vapor around the comburent, which can keep out air, reduce the oxygen content, and make the combustion anoxic and asphyxiated.

When using water to put out a fire, pressurized fire hose nozzles can spray water far away; thus it is safer for firefighters. Moreover, the high water pressure has the impact effect, which can rush through the burning surface into the interior, so as to isolate the parts not on fire from those already on fire, and enhance the fire extinguishing effect, as shown in Fig. 9-9.

Fig. 9-9　Firefighters using fire hose nozzles to put out the fire

However, subject to water properties, some fires cannot be put out by water extinguishing agents, including:

(1) The fire caused by combustible liquids less dense than water and insoluble in water. For example, in case of the combustion of gasoline, kerosene, diesel, benzene series, alcohols, ketones and other liquids not completely soluble in water and less dense than water, if used, water will, first, settle to the bottom layer and cannot isolate air; second, the highly mobile water will drive the burning liquid to flow and thus spread the fire; third, when heated, the water underneath will cause explosive boiling, resulting in a splash of combustible liquids and spreading the fire.

(2) Fire caused by chemical substances that burn in water, such as potassium, calcium, sodium, and calcium carbide, will react violently with water, and even explosion. As a result, water extinguishing agents cannot be used; instead, sand or special extinguishing agent for metals is preferred. Sodium metal reacts with water, as shown in Fig. 9-10.

(3) If caused by sulfuric acid, hydrochloric acid, and nitric acid, the fire cannot be extinguished with water, otherwise the acid will splash, resulting in secondary fire or explosion after encountering combustible materials, and even burn the human body.

(4) Electrical fires cannot be extinguished with water extinguishing agents before cutting off the power supply, since water is a good conductor, which is prone to electric shock injury.

(5) In case of fire, chemical equipment at high temperature cannot be extinguished by water extinguishing agents, in order to prevent the high-temperature equipment from sudden cooling after encountering water, thus causing deformation and burst.

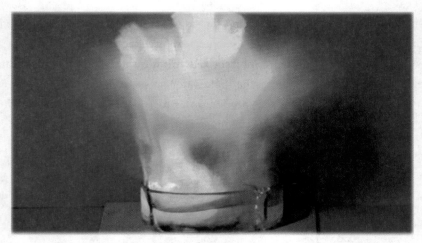

Fig. 9-10   Sodium metal reacting with water

2. Gaseous extinguishing agent

Gaseous extinguishing agents have been used since the end of the 19th century. Gaseous extinguishing agents have no damage to equipment after release, making their protected objects gradually expanded to different fields since the 20th century. With a wide range of sources, carbon dioxide (CO_2) gas can contain a fire successfully using the smothering action of isolating air. Therefore, the early gaseous extinguishing agent is mainly carbon dioxide gas.

Carbon dioxide gas is free from water, conductivity, corrosion, and pollution, having no destructive effect on most substances. It can be used to extinguish precision instruments and general electrical fires. It is also suitable for fighting fires with combustible liquids or solid substances. However, some fires cannot be put out using carbon dioxide gas:

(1) The fire caused by active metals such as potassium, calcium, sodium, and aluminum.

(2) The fire caused by metal peroxides such as potassium peroxide and sodium peroxide.

(3) The fire caused by organic peroxides such as benzoyl peroxide, benzoyl peroxide and other chemical materials.

(4) The fire caused by oxidizing agents such as chlorate, nitrate, and permanganate.

When carbon dioxide is spewed out, the surrounding temperature will drop, and water in the air will condense into droplets. The above substances will immediately react violently with water, releasing a lot of heat and even oxygen, which may affect the asphyxiation of carbon dioxide. Therefore, for the fires caused by the above substances, the carbon dioxide extinguishing agent may produce poor effect.

While studying the carbon dioxide extinguishing agent, Western developed countries are developing new gaseous extinguishing agents, halon 1211 and halon 1301 extinguishing agents are known for excellent extinguishing performance, so halon fire extinguishers had dominated the field for a while. Later, it was found that halon would react with ozone, resulting in a hole in the ozone layer, destroying the global environment and deteriorating the human living environment. Therefore, the State Environmental Protection Administration issued a notice in 1994 to stop the provision of halon fire extinguishers in non-essential places.

After the elimination of halon extinguishing agents, a variety of environmentally friendly gaseous extinguishing agents were found as substitutes. Heptafluoropropane (CF_3CHFCF_3) has the most value for popularization and application. It is an fluorinated extinguishing agent, abbreviated as FM-200. This gas is colourless, odourless, low toxic, non-conducting, pollution-free, having no damage to property and precision facilities, and high extinguishing efficiency. It's in heavy use now.

Some gaseous extinguishing agents, such as IG-541 mixed gas extinguishing agent, are used, too. It is a mixture of nitrogen (N_2), argon (Ar) and carbon dioxide (CO_2) in gaseous form, and will not cause blurred vision when sprayed. Besides, it is completely non-toxic to humans.

3. Foam extinguishing agent

Two types are available: chemical foam and air foam extinguishing agents. Chemical foam extinguishing agent is formed by the chemical reaction of aluminum sulfate [$Al_2(SO_4)_3$] and sodium bicarbonate ($NaHCO_3$) in aqueous solution to produce carbon dioxide and foams. Air foam extinguishing agent is produced by the aqueous solution containing surfactant in the foam generator through mechanical action, the gas contained in the foam is air, also known as mechanical foam extinguishing agent.

There are many kinds of air foam extinguishing agents. They can be fallen into low-expansion foam, medium-expansion foam and high-expansion foam extinguishing agents by the foam expansion ratio. It is the current development trend to replace low-expansion foam extinguishing system with high-expansion one. With high foam expansion ratio (201–1,0000 times), the latter can fill the whole fire space with foam in a short time, especially applicable to large space fires, such as warehouses, powerhouses, oil tankers, and refrigeration houses. Furthermore, it is known for fast extinguishing speed, as shown in Fig. 9-11. Unlike the high-expansion foam extinguishing system, the low-expansion system is mainly to isolate the air and extinguish the fire by covering the surface of combustibles with foam, and with water stains. It can do nothing to liquefied hydrocarbon flow, vessel, instruments, high-precision equipment, electrical fires, etc.

Fig. 9-11　High-expansion fire foam

4. Dry powder extinguishing agent

The dry powder extinguishing agent is composed of one or more fine inorganic powders with fire extinguishing capability, mainly including active fire extinguishing components, hydrophobic components, and inert fillers. The size and distribution of powder particles have a great influence on the fire extinguishing effect. The dry powder extinguishing agent has the functions of asphyxiation, cooling and chemical inhibition on the flame and combustibles. The chemical inhibition is the basic principle of fire extinguishing and plays the main role in extinguishing fire. The fire extinguishing component in the dry powder extinguishing agent is the inactive substance of the combustion reaction. When entering the combustion zone, the free radical of the combustion reaction is captured and terminated to reduce the combustion reaction rate. When the dry powder concentration is high enough and the contact area with the flame is large enough, the chain combustion reaction will be terminated, thus extinguishing the flame.

Compared with water extinguishing agents, foam extinguishing agents and carbon dioxide extinguishing agents, dry powder extinguishing agents have certain advantages in terms of extinguishing rate, area and cost. The dry powder extinguishing agent is characterized by fast extinguishing rate, simple production process, wide operating temperature range, no special requirements for the environment, user-friendliness, no external power and water source required, non-toxicity, non-pollution, and safety. At present, it is widely used in portable fire extinguishers and fixed extinguishing systems.

5. Other special extinguishing agents

7150 extinguishing agent: It is an extinguishing agent specially used to fight class D incipient fire, that is, combustible metal fire. Its main component is trimethoxyboroxine, a colorless transparent flammable liquid with poor thermal stability, which can decompose or burn under the action of flame temperature. Its combustion is so oxygen-intensive that it can quickly consume oxygen in the space around the metal, and the resulting boron anhydride melts at high temperatures into a glassy liquid, covering the metal surface to block the contact between metal and oxygen, so as to extinguish the fire.

Carbon tetrachloride extinguishing agent: Carbon tetrachloride is a colorless transparent liquid, neither combustion-promoting, spontaneous combustion, nor conductive, with low boiling point (80 °C). When falling into the fire zone, carbon tetrachloride evaporates rapidly; because of its high vapor density (about 5 times that of air), it can densely surround the burning material around the fire source to isolate the air. Carbon tetrachloride extinguishing agent is an extinguishing agent with strong flame retardant ability, especially suitable for the extinguishing of class E incipient fire of live equipment. Carbon tetrachloride extinguishing agent is not allowed to fight the fire caused by calcium carbide, potassium, sodium, aluminum, magnesium, etc.

In situ expanded graphite extinguishing agent: the treated variant of graphite, a new type of metal extinguishing agent. Graphite is a carbon isomer, free from toxicity and corrosion. When the temperature is below 150 °C, its density is basically stable; when the temperature exceeds 150 °C, it becomes less dense and begins to expand; when the temperature reaches 800 °C, its volume expands up to 50 times. This extinguishing agent is specially used to fight class D incipient fire. In case of fire to alkali metal or light metal, expandable graphite extinguishing agent is sprayed on the surface of the combustion material. Under the action of high temperature, graphite expands rapidly in volume, forming a sponge foam on the surface of the combustion material. Meantime, part of the metal in contact with the burning metal is moistened by the liquid metal, forming metal carbides or partial graphite interlayer compounds as well as an air insulating diaphragm to stop combustion.

Fire sand and gray iron dust: These are two extinguishing agents non specially made. They are applied alone to small scale phosphorus, magnesium, sodium and other fires, play the role of insulating air or absorbing heat from the flame, and can extinguish or control the development of fire. Fire sand can be even used to stop the spread of class B fires.

(Ⅱ) Fire extinguishers

The fire extinguisher is composed of a cylinder, a head, a nozzle, a handle and other components. The driving pressure can eject the filled extinguishing agent out from the nozzle to extinguish the fire. Thanks to simple structure, low price, convenient operation, portability and flexibility, fire extinguishers are important fire-fighting facilities for incipient fire.

There are many types of fire extinguishers. They can be classified into portable, trolley and hanging types by way of movement; gas cylinder type, accumulation type and chemical reaction type by the driving power source. By the filled extinguishing agent, they can be mainly divided into the following 6 categories.

1. Water extinguisher

Water extinguishers, also known as plain water extinguishers, use clear water and certain additives as extinguishing agents. Stored in the gas cylinder, the gas in the carbon dioxide cylinder is used as the power to spray the extinguishing agent onto the combustibles for the purpose of smothering. The clear water extinguishing agent is suitable for extinguishing fires caused by flammable solid substances, that is, class A fires. It should be noted that due to the conductivity of water, the water extinguisher must not be used to extinguish electrical fires and class E fires. Water extinguishers have a narrow range of application and are less used. The clear water extinguishing agent is usually used in fire-fighting equipment, but not fire-fighting facilities.

2. Foam extinguisher

Foam extinguishers are fallen into chemical foam and mechanical foam extinguishers. Foam extinguishers are suitable for fighting class B incipient fire such as grease and petroleum products and Class A incipient fire such as wood. They cannot fight class B water-soluble fires, such as fires caused by ethanol, nor extinguish class C, class D, and class E fires.

The chemical foam extinguisher is filled with acidic and alkaline chemical agents; when used, the two solutions are mixed to cause a chemical reaction and produce foam, which is sprayed out to extinguish the flame. It can be fallen into the portable (as shown in Fig. 9-12), wheeled, and hand-push types (as shown in Fig. 9-13) by the mode of movement.

The air foam extinguisher, also known as mechanical foam extinguisher, boasts good thermal stability, long time of fire resistance, extinguishing ability 3 to 4 times higher than chemical foam, excellent performance, long shelf life, and user-friendliness, making it an upgrading product of the chemical foam extinguisher. It is filled with protein foam or fluorine nitrogen foam as needed to extinguish the incipient fire caused by oils and polar solvents. The fire company has launched an air foam extinguisher, commonly known as the "water-based fire extinguisher". It can be used for class E electrical fires with voltages below 36 kV, and also available for class C gas fires. This fire extinguisher is very resistant to rekindling, but more expensive than other fire extinguishers.

Fig. 9-12   Portable water-based fire extinguisher   Fig. 9-13   Hand-pushed foam extinguisher

3. Soda-acid fire extinguisher

The soda-acid fire extinguisher contains an aqueous solution of 65% industrial sulfuric acid and sodium bicarbonate as the extinguishing agent. When used, the two reagents are mixed to react and produce carbon dioxide, which is sprayed under its pressure to extinguish the fire. This fire extinguisher is suitable for fighting the incipient fire of class A substances, but not available for class B, class C, class D, and class E fires.

4. Carbon dioxide fire extinguisher

Carbon dioxide fire extinguishers are externally filled with liquid carbon dioxide. When used, open the valve, carbon dioxide is ejected under its own pressure to fill the space around the combustibles and isolate the air for fire suppression.

Generally, when the oxygen content is less than 12% or the carbon dioxide concentration reaches 30% to 35%, the combustion can be terminated. 1kg of carbon dioxide can generate 500 L of carbon dioxide gas at room temperature, enough to extinguish the flame in a space of 1 m^3. Carbon dioxide fire extinguishers are non-toxic, pollution-free, leaving no traces, and insulated. They are often used to extinguish the class E incipient fire of live electrical appliances below 600 V, precision instruments, and valuable equipment, and also used to extinguish class A and class B incipient fires in small areas and enclosed spaces.

When the carbon dioxide fire extinguisher is sprayed, the temperature of the surrounding space will plummet; thus, the extinguisher handle usually uses a large nozzle made of thermal insulation material. Compared with dry powder, foam and other fire extinguishers, the carbon dioxide fire extinguisher features a higher internal pressure. As a result, the fire extinguisher cylinder is of higher sealing process, making the fire extinguisher proper more expensive. Fig. 9-14 shows a portable carbon dioxide fire extinguisher.

Fig. 9-14 Portable carbon dioxide fire extinguisher

5. Halon fire extinguisher

Those fire extinguishers filled with halon extinguishing agents are collectively referred to as halon fire extinguishers. They mainly inhibit the chemical reaction process of combustion, so that the combustion is interrupted to achieve the purpose of fire suppression. There are many kinds of halon fire extinguishers, including halon 1211, halon 1301, halon 2402, and halon 1202 fire extinguishers. China used to apply halon 1211 fire extinguishers, mainly used to put out some class A, class B, and class C incipient fires, especially applicable to precision instruments, computers, cultural relics, etc. It's about five times as efficient as a carbon dioxide fire extinguisher. The halon fire extinguisher can neither put out metal fires, nor extinguish fires that burn nitrocellulose and gunpowder with their own oxygen supply in inert media. It cannot extinguish the fire of metal hydrides such as potassium hydride and sodium hydride and the chemical substances of spontaneous decomposition, such as peroxide and hydrazine. Allowing for the serious air pollution and somewhat toxic steam, the halon extinguishing agent has been gradually eliminated. It is little produced and used in China.

6. Dry powder extinguisher

The dry powder extinguisher is powered by liquid carbon dioxide or liquid nitrogen to spray the dry powder extinguishing agent out of the fire extinguisher for fire extinction. This extinguisher mainly extinguishes fire by inhibiting the flame, and can be classified into ordinary dry powder and multi-purpose dry powder extinguishers by the range of application.

The ordinary dry powder, also known as BC dry powder, contains sodium bicarbonate dry powder, modified sodium salt, and/or amino-group powder, mainly used to extinguish class B, class C, and class E incipient fires. The multi-purpose dry powder, also known as ABC dry powder, refers to ammonium phosphate powder, ammonium polyphosphate powder, etc. It is the most widely used fire extinguisher. It is applicable to class B, class C, and class E incipient fires, and even to general class A fires, but cannot extinguish class D metal fires. Fig. 9-15 shows a portable ABC dry powder fire extinguisher.

Fig. 9-15　Portable ABC dry powder fire extinguisher

(Ⅲ) Fire detector

During the process of combustion, substances usually produce smoke, release the aerosol-like combustion gas, and form a lot of infrared and ultraviolet light, resulting in an increase in ambient temperature. These smoke, temperature, flame, and combustion gas are collectively referred to as fire parameters.

The basic function of the fire detector is to effectively respond to fire parameters such as smoke, temperature, flame, and combustion gas, and convert the fire parameters into digital electric signals to the alarm controller through the sensitive element.

Depending on the response to different fire parameters, it can be divided into the optical flame fire detector, smoke-sensing fire detector (see Fig. 9-16), heat fire detector, combination type fire detector, combustible gas fire detector, etc.

Fig. 9-16　Smoke-sensing fire detector

(Ⅳ) Fire hose

The fire hose is an essential device for the supply of fire water or the delivery of foam extinguishing agent. It is widely used on fire trucks, fire pumps, and fire hydrants, as shown in Fig. 9-17.

Fig. 9-17 Fire hose

(Ⅴ) Fire hose nozzle

The fire hose nozzle is a tool used to shoot water while extinguishing the fire. Its role is to speed up the flow rate of water and change the shape of water flow. Different calibers are available for fire trucks, fire pumps, and fire hydrants, as shown in Fig. 9-18.

Fig. 9-18 Fire hose nozzle

(Ⅵ) Fire truck

The following fire trucks are available in China: fire-extinguishing water tanker, fire-extinguishing foam tanker, dry powder fire-engine, carbon dioxide fire vehicle, dry powder, foam and water fire fighting vehicle, lighting fire fighting vehicle, articulating aerial platform apparatus, etc.

(Ⅶ) Fire sandbox (bucket)

The yellow sand box (bucket) for fire protection is a box (bucket) made of pure iron sheet, with simple structure. It contains fire sand, which is mainly used to prevent the spread of class B fires, or to extinguish small incipient fires of phosphorus, magnesium, sodium, etc. It is simple to use, as shown in Fig. 9-19.

Fig. 9-19 Fire sandbox

任务三 防火、灭火、逃生与自救

据统计,全世界每年至少有十万人死于火灾,约有三分之一的人本可获救,却死于自身的惊慌失措,如果人们能掌握一定的火灾逃生以及消防常识的话,应该有大量的人能够"火"口逃生。

一、火灾死亡原因

(一)化学中毒死亡

吸入一氧化碳、硫化氢及氰化物等有毒气体后会出现化学窒息死亡。

一氧化碳(CO)是一种无色、无嗅、无味、难溶于水的有毒气体。它与血红蛋白的亲和力要比氧大210倍,当空气中一氧化碳含量达0.1%时,血液中将形成50%碳氧血红蛋白和50%氧合血红蛋白,造成一氧化碳重度中毒,其中,当一氧化碳浓度达到1.28%时,人在这样的环境里3 min即可死亡。而火灾现场中,特别是室内火灾,氧气含量有限,可燃物燃烧不充分,几分钟内烟气中的氧含量就会远低于21%,一氧化碳含量会远高于0.1%,能使人短时间内死亡。

硫化氢(H_2S)是一种有刺激性气味、无色、易溶于水的气体,低浓度的接触对呼吸系统和眼睛有刺激作用。当人处于高浓度硫化氢气体内,察觉到硫化氢气体时,就会立刻丧失嗅觉,然后出现头晕、头痛、意识模糊甚至抽搐痉挛,数分钟内就可使心脏呼吸停止而死亡。

氰化物是指带有氰基(CN)的化合物,具有极强的细胞毒作用,进入体内后会迅速与细胞线粒体内的色素氧化酶结合,使细胞色素丧失传递电子的能力,阻止细胞正常呼吸能力,使呼吸链中断,细胞死亡,造成人短时死亡。

(二)单纯窒息死亡

人体与外界的气体交换过程是吸入氧气和呼出二氧化碳,是通过肺实现的,这包括肺通气和肺换气两个方面。肺通气量不足时会使肺部氧气含量降低,二氧化碳含量增高;肺换气不足时会使血液中氧气含量降低,二氧化碳含量增高,会出现窒息死亡。

火灾发生时,可燃物燃烧过程要消耗大量氧气,其氧含量会降低当空气中含氧量低于6%时,短时间内人就会因缺氧而窒息死亡。因此,在缺氧环境中,大脑首先受影响,产生功能障碍,即使含氧量在6%~14%,虽然不会因缺氧而短时死亡,但也会因活动能力下降或丧失活动能力不能顺利逃离火场,最终因其他原因死亡。

(三)烟尘堵塞窒息死亡

当含有大量烟尘(空气中的固体粉末)的烟气被火灾现场中人员吸入后,会黏附在鼻腔、口腔和气管内,进入支气管、细支气管和小支气管,甚至扩散进入肺部黏附在肺泡上,所以火灾中死者的鼻腔、口腔、舌体上表面和气管处会发现大量烟尘,有时会是厚厚的一层,严重时会堵塞鼻腔和气管,致使肺通气不足,最终窒息死亡。

（四）热力损伤窒息死亡

发生火灾后，从火场中扩散出来的烟气温度可高达几百摄氏度，火焰的温度可达 1 000 °C，醚类和一些可燃气体火灾的火焰温度可达 2 000 °C。当人吸进高温烟气后，烟气流经鼻腔、咽喉、气管进入肺部的过程中，会灼伤鼻腔、咽喉、气管甚至肺，致使其黏膜组织出现水泡、水肿或充血，使伤者窒息死亡。另外，火灾中含有水蒸气的烟气造成的热力损伤更为严重。

（五）麻醉导致死亡

有些气体，如笑气（N_2O）、醚类气体吸入不会使人窒息死亡，但对人体有麻醉使用，人因被麻醉神志不清、活动能力下降而不能及时逃离现场被火烧死或其他因素致死。

（六）高温导致死亡

火灾烟气温度可高达几百摄氏度，在密闭性高的空间内（如地下室）烟气的温度可高达 1 000 °C。人对高温烟气的忍耐是有限的，在 65 °C 时，可短时忍受；在 120 °C 时，15 min 内可产生不可恢复的损伤；140 °C 时，可忍受 5 min；170 °C 时可忍受 1 min；温度再高些 1 min 也忍受不了，会有强烈的疼痛感，心率加快，肌肉痉挛，出现休克，不能及时逃离火场而被烧死。

（七）其他类型死亡

在火灾中，除了上述常见原因外，还有很多不常见但是也不能忽略的原因会导致人的死亡，例如：因为火灾的发生导致建筑、设备等坍塌或倒下，造成人员伤亡；人处于着火点，或者在扑灭初起火灾的时候因为不小心或者其他原因直接被大火烧死；高层建筑着火，人惊慌失措跳楼摔死；压力设备或者其他设备等受热爆炸将人炸死等等。

二、基本防火、灭火方法

（一）防火基本方法

根据物质燃烧的条件，防火的基本方法是控制可燃物、控制氧化剂、消除着火源、阻止火势蔓延，基本方法如表 9-3 所示。

表 9-3　防火基本方法

防火方法	防火原理	方法举例
控制可燃物	破坏燃烧的基础，缩小燃烧范围	限制可燃物储运量； 加强通风，降低可燃气体、粉尘的浓度； 用防火漆涂料浸涂可燃物； 及时清除撒漏的可燃物
控制氧化剂	破坏助燃条件	采用密封容器储存可燃物质； 将金属钠存放于煤油中，白磷存放于水中，二硫化碳用水封存，镍储存在乙醇中等

续表

防火方法	防火原理	方法举例
消除着火源	破坏燃烧的激发能源	危险场所禁止吸烟，禁止穿铁钉鞋，采用防爆灯等； 常润滑轴承，防止摩擦生热； 室外设备涂白漆或银漆，减少阳光辐射升温； 油罐防静电接地； 避雷针防雷击
阻止火势、爆炸蔓延	不使新的燃烧条件形成，防止火灾扩大范围，减少损失	在可燃气体管路上安装阻火器、水封； 压力的容器安装防爆膜、安全阀； 在建筑物之间留防火间距，筑防火墙； 危险货物与机车隔离； 变压器设油池

（二）灭火基本方法

在火灾中，阻断火的三要素中的任何一个要素，即可扑灭火灾。具体方法见表9-4。

表9-4 灭火基本方法

灭火方法	灭火原理	具体措施举例
隔离法	使燃烧物和未燃烧物隔离，限定灭火范围	搬迁未燃烧物； 拆除邻近燃烧处的建筑、设备； 断绝燃烧气体、液体的来源； 放空未燃烧的气体； 堵截流散的未燃烧液体
窒息法	稀释燃烧区的氧化剂含量，隔绝可燃物与氧化剂	往燃烧物上喷射氮气、二氧化碳； 往燃烧物上喷洒水、泡沫； 用沙土掩埋燃烧物； 封闭着火建筑、设备
冷却法	降低燃烧物温度于燃点之下，从而停止燃烧	往燃烧物上喷射水、泡沫； 用沙土掩埋燃烧物； 往燃烧物上喷射二氧化碳
抑制法	采用化学灭火剂参与燃烧，终止燃烧的链反应，从而阻止燃烧	往燃烧物上喷射干粉； 往燃烧物上喷射卤代烷

（三）灭火剂与灭火器

不同类别的火灾，应当选择不同类别的灭火剂，见表9-5。

表 9-5 火灾类型与灭火剂

火灾种类	水型（清水、酸碱）	ABC 干粉	泡沫	1211卤代烷	二氧化碳	其他
A 类（一般固体火灾）	最适用，水能冷却并穿透可燃物，有效防止复燃	适用，ABC 干粉附着在燃烧物表面，能隔绝空气，还能参与反应，起抑制作用	适用，具有冷却和窒息作用	适用，能参与反应，起抑制作用，但污染大气	密闭空间内或火焰较小时，可用，能够迅速降低空间氧气含量，使燃烧窒息	特殊情况可使用消防沙，灭火速度较慢
B 类（可燃液体火灾）	不适用	最适用，干粉在燃烧液体中参与反应，起抑制作用	最适用，覆盖燃烧物，将其与空气隔离，防止复燃（化学泡沫灭火剂不能扑救 B 类极性水溶性火灾）	适用，能参与反应，起抑制作用，但污染大气	适用，二氧化碳包裹燃烧物，使其与空气隔绝，并且有降温作用	特殊情况可使用消防沙，灭火速度较慢
C 类（可燃气体火灾）	不适用	最适用，喷射的干粉能迅速在气体中扩散，参与反应，抑制燃烧	仅部分机械泡沫（水基式）可用	适用，能够迅速在气体中扩散，参与反应，阻止燃烧，并且不留任何痕迹，但是污染大气	密闭空间内或火焰较小时可用，能够迅速降低空间氧气含量，使燃烧窒息	无
D 类（金属火灾）	不适用	不适用	不适用	不适用	不适用	只能使用膨胀石墨、7150、消防沙、铸铁粉等灭火剂灭火
E 类（带电火灾）	不适用	适用，ABC 干粉不导电，能用于带电灭火，但是会形成残留，造成设备损坏	仅部分机械泡沫（水基式）可用于36kV以下设备，但会有残留，造成设备损坏	最适用，卤代烷不导电，并且灭火后无残留，不损坏仪器设备，但是污染大气	封闭空间内最适用，能够迅速窒息扑灭火焰，无残留，不损坏仪器设备，无污染	无

三、电气火灾的扑救

人类的生存离不开电，但是由于一些不确定的或人为的因素，电气火灾不断发生，使人们生活、生产受到严重影响。

设备或充气设备有化学、物理爆炸风险。

1. 切断电源

电气火灾发生后，切断电源，就可按照常规方法扑救。切断电源的安全要求如下：

（1）火灾时，由于烟熏火烤，开关电器设备绝缘降低，操作时应使用绝缘工具或物品代替人体，以防止人体直接接触开关设备。

（2）对于变电站，应严格遵守倒闸操作顺序，防止误操作，扩大事故范围。

（3）切断低压电源必要时可用带绝缘套的斜口钳等工具间断电线，剪断时应当穿绝缘靴或站在绝缘垫上。剪断后断点应当有支撑，以防接地短路。剪断时应当分相剪，不能一起剪断，防止相间短路产生高温火花。

（4）夜间扑救时，应当有备用照明措施，避免断电后影响灭火工作。

2. 带电灭火

有时候为了争取有利的灭火时机，来不及断电，或者因某些原因不能切断电源时，需要带电灭火，注意事项如下：

（1）选用不导电的灭火器，二氧化碳灭火器、1211卤代烷灭火器、ABC干粉灭火器、部分机械泡沫（水基）灭火器都是不导电的，可用于带电灭火。但是应注意其适用范围，并且保证人员与带电体的安全距离。普通机械泡沫或化学泡沫灭火器具有导电性，不能用于带电灭火。

（2）使用消防栓的清水扑灭带电火焰时，应注意水导电性较强，不能直接用直流式水枪，以避免喷出的水柱导电导致灭火人员触电，应用喷雾水枪，在水压足够大时，能够使水充分雾化，并且使用抛物线的方式灭火，可大大减小水中的泄漏电流。除此之外，使用喷雾灭火时，还应与带电设备保持安全距离，110 kV及以下应当保持3 m安全距离，220 kV及以上保持5 m安全距离，同时，应当将水枪喷嘴接地，操作人员穿戴绝缘靴与绝缘手套。

（三）电气火灾安全注意事项

电气设备和线路的种类很多，结构错综复杂，必须根据其特点进行灭火，以保证安全性、可行性和有效性。

（1）火场上有架空线路经过时，人不能站立在导线下方，以防断线落地造成触电。

（2）充油设备着火，有爆炸危险。若充油设备外部着火，确保安全的情况下，可用不导电灭火器灭火。若充油设备内部故障起火，则必须先断电，确保安全的情况下，再用冷却法扑救，即使火焰熄灭后，还应继续喷洒降温。若油箱出现油外泄，而周围地面无油池，应用消防沙扑灭地面的燃油，防止油蔓延。若发现火焰较晚，充油设备温度已经过高，有较大爆炸风险时，应当在断电后立即撤离到安全地点，拨打消防电话，不得靠近，防止爆炸伤人。

（3）发电机和电动机等旋转电机着火时，为防止机械部分变形，采用喷雾冷却法扑救时，应注意电机应当均匀冷却。不宜使用消防沙或干粉灭火，以免损坏电机。

（4）变配电装置着火，必须先切断电源，然后用水枪喷雾灭火，室内也能使用1211卤代烷灭火器或者二氧化碳灭火器，但不得使用消防沙或干粉灭火，以免损坏设备。

四、火场的逃生与自救方法

（一）火场逃生与自救原则

（1）要了解和熟悉周围环境：包括留意逃生通道、安全出口、消防电梯、消防栓、灭火器等位置，以便在火灾浓烟中找到逃生的路线。

（2）迅速撤离并报警：一旦听到火灾警报响起，或者发现浓烟、火焰，意识到自己处于火场时，不要犹豫，立刻想方法撤离，同时打 119 报警。

（3）保护呼吸系统：火灾时会有大量有毒、高温度浓烟，防止吸入浓烟和有毒气体，可以使用湿毛巾、湿衣服捂住口鼻，以降低呼吸系统中毒、灼伤的程度；逃生时注意要弯腰甚至匍匐前进，防止上层烟雾的伤害。

（4）防止高温灼伤：如果人离着火点较近，或者是逃离过程中会路过着火区域附近，除了保护呼吸系统，还要防止被高温灼伤，最简单的方法是利用湿的纯棉衣服、被褥裹在身上逃离，切记不能穿着涤纶、锦纶、腈纶、维纶等合成纤维制成的衣物，因为相比棉、毛类织物，它们容易受热熔融覆在皮肤上，造成大面积烫伤。

（5）暂时避难措施：在无路可逃的情况下，可以利用卫生间、浴室等可燃物非常少，并且有水的空间来避难。避难时要用水喷淋迎火门窗，把房内一切可燃物淋湿，包括自身，使用湿毛巾堵住门缝，防止烟气进入，以延长生存时间。在避难期间还应主动与外界联系，以便尽早获救。

（6）注意指示牌：户内火场中，建筑中上部可能会布满浓烟，而脚底通常不会被浓烟覆盖，这时候应当匍匐前行，而很多"安全出口"标示牌（图 9-20，断电后也亮灯）都是布置在墙角，匍匐前进的过程中应能够比较清晰地看到，这时候应当跟随标示牌的引导逃离火场。

图 9-20 "安全出口"标示牌

（7）利用硬件条件：充分利用火场的硬件条件，例如消防电梯、室外疏散楼梯、屋顶、阳台、煤气管道等硬件逃生；或者利用绳子、窗帘被单撕成条，拴在牢固构架上，沿绳索下楼逃生。

（8）利人利己：在火场中逃生不能不顾他人死活，做损人利己的事，应当自救与互救相结合，利人利己才能最大程度地减少伤亡。

（二）火场中的典型错误做法

（1）贪恋财物：火场逃生切忌贪恋财物，逃出再折回，往往会导致当事人的死亡。

（2）盲目跳楼：火灾发生不冷静沉着，盲目跳楼，往往会导致死亡或者残疾。不到关键时刻，不能保证自身安全，绝不采用跳楼的方法逃避火灾。

（3）着火乱跑：身上衣物着火后乱跑，会使燃烧更充分、更猛烈，更不容易灭火。若附近无灭火剂，可以就地打滚或使用厚重的衣物覆盖窒息灭火。

（4）大喊大叫：火场中有大量浓烟，大喊大叫会吸入更多烟气，导致窒息伤亡。

（5）乘坐电梯：非消防电梯，没有防烟、绝热功能，在火灾中都不能使用，假如中途停电，乘客都会因为浓烟而窒息死亡。

（6）直立行走：直立行走会导致吸入过多浓烟而窒息死亡。

（三）注意日常生活中的细节

（1）熟悉身边的消防设施和消防器材：高层建筑，都会配备各种消防设施和消防器材，例如烟雾警报、封闭楼梯、防烟楼梯（32层以上）、消防电梯、消防栓、楼梯口的防火门、灭火器等。在熟悉这些消防设施和消防器材的情况下，沉着冷静地扑灭初起火灾并不困难。

（2）日常生活中在楼梯间堆放杂物，会影响逃生，甚至会引发封闭楼梯或者防烟楼梯出现火灾，所以日常生活中楼梯间一定不能堆放杂物。

（3）常闭式防火门（图9-21）只有在处于关闭状态时，在发生火灾后才能有效地阻挡浓烟烈火的侵袭，能够在发生火灾时，有效延缓火势和浓烟向楼梯间、消防通道蔓延，为人员疏散、火灾救援赢得宝贵时间。因此，要保持常闭式防火门处于常闭状态，决不能图方便将其打开，或为了防盗将防火门锁上。

（4）在开放式阳台上不能堆放易燃易爆物品，否则外侧飘来的点火源有可能将其引燃，并且阳台堆放可燃物也容易成为火灾外部垂直蔓延的途径。

（5）现在市面上有大量电动车锂电池电瓶是不符合标准的三无产品，充电过程中出现鼓包、爆炸、燃烧的可能性较大，所以电动车不要过夜充电，也不要在楼道、安全出口等地方进行充电，万一电动车充电引发火灾，堵住疏散通道，再加上楼道堆放了可燃物，后果不堪设想。

图9-21　常闭式防火门

图9-22　电动车楼道着火

（四）高层建筑火灾

我国《建筑设计防火规范》（GB 50016—2014）对高层建筑防火规范有明确规范和要求，但我国高层建筑火患依然突出。

高层建筑火灾具有火势蔓延快、疏散困难和扑救难度大的特点，由于高楼结构复杂、人员密集，一旦失火难以控制和逃离。高层建筑火灾案例也表明，现有的火灾救援手段很难适应快速的高层建筑的建设发展。所以我国对于高层建筑的防火防爆要求非常高，往往都配备了很齐全的消防设施和消防器材。高层建筑火灾如图 9-23 所示。

研究高层建筑火灾的特点，了解消防装备的技术状况，对于指导人们正确选择救援逃生设施，正确应对高层建筑火灾进行应急逃生，具有现实的意义。当身处高层建筑中，发生火灾时，应当注意以下几点：

图 9-23　重庆 2020 年 12 月 26 日高层建筑火灾

（1）当某一层楼起火，已经开始蔓延了，无法立刻扑灭，并且浓烟滚滚，不清楚着火楼层时，应注意听广播、警报通知、观察窗外情况，判断着火楼层，不要一听见警报，看见浓烟就惊慌失措，盲目行动。

（2）当房内起火，且门已经被火焰阻断、封闭，无法顺利逃离时，可以另寻其他通道，如通过阳台、走廊等暂时转移到其他未着火的房间，切记不能穿越大火开门或者在没有安全措施的情况下跳楼。

（3）如果是晚上听到警报，应当用手背去触碰房门，如果房门变热，门不能打开，否则火焰立刻会冲进房间。这时候只能用湿毛巾、湿被褥等堵住门缝，向门上泼水以增强门的抗火能力。如果房门不热，则表示火焰不大，打开房门离开后，一定随手关门，以防火势蔓延。

（4）如果在逃生过程中出现浓烟扑面而来，不能试图冲过浓烟，因为浓烟来的方向，必定有大火在快速蔓延，而且人冲进浓烟如果不能及时冲出，几分钟内即可窒息而死。

（5）如果楼层着火已经将所有向下的通道堵死，不能向下疏散，则可以先到屋顶。若有楼顶相邻楼梯间相连，则可以从相邻楼梯间撤离，否则只能在楼顶向救援人员发出信号，等待救援。

（6）在充满浓烟的地方，无法迅速冲出浓烟时，应当弯腰甚至匍匐前进，因为离地板越近，浓烟就相对更少，可以延缓窒息死亡的时间。有条件的情况下可以用湿毛巾捂住口鼻。

（7）如果在高层建筑的低层被困，当人员被火灾危及生命又无其他方法自救时，可以将床垫、被子等软物抛至楼底，跳楼逃生；或者将床单、绳索等淋湿，编制成长绳放至楼下，利用长绳逃生。

（8）封闭楼道着火，不能从楼道逃生的情况下，也不能盲目乘坐电梯逃生，乘坐普通电梯过程中可能会断电，出现浓烟、高温。应认准消防电梯，只有消防电梯才有防烟隔热功能。使用消防电梯应注意需要砸碎消防开关外面的玻璃，按下"ON"键，这时候会出现警报，电梯的消防功能才能启动，见图 9-24。

模块九　防火与防爆（Module Ⅸ　Fire and explosion protection）　　349

图 9-24　消防电梯开关

（9）高层建筑每一层楼道都配备有手提式灭火器，可以用于扑灭室内初起火灾。除此之外还会配备室内消防栓，如果手提式灭火器无法扑灭火焰，则可以使用消防栓喷水灭火，但是消防栓使用的是清水灭火剂，是电的良导体，使用前必须保证全屋断电，否则会导致触电伤亡。

（五）中低层住宅楼火灾

我国很多老旧房屋都属于中低层住宅楼，这类房屋在消防方面有三个特点：

第一，住房老旧，存在电力线路老化、厨房产生的油烟在墙壁附着等现象，加上楼道杂物堆积等问题，着火可能性远高于高层建筑。

第二，这类住宅楼往往没有消防电梯，没有防火门，没有警报装置，甚至也没有消防栓、灭火器，所以在发生火灾后，消防员来之前，住户的灭火能力和自救能力将大大受到限制。

第三，由于老旧小区未规划地下停车场，地面有大量汽车，很可能堵塞消防通道，增加消防员救援时间。

由于以上三点，当中低层住宅发生火灾，住户的自救能力就显得尤为重要。在中低层老旧住宅中出现火灾时，应当特别注意以下几点：

（1）老旧小区没有消防联动装置，所以发现火情，立刻拨打 119 报警，越快越好。

（2）发现火情应当立即断电，因为火焰中有大量的离子，是电的良导体，老旧房屋线路复杂，很可能由于漏电造成火灾蔓延或者触电伤亡。

（2）最好在家中自行配备手提式灭火器，当出现初起火灾时，能够快速扑灭。

（3）所处楼层较低时，当人员危及生命又无其他方法自救时，可以将床垫、被子等软物抛至楼底，跳楼逃生；或者将床单、绳索等淋湿，编制成长绳放至楼下，利用长绳逃生。

（4）低层住户为了防盗，往往增设了防盗窗，普通防盗窗会阻碍火灾中的逃生，所以尽量装设有消防设计的防盗窗，必要时可以打开。

（5）逃生时切勿贪恋财物，特别是逃离后折返回去拿财物，非常危险。

（6）逃生时使用湿毛巾捂住口鼻，延缓窒息死亡时间。

（7）在浓烟中无法逃生时，可以将半个身子探出窗外，呼吸新鲜空气，等待救援，切记不得随意跳楼。

（8）中低层建筑物高度低，消防梯容易达到，若有阳台，则可在阳台呼救，等待消防员救援。

（9）老旧住宅室的煤气管道、下水管道等可能布置在室外，在无路可逃的时候也可以利用管道逃生，但这种方法一般不适用于老人和小孩。

（六）公共场所火灾

公共场所，特别是电影院、KTV、餐馆、百货商场等人口密集，相比之下疏散通道较少，给人员逃生带来一定的困难。对于有一些公共场所，有一些独特的特点，也使得在不同公共场所发生火情时的逃生手段也有独特的特点。

（1）电影院：由于电影播放时，没有灯光，如果出现火灾，将会出现找不到出口的情况，所以在进入影院时应当仔细观察"逃生通道""安全出口"的方位，以便紧急撤离。

（2）大型商场、超市：商场通常使用自动扶梯上下，而自动扶梯非常窄，在发生火灾时无法作为逃生通道，所以在进入商场时应当观察封闭楼梯的方位。有一些商场因为各种原因，将通向消防通道的防火门上锁（图9-25），这是违反《中华人民共和国消防法》的行为，发现后可以拨打当地消防局电话投诉，或者拨打96119火灾隐患举报投诉电话进行投诉。

图 9-25　防火门违规上锁

（3）酒吧等：此类场所通常晚上人较多，进出顾客随意性较大，密度很高，灯光暗淡，出现火情时很容易出现踩踏事故；还有就是这类场所房屋结构中使用大量的塑料、泡沫作为隔音材料，火灾发生时极易燃烧，并且产生大量有毒气体，所以在逃生时注意弯腰甚至匍匐前行，并且用湿毛巾、湿衣物捂住口鼻。

五、家庭日常防火措施

（一）日常火灾发生的原因

现代家庭中一般都具备火的三要素。家庭中有可燃的装潢材料，如木材、壁纸、密度板等；有易燃的清洁剂，如汽油、乙醇、香蕉水等。而空气中20%的氧气都是属于氧化剂，所以只要上述的可燃物遇到点火源，即可出现火焰，若不能及时控制，则会演变成火灾。目前家庭中出现火灾的主要原因有以下几点：

（1）煤气、天然气使用不当造成的火灾，即厨房火灾。虽然目前城镇住户基本上都已经使用天然气或者煤气做饭，但是天然气或者煤气相比原始的木柴、煤炭，更容易由于使用不当造成火灾，例如燃气灶故障后继续使用、干烧等。

（2）电气火灾。现代家庭都有大量的家用电器，当线路老旧，或者电器线路破损、插线板过载等，都有可能造成火灾。电气火灾也是家庭火灾中最常见的类型。

（3）吸烟酿成的火灾。随意扔烟头，或者未熄灭烟头即扔进垃圾桶引燃垃圾，都有可能造成火灾。

（4）儿童玩火。小孩的好奇心强，常把玩火作为一种乐趣，如果家长没有及时制止，往往会引发火灾。

（二）厨房防火

（1）使用天然气、煤气必须遵循"先点火，后开气"的操作。否则会让可燃气体与空气充分混合，出现前文所述的混合燃烧，表现为"爆燃"，若旁边放有易燃物，如白酒等，就会立刻引发燃烧。

（2）天然气或煤气本身无色无味，但是天然气公司、煤气公司都添加了叫作四氢噻吩的带有臭味的气体，以便人们察觉到它的泄漏，当闻到家中有天然气、煤气泄漏时，应打开门窗通风，然后再关闭阀门。切记不能使用明火检查，不能开灯，否则可能会引发混合燃烧甚至爆炸。

（3）如果阀门本来就是关闭的，却能闻到四氢噻吩的臭味，那么证明管道已经泄漏，应当打开门窗通风、关闭总阀后打电话通知天然气公司前来检查。

（4）严禁个人更改、拆卸、迁移天然气、煤气管道中的阀门、计量表等装置。

（5）天然气、煤气软管是燃气专用软管，不能使用水管或者其他塑料管代替，应遵循天然气、煤气公司工作人员指导购买和更换。

（6）如果使用液化煤气罐，出现阀门损坏，无法关闭，分为两种情况：一种是没有点火，这时候应当将煤气罐放置在空旷的室外，不能使其接触点火源，并拨打119；第二种情况是已经点燃火，如果火焰不大，不至于点燃其他物体，应任其燃烧不可扑灭（若扑灭则会出现煤气、空气混合，造成混合燃烧风险），并且让其他可燃物远离火焰，拨打119。

（7）如果出现液化煤气罐本体着火，不用惊慌失措，因为煤气罐中只有可燃气体，没有氧化剂，只要不过热，是不会发生爆炸的，这时候应当带上湿手套关闭气瓶角阀，切断气源，然后用湿抹布盖住气瓶或使用干粉灭火器灭火。

（8）在使用天然气、煤气做饭的时候，炉灶不得离开人，否则可能会由于遗忘而导致厨房发生火灾。

（9）如果发生油锅内着火，切忌惊慌失措地使用水来扑灭，这样只会增长火势，见图9-26。最直接的办法是盖上锅盖，窒息灭火。

图 9-26 油锅着火

（三）电器防火

随着电能的越来越普遍，电所带来的火灾比例在逐年增加，如果家用电器等带电物品使用不规范，则会带来较大火灾隐患。无论是哪一种家用电器发生火灾，第一时间不是灭火，而是先断电再灭火。

（1）洗衣机：洗衣机通常着火风险较小，但是有一些特殊情况需要注意，某些衣物沾染特殊污渍如油漆等，在使用乙醇、汽油等易燃液体清洁后，应当用清水清洁干净后再放入洗衣机。老式洗衣机一次投入洗衣机内衣物不得过多，否则容易导致电机负荷过大，甚至堵转，而又因老式洗衣机没有过热保护，会导致电机烧毁甚至造成火灾。

（2）空调：空调一般分为内机和外机，内机着火可能性非常小，而如果外机朝向太阳，没有遮阳措施，在夏季时，长时间开启空调可能会导致外机过热，出现火灾隐患。

（3）电熨斗：对于无过热保护的电熨斗而言，使用时人一定不能离开，否则一旦遗忘，很容易造成火灾。

（4）电热毯：不能购买粗制滥造、无安全保护措施的三无产品。三无产品容易出现电热毯内匝间短路，导致部分过热甚至烧坏被褥导致火灾，更严重的是出现漏电造成人员的触电伤亡，见图 9-27；在人员离开时，电热毯也应当随时关闭。

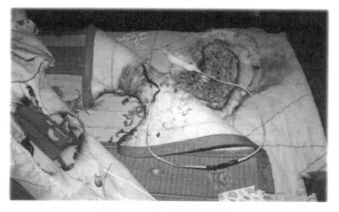

图 9-27 劣质电热毯自燃

（5）电水壶：电水壶烧水时可能会有水溢出，所以不能摆放于插座旁或者其他带电设备旁，要保证即使有水溢出，也不会出现漏电、短路的情况。

（6）电吹风：使用通电的电吹风时，人员不得离开，不能随意放置在沙发、被褥等可燃物上，更不能用毛巾等物品堵塞电吹风出风口或者进风口，见图 9-28，这对于老式没有过热保险的电吹风尤为重要。通电的电热丝是由空气绝缘，无其他绝缘，属于明火管理范围，如果空间中存在大量乙醇气体、悬浮面粉等可燃气体或者可燃固体粉末，禁止使用电吹风。

图 9-28　电吹风进、出风口堵塞导致起火

（7）电冰箱：对于新买回来的电冰箱，用于包装的塑料板、泡沫等一定要抽掉，以保证冰箱的正常散热；冰箱背面是压缩机散热管，夏季温度较高，所以冰箱背面不能放置低燃点易燃易爆物体，比如白酒、汽油、香蕉水等。

（8）电热取暖器：电热取暖器有很多种，如油酊类、热辐射类、热风类等，无论哪一种，其加热部位都不能靠近易燃易爆物体，也不能为了快速烤干衣物，用衣物阻挡其散热，否则易引起火灾，见图 9-29。

图 9-29　电暖气烤焦毛巾

（9）插座、插线板、开关等：首先应当购买正规产品，正规产品大多使用铜片导电，夹紧力度较大，而三无产品很多采用铁片，夹紧力也不够，造成接触电阻大，发热严重，重载易引起火灾；使用时应当注意其额定功率，用电器功率不应大于插座、插线板、开关额定功

率，否则长时间过载会因过热而引发火灾，特别是应当避免插线板外接插线板的形式，该形式容易引起过载。

（10）抽油烟机：抽油烟机导致的火灾通常不是由于电引发，而是明火点燃油烟。抽油烟机表面很容易沾染油污，油污遇明火容易被点燃，而它的工作环境正好在炉灶上方，环境非常恶劣，特别是发生锅内着火，火焰可能会点燃抽油烟机表面油烟，导致火灾，所以应当定时清洁油烟机表面。

（四）家庭吸烟的防火

燃烧的烟头温度一般为 200～300 ℃，中心温度达到 700～800 ℃。而一般的纸张、棉花、布等可燃物，燃点大多在 300 ℃ 以下，卫生纸之类的燃点更低。这说明烟头的温度实际超过了很多可燃物的着火点，所以在家庭中吸烟时一定要注意防火。

（1）不要躺在沙发、床上吸烟，掉落的高温烟灰可能会引燃沙发、被褥等。

（2）不要在喝醉了或者过度疲劳的情况下在房间吸烟，可能会发生一支烟还未吸完，人已入睡，导致烟头掉落在可燃物上，发生火灾。

（3）不要随意乱放点燃的烟头，如放在写字台、窗台、餐桌上等。当人暂时离开做其他的事，可能会忘记燃烧的烟头，导致其点燃其他的可燃物。

（4）处理好吸完的烟头，一定完全灭火后再扔垃圾桶，否则可能会引燃垃圾桶中其他垃圾。

Task III Fire prevention, extinguishing, escape and self-rescue

At least 100,000 people die in fires around the world every year, and about one-third of those who could have been saved died of their own panic, according to statistics. With certain fire escape and fire-fighting knowledge, a large number of people may be able to escape from the fire.

I. Cause of loss of life

(I) Chemical poisoning

Chemical asphyxia occurs after inhaling toxic gases such as carbon monoxide, hydrogen sulfide, and cyanide.

Carbon monoxide (CO) is a toxic gas that is colorless, odorless, tasteless and insoluble in water. It has a 210 times greater affinity with hemoglobin than oxygen. When the carbon monoxide content in the air reaches 0.1%, 50% carboxyhemoglobin and 50% oxyhemoglobin will be formed in the blood, resulting in severe carbon monoxide poisoning. People will die immediately in 3 minutes when exposed to 1.28% of CO concentration. At the scene of fire, especially indoor fire, the oxygen content is limited, giving rise to insufficient combustion of combustibles; within minutes, the oxygen content in the smoke will be well below 21%, and the carbon monoxide level will be well above 0.1%, causing a person to die in a short time.

Hydrogen sulfide (H_2S) is a colorless, water-soluble gas with a pungent odor that can irritate the respiratory system and eyes at low concentrations. When people are exposed to a high concentration of H_2S gas and perceive the H_2S gas, they will immediately lose their sense of smell, followed by dizziness, headache, confusion, and even convulsive spasms. Cardiac arrest and death may occur within a few minutes.

Cyanide refers to compounds with cyanogroup (CN), which are extremely cytotoxic. After entering the body, it will quickly bind to the cytochrome oxidase in the cell mitochondria, making the cytochrome lose the ability to transfer electrons, and preventing cells from breathing properly. In this case, the respiratory chain is interrupted, resulting in cell death, and then human death in a short time.

(II) Simple asphyxia

The process of gas exchange between the human body and the outside world is taking in oxygen and exhaling carbon dioxide through the lungs, including pulmonary ventilation and pulmonary exchange. Insufficient pulmonary ventilation reduces oxygen levels in the lungs and increases carbon dioxide levels. Insufficient pulmonary exchange reduces oxygen levels in the blood and increases carbon dioxide levels, resulting in death by suffocation.

In case of fire, the combustion of combustibles consumes a lot of oxygen, thus decreasing the oxygen levels. When the oxygen content in the air is less than 6%, people will die of asphyxia due to lack of oxygen in a short time. Therefore, in a hypoxic environment, our brain is first affected, resulting in dysfunction. Even if the oxygen levels are between 6% and 14%, victims may not die temporarily from lack of oxygen, but will eventually die from other causes since they cannot get out of the fire scene due to their decreased mobility or the loss of mobility.

(Ⅲ) Smoke clogging

When inhaled by those at the fire scene, smoke containing a large amount of soot (solid powder in the air) will stick to their nasal cavity, mouth and air duct, enter the bronchi, bronchioles and bronchia, and even spread into the lungs and stick to the alveoli. Therefore, a lot of smoke is found in the nasal cavity, mouth, upper surface of the tongue, and trachea of the fire victim, sometimes a thick layer. In severe cases, it will block the nasal cavity and trachea, resulting in insufficient pulmonary ventilation and eventually suffocation.

(Ⅳ) Thermal damage

In case of fire, the smoke diffused from the fire scene can be up to several hundred degrees Celsius, the flame more than 1,000°C, and the flame temperature of the ethers and some combustible gas fires can reach more than 2,000°C. When people breathe in hot smoke, the smoke flows through the nasal cavity, throat, and trachea into the lungs. This will burn the nasal cavity, throat, trachea and even lungs, causing blisters, edema or congestion in their mucosal tissues. In the end, the victims suffocate and then die. In addition, the smoke containing water vapor in the fire may cause even more severe thermal damage.

(Ⅴ) Anesthesia

Some gases, such as nitrous oxide (N_2O) and ether gases, will not cause us to suffocate when inhaled, but they have an anesthetic effect on the human body. Those unconscious and unable to escape timely from the fire scene will be killed by fire or other factors.

(Ⅵ) High temperature

Fire smoke can reach hundreds of degrees Celsius, and the smoke can be as high as 1,000°C in a highly sealed space (such as basement). Our tolerance for high temperature smoke is limited: At 65 °C, we can tolerate for a short time; at 120 °C, we may suffer irreversible damage within 15 min; at 140 °C, we can tolerate for 5 min; and at 170 °C, we can tolerate for 1 min. We cannot stand for 1min at a higher temperature. We will suffer intense pain, increased heart rate, muscle spasm, and even shock, which prevents us from escaping the fire scene timely. In the end, the victims may be burned to death.

(Ⅶ) Other types of death

In addition to the above common causes, there are many less common but not ignorable causes, for example, casualties as a result of the collapse or dilapidation of buildings/equipment because of the fire; killed in the fire directly because of carelessness when putting out the incipient fire or other reasons at the fire point; panicked and jumped to his/her death when high-rise buildings caught fire; killed in the heated explosion of pressure facility or other equipment.

II. Basic fire prevention and extinguishing methods

(I) Basic methods of fire prevention

Depending on the conditions under which the substance burns, the basic methods of fire prevention are to control combustibles, control oxidizing agents, eliminate fire sources, and arrest the spreading of fire, as shown in Tab. 9-3.

Tab. 9-3 Basic methods of fire prevention

Method	Principle	Examples
Control the oxidizing agent	Destroy combustion-supporting conditions	Use sealed containers to store combustible materials; Deposit sodium metal in kerosene, white phosphorus in water, carbon disulfide in water, and nickel in ethanol
Eliminate the ignition source	Destroy the excitation energy of combustion	No smoking and iron shoes in dangerous places, and use explosion-proof lights; Lubricate the bearing regularly to prevent the generation of heat by friction; Apply white or silver paint to outdoor equipment to reduce the temperature rise of sunlight radiation; Electrostatic grounding of oil tank; Lightning rod to protect against lightning strikes
Arrest the spreading of fire and explosion	Do not enable new burning conditions, prevent the spread of fire, and cut losses	Install the flame arrester and water seal on combustible gas piping; Install the explosion-proof membrane and safety valve on the pressure vessel; Keep fire separation between structures, and build firewalls; Isolate hazardous goods from locomotives; Provide the transformer with an oil pool

(II) Basic methods for fire extinguishing

Blocking any of the three elements of fire can extinguish the fire. See Table 9-4 for details.

Table 9-4 Basic methods for fire extinguishing

Method	Principle	Examples
Isolation method	Segregate the comburent from unburned materials to limit the fire suppression range	Remove unburned materials; Demolish buildings and equipment near the burning site; Cut off the source of combustion gas and liquid; Vent the unburned gas; Block loose unburned liquid
Smothering method	Dilute the oxidizer content in the combustion zone to isolate the combustible and oxidizer	Spray nitrogen and carbon dioxide onto the comburent; Spray water and foam onto the comburent; Bury the comburent with sand; Enclose burning structures and equipment

Continued

Method	Principle	Examples
Cooling method	Stop combustion by lowering the combustion temperature below the fire point	Spray water and foam onto the comburent; Bury the comburent with sand; Spray carbon dioxide onto the comburent
Suppression method	Use chemical extinguishing agents to participate in combustion and terminate the chain reaction of combustion, thus stopping combustion	Spray dry powder onto the comburent; Spray haloalkane onto the comburent

(Ⅲ) Extinguishing agent and fire extinguisher

Different extinguishing agents should be selected for different types of fires, as shown in Tab. 9-5.

Tab. 9-5 Fire types and extinguishing agents

Fire type	Water-based (plain water, soda-acid)	ABC dry powder	Foam	Halon 1211	Carbon dioxide	Miscellaneous
Class A (general solid fire)	Most applicable, water cools and penetrates the combustibles, effectively preventing rekindling	Applicable, ABC dry powder attaches to the surface of comburent, can isolate the air, and even inhibit the reaction	Applicable, with cooling and smothering actions	Applicable, able to inhibit the reaction, but pollute the atmosphere	Available for confined space or little flame, able to quickly reduce the oxygen content in the space to suffocate the combustion	Fire sand can be used under special circumstances, subject to low extinguishing speed
Class B (combustible liquid fire)	Not applicable	Most applicable, the dry powder participates in the reaction in the burning liquid and acts as an inhibitor	Most applicable, to cover the comburent, isolate it from air, and prevent rekindling (chemical foam extinguishing agent can not fight class B polar water-soluble fire)	Applicable, able to inhibit the reaction, but pollute the atmosphere	Applicable, carbon dioxide wraps the comburent to isolate it from air, and has a cooling effect	Fire sand can be used under special circumstances, subject to low extinguishing speed
Class C (combustible gas fire)	Not applicable	Most applicable, the dry powder can quickly diffuse in the gas, participate in the reaction, and inhibit the combustion	Available to some mechanical foam (water-based) only	Applicable, able to quickly diffuse in the gas, participate in the reaction, prevent the combustion, and leave no trace, but pollute the atmosphere	Available for confined space or little flame, able to quickly reduce the oxygen content in the space to suffocate the combustion	None

Continued

Fire type	Water-based (plain water, soda-acid)	ABC dry powder	Foam	Halon 1211	Carbon dioxide	Miscellaneous
Class D (metal fire)	Not applicable	Not applicable	Not applicable	Not applicable	Not applicable	Only expanded graphite, 7150, fire sand, cast iron powder and other extinguishing agents available
Class E (live fire)	Not applicable	Applicable, the non-conducting ABC dry powder can be used for live fire extinguishing, but will give residues, resulting in equipment damage	Only part of the mechanical foam (water-based) can be used for equipment below 36kV, but will give residues, resulting in equipment damage	Most applicable, haloalkane is non-conducting, and free from residue after fire extinguishing, without equipment damage, but pollutes the atmosphere	Most applicable in the enclosed space, able to put out the flame quickly, free from any residue, equipment damage and pollution	None

III. Electrical fire fighting

Human beings cannot live without electricity. However, due to some uncertain or human factors, Electrical fires keep breaking out, seriously affecting our lives and production.

(I) Causes of electrical fire

1. Dangerous temperature

It can be known from the electrical principle that when the current flows through the conductor, the resistance will cause the conductor to heat up, raising the temperature of the conductor and the surrounding insulating media. In addition to the heat generated by the current flowing through the conductor, the alternating magnetic field in the permeability magnetic material such as the transformer core will cause eddy current loss and hysteresis loss in the iron core and other ferromagnetic materials, and also generate heat to raise the temperature. Dielectric loss in the insulating electrolyte will also raise the temperature of the insulating material.

In short, the electrical equipment will heat up as long as it is running. For normal operation of electrical equipment, the heat generation and heat dissipation are in equilibrium, and the temperature will not exceed the specified allowable value. This is the thermal equilibrium. When the normal operation of equipment is broken, there may be excessive heat, and the temperature is above the specified allowable value, eventually leading to fire. The high temperature of equipment arising from the breaking of thermal equilibrium is mainly ascribed to the following reasons.

(1) Overload: The equipment or conductor runs overload for a long time, resulting in excessive

heat due to overcurrent.

(2) Short circuit: The equipment or conductor is short-circuited, the current suddenly increases several or even dozens of times, far higher than the rated current. A lot of heat is generated in a short time, causing the temperature to rise sharply.

(3) Poor contact: For some reasons (corrosion, insufficient pressure, loose bolts, etc.), the conductor joint points and equipment contact terminals are of excessive contact resistance, and the contact heats up seriously when the normal current is flowing.

(4) Core heating: The transformer core silicon steel sheet is insulation damaged or multi-point earthed, resulting in increased eddy current loss and local overheating.

(5) Poor heat dissipation: The electrical equipment suffers poor heat dissipation condition, and the thermal equilibrium cannot be reached even in normal operation, resulting in local or global overheating.

(6) Improper use of electric heaters: Resistive electrical equipment uses resistance to generate heat and emit light, too close to flammable substances, causing flammable substances to burn, or overheating as a result of poor heat dissipation.

2. Electric spark or arc

The electric spark or arc features very high temperature, especially the arc, to the extent that the stabilized arc pole area can be as high as 5,000–10,000 °C. This temperature can ignite combustibles, and even melt and splash metals, thus forming a secondary ignition source. In places with potential risks of fire or explosion, electric sparks and arcs are dangerous ignition sources.

(1) Working spark and arc: Sparks produced during normal running or operation of electrical equipment, such as switching, operating of motor carbon, welding machine, and stove ignition.

(2) Accident spark and arc: Abnormal sparks generated by electrical equipment accidents, such as arc produced by short circuit, spark from a broken line, flashover of insulating media, and lightning discharge.

(3) Other types: Sparks produced by causes other than electricity, such as sparks generated by metal collisions.

(Ⅱ) Electrical fire fighting

The electrical fire differ from ordinary fire in the following ways: First, the fire source is electrically charged, and there is a risk of electric shock for those present; second, the oil-filled or inflation equipment has the risk of chemical and physical explosion.

1. Power off

In case of an electrical fire, cut off the power supply, and then do it the usual way. The safety requirements for cutting off the power supply are as follows:

(1) In case of fire, the insulation of the switchgear is reduced as a result of smoke and fire; we should use insulation tools or items instead of the human body during operation to prevent the direct human contact with switchgear.

(2) For substations, the switching operation should be strictly observed to prevent misoperation and expand the scope of accidents.

(3) Cut off the low-voltage power supply. If necessary, use tools such as diagonal pliers with insulation covering to cut the wires. Wear insulating boots or stand on insulation pads when cutting the wires. The breakpoint should be supported after cutting to prevent grounding short circuit. Cutting should be done in phases, not together, to prevent the high temperature sparks from interphase short circuit.

(4) When fighting at night, there should be standby lighting measures in order not to affect the fire extinguishing after power failure.

2. Live-wire fire fighting

Sometimes it's too late to turn off the power in order to get a good time to put out the fire, or the power supply cannot be cut off for some reason, the precautions for live-wire fire fighting are as follows:

(1) Non-conductive fire extinguishers, including carbon dioxide, halon 1211, ABC dry powder, and some mechanical foam (water-based) fire extinguishers, can be used for live-wire fire fighting. However, attention should be paid to its scope of application, and ensure the safe distance between operators and charged bodies. Ordinary mechanical foam or chemical foam fire extinguishers are electrically conductive and cannot be used for live-wire fire fighting.

(2) When the clear water from a fire hydrant is used to extinguish a charged flame, we should note the high conductivity of water, and the straight stream nozzle cannot be used directly to avoid the electric shock to firefighters. Instead, spray nozzles must be used, which can fully atomize water when the water pressure is large enough; moreover, the parabolic fire extinguishing can greatly reduce the leakage current in water. In addition, while spraying to extinguish fire, be sure to keep a safe distance from live equipment: 3 m for 110 kV and below, 5 m for 220 kV and above. Meantime, the fire hose nozzle should be grounded, and the operator must wear insulating boots and gloves.

(Ⅲ) Safety precautions for electrical fire

Electrical equipment and lines are known for many types and complicated structures, and fire suppression must be carried out according to their characteristics to ensure safety, feasibility and effectiveness.

(1) When overhead lines are above the fire scene, please do not stand under the wires to prevent the broken line from falling and causing electric shock.

(2) When caught fire, the oil-filled equipment is prone to explosion. If the oil-filled equipment is on fire outside, non-conductive fire extinguishers can be used to put out the fire as long as it is safe. If the oil-filled equipment catches fire as a result of internal fault, you must first power off to ensure safety, and then put out the fire using the cooling method. Even if the flame is extinguished, please keep spraying to cool down. If oil leaks from the tank and there is no oil pool on the ground nearby, fire sand must be applied to extinguish the fuel on the ground to prevent the spread of oil. If

the flame is found late, and the temperature of the oil-filled equipment is so high that there is a greater risk of explosion, please evacuate to a safe place immediately after power failure, call the fire department, and do not get close to prevent any injury by the explosion.

(3) When the electric rotating machinery such as the generator and motor is on fire, the motor should be evenly cooled in the spray cooling process to prevent the deformation of mechanical parts. Fire sand or dry powder fire extinguishers should not be used to avoid any damage to the motor.

(4) When the transformer and power distribution unit catch fire, first cut off the power supply, and then spray the fire with fire hose nozzles. Halon 1211 or carbon dioxide fire extinguishers can be used indoors, but not fire sand or dry powder, so as to avoid equipment damage.

IV. Escape and self-rescue methods from fire scene

(I) Principles

(1) Know and get familiar with the surroundings, including the locations of escape passages, safety exits, fire elevators, fire hydrants, fire extinguishers, in order to find escape routes in the fire smoke.

(2) Evacuate quickly and call the police: Once you hear the fire alarm, or see the smoke and/or flames, and realize that you are in the fire scene, do not hesitate to evacuate immediately and call 119.

(3) Protect the respiratory system: Fire will produce a lot of smoke, toxic gas, and high temperature gas. To prevent the inhalation of smoke and toxic gas, please cover your mouth and nose with wet towels or clothes, in an effort to minimize the degree of respiratory poisoning and burns. When escaping, bend or even crawl forward to prevent the damage from upper smoke.

(4) Prevent high temperature burns: If you're close to the fire, or will pass near the fire zone on your way out, be sure to prevent high temperature burns in addition to protecting your respiratory system. The easiest way is to wrap wet cotton clothes and bedding around you. Remember not to wear clothes made of synthetic fibers such as lacquer, polyamide, acrylic, and vinylon. Compared with cotton and wool fabrics, they are prone to heat and melt on our skin, causing extensive scalding.

(5) Temporary refuge measures: With no way out, please shelter in toilets, bathrooms and other spaces with few combustibles and water available. When taking refuge, it is necessary to spray the fire doors and windows with water, wet all combustibles in the room, including yourself, and plug the door with wet towels to prevent smoke from entering and prolong the survival time. During the refuge, take the initiative to contact the outside world for early rescue.

(6) Notice the sign boards: In an indoor fire, the middle and upper parts of the building may be full of smoke, and the bottom is usually not covered by smoke. We should crawl forward at this time. Many "Fire Exit" sign boards (see Fig. 9-20, still lit after power failure) are placed in the corners. You'll see them clearly while crawling forward, and then follow the signs to get out of the fire scene.

Fig. 9-20 "Fire Exit" sign board

(7) Take advantage of hardware: Take full advantage of hardware at the fire scene, such as fire elevators, outdoor escape stairs, roofs, balconies, gas pipes and other facilities; or tie the rope, curtain or sheeting (torn into strips) to a strong frame, and follow the rope down the stairs to escape.

(8) Benefit other people as well as oneself: We cannot escape a fire at the expense of others. Both self-rescue and mutual rescue are a good solution to minimize casualties.

(Ⅱ) Typical mistakes in a fire scene

(1) Being greedy for property: Fire escapers should not be greedy for property, otherwise it will often lead to the death of those involved.

(2) Jumping off a building: Failure to keep cool and calm in case of fire, and blind jumping will often result in death or disability. Never escape a fire by jumping from a building until the critical moment and you can keep yourself safe.

(3) Running on fire: Running around with your clothes on fire will make the burning more full and violent, and it is less likely to be extinguished. In the absence of extinguishing agents nearby, you can roll on the spot or cover yourself with heavy clothing for fire suppression.

(4) Shouting: There is a lot of smoke in the fire scene. Shouting will make the victims inhale more smoke, resulting in asphyxia.

(5) Taking the elevator: Non-fire elevators have no smoke and heat insulation functions, and cannot be used in the fire; in case of power failure, passengers will suffocate from the smoke.

(6) Walking upright: Walking upright will lead to suffocation by breathing in too much smoke.

(Ⅲ) Noticing the details of everyday life

(1) Familiar with the fire-fighting equipment and facilities around you: High-rise buildings will be equipped with fire-fighting equipment and facilities, such as smoke alarms, enclosed stairs, smoke prevention stairs (more than 32 floors), fire elevators, fire hydrants, fire doors at the stair exit, and fire extinguishers. Having known these fire-fighting equipment and facilities, you'll not find it difficult to put out the incipient fire calmly.

(2) Stacking debris in the stairwell in daily life will affect escape, and even cause fire in enclosed stairs or smoke prevention stairs. Therefore, be sure to keep debris away from the stairwell in daily life.

(3) Only in the closed state can the normally closed fire door (see Fig. 9-21) block the smoke and fire and delay the spread of fire and smoke to the stairwell and fire fighting access effectively in the event of a fire, so as to gain valuable time for personnel evacuation and fire rescue. Therefore,

the normally closed fire door must be kept normally closed. Never open it for convenience, nor lock it for the purpose of guard against theft.

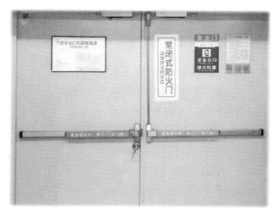

Fig. 9-21　Normally closed fire door

(4) Flammables and explosives cannot be stacked on the open balcony, otherwise they may be ignited by the ignition source floating from the outside. Furthermore, the stacking of combustible materials on the balcony is prone to the external vertical spread of fire.

(5) Now, a large number of electric vehicle lithium batteries on the market do not meet the product standards, and there is a high possibility of bulging, explosion and burning during the charging process. Therefore, electric vehicles must not be charged overnight, nor charged in places such as corridors and safety exits. Otherwise, the fire will block the escape route, coupled with the combustible materials stacked in the corridor, giving rise to unimaginable consequences.

Fig. 9-22　Electric vehicle on fire in the corridor

(Ⅳ) High-rise building fire

The *Code for Fire Protection Design of Buildings* (GB 50016—2014) has clear norms and requirements for high-rise buildings and fire codes. However, the fire hazards in high-rise buildings are still prominent in China.

The high-rise building fire is characterized by fast spread of fire, difficult evacuation and great rescue difficulty, since high-rise buildings are structurally complex and densely populated, making them difficult to control and escape. High-rise building fire cases show that the existing fire rescue means are difficult to adapt to the rapid construction and development of high-rise buildings. Therefore, China has set very high requirements for fire and explosion protection of high-rise buildings, which are often equipped with a complete set of fire-fighting equipment and facilities. Fig. 9-23 shows a high-rise building fire.

Fig. 9-23 High-rise building fire in Chongqing on December 26, 2020

It is of practical significance to study the characteristics of high-rise building fire and understand the technical status of fire-fighting equipment. In this way, people will be guided to choose the correct rescue and escape facilities and respond to the high-rise building fire properly for emergency escape. When caught in a high-rise building fire, you should note the following points:

(1) When any floor catches fire, the fire has spread and cannot be extinguished immediately, and a surge of smoke prevents us from judging the correct floor, please listen for the radio and alarm notice, look out the window, and judge the floor on fire. Don't panic and act blindly when you hear the alarm and see the dense smoke.

(2) When the house catches fire, and the door has been blocked and sealed by the flame, you may find other escape routes, for example, temporarily moving to other non-burning rooms through balconies, corridors, etc. Never open a door through the fire or jump off a building in the absence of safety measures.

(3) If you hear the alarm at night, touch the door with the back of your hand. If the door gets hot, do not open the door, otherwise the flames will rush into the room immediately. At this time, you can plug the door with wet towels or bedding, and pour water over the door to make it more resistant to fire. If the door is not hot, the flame is not large. Then open the door and leave the room. Do keep the door closed to prevent the spread of fire.

(4) In case of heavy smoke in your escape, do not try to rush through the smoke, because a fire must be in the rapid spread behind the smoke. Otherwise, you may suffocate in a few minutes in case of failure to rush out timely.

(5) If the fire has blocked all the way down, you may go to the roof first. If there is an adjacent stairwell on the roof, just evacuate from that stairwell; otherwise you can only signal to the rescue workers on the roof and wait for rescue.

(6) In places full of smoke, when it is not possible to rush out of the smoke, you should bend or even crawl forward. The closer you are to the floor, the less smoke there is. This can delay the time of your death from asphyxia. Cover your mouth and nose with a wet towel where applicable.

(7) If trapped in the lower levels of a high-rise building, you can throw soft objects such as mattresses and quilts downstairs and jump to escape, when your life is in danger and there is no other way out. Alternatively, you can wet the sheets and ropes, weave them into a long rope and put it downstairs for escape.

(8) If the enclosed corridor catches fire and you cannot escape from the corridor, please do not take the elevator blindly. Power outage, heavy smoke and high temperature may occur during ordinary elevator rides. Take fire elevators only, which features the function of smoke and heat insulation. When using the fire elevator, please smash the glass outside the fire switch and press ON. At this time, an alarm will sound and the elevator will enable the fire protection function, as shown in Fig. 9-24.

Fig. 9-24　Fire elevator switch

(9) Each floor of the high-rise building is equipped with portable fire extinguishers, which can be used to extinguish the incipient fire in rooms. There will also be indoor fire hydrants, which can be used to spray water if the portable fire extinguisher cannot put out the flame. However, the fire hydrants are filled with the plain water extinguishing agent, which is a good conductor of electricity. Make ensure that the whole house is powered off before use, otherwise it will lead to electric shock casualties.

(Ⅴ) Fire in low- and mid-rise residential buildings

Many ageing buildings in China are low- and mid-rise residential buildings, which have three characteristics in terms of fire protection:

First, the old housing is fraught with aging power lines, kitchen fumes clinging to the walls, accumulation of debris in the corridors, etc. The possibility of fire is much higher than that of high-rise buildings.

Second, such residential buildings are often not equipped with fire elevators, fire doors, alarm devices, even fire hydrants and fire extinguishers. Therefore, in the event of fire, the fire-fighting capacity and self-rescue capacity of households will be greatly limited before the firefighters come.

Third, since no underground parking lot is planned for the old compound, a large number of cars are parked on the ground, which is likely to block the fire escape and increase the rescue time of firefighters.

Given the above three points, the self-rescue ability of residents is particularly important in case of fire in the low- and mid-rise residential buildings. When a fire occurs in low- and mid-rise old residential buildings, special attention should be paid to the following points.

(1) There is no fire linkage device in the old compound. In case of a fire, call 119 immediately, the sooner the better.

(2) In case of fire, cut off the power supply immediately. There are a lot of ions in the flame, which are a good conductor of electricity. Old houses have complicated wiring, and are prone to fire spreading or electric shock due to leakage.

(2) It is best to equip your own portable fire extinguisher at home. It can quickly extinguish any incipient fire.

(3) If trapped in the lower levels, you can throw soft objects such as mattresses and quilts downstairs and jump to escape, when your life is in danger and there is no other way out. Alternatively, you can wet the sheets and ropes, weave them into a long rope and put it downstairs for escape.

(4) In order to prevent theft, low-rise residents often install security windows, and ordinary security windows will hinder escape in the fire. Please try to choose security windows with fire design, which can be opened if necessary.

(5) While escaping, do not be greedy for property. It is very dangerous to go back and retrieve your belongings.

(6) While escaping, cover your mouth and nose with a wet towel to delay the time of death from asphyxia.

(7) When you cannot escape through the smoke, lean half your body out of the window, breathe fresh air, and wait for rescue. Remember not to jump off the building.

(8) The low- and mid-rise buildings are of low height, and accessible by the fire ladders. You can call for help on the balcony (if any), and wait for firefighters.

(9) The gas pipes and sewer lines of the old dwellings may be arranged outdoors, and can be used to escape routes when there is no way out. This method is generally not suitable for the elderly and young children.

(Ⅵ) Fire in public places

Public places, especially cinemas, KTV, restaurants, and department stores, are densely populated, but there are fewer escape routes, which brings certain difficulties to personnel escape. Some public places feature unique characteristics, making the means of escape in different public places have unique characteristics.

(1) Cinemas: Since there is no light when the movie is playing, there will be no way to find the exit in case of fire. Therefore, when entering the cinema, please carefully observe the location of the "escape route" and "fire exit" for emergency evacuation.

(2) Large shopping malls and supermarkets: Shopping malls usually use escalators up and down, and the escalators are too narrow to serve as an escape route in the event of a fire. Please

observe the location of enclosed stairs when entering the shopping malls. Some shopping malls lock the fire door leading to the fire fighting access for various reasons (as shown in Fig. 9-25). This goes against the *Fire Law of the People's Republic of China*. Once found, please call the local fire station, or call 96119 to make a complaint.

Fig. 9-25　Fire door locked improperly

(3) KTV and bars: Such places are usually more crowded at night, coupled with dim lighting, prone to stampede accidents in case of a fire. Besides, a large amount of plastic and foam are used as sound insulation in the housing structure of such places. They are highly combustible in case of fire, producing a large number of toxic gases. Therefore, please bend or even crawl forward when escaping, and cover your mouth and nose with wet towels and clothes.

V. Daily fire prevention measures at home

(I) Causes of daily fire

Modern homes generally have the three elements of fire. There are combustible upholstery materials in our houses, such as wood, wallpaper, and density board, and flammable cleaning agents, such as gasoline, ethanol, and banana oil. 20% of oxygen in the air is an oxidizing agent. As long as the above combustibles meet the ignition source, a flame can appear; if not controlled in time, it will evolve into a fire. At present, the occurrence of home fire is ascribed to the following reasons.

(1) Kitchen fire is caused by improper use of coal gas and natural gas. Urban households have basically used natural gas or coal gas for cooking. The natural gas or coal gas is more likely to cause fire due to improper use than firewood and coal, such as continued use of gas stoves after failure, and heating up without water.

(2) Electrical fire: Modern families have a large number of household appliances. The old line, damaged electrical wiring, and overloaded plug board are prone to fire. Electrical fire is the most common type of home fires.

(3) Fires caused by smoking: Throwing cigarette butts at random, or throwing them into garbage cans without extinguishing them can cause fires.

(4) Children playing with fire: Children are so curious that they often play with fire as a pleasure. If their parents do not stop it in time, it will often lead to fires.

(Ⅱ) Kitchen fire protection

(1) The natural gas and coal gas must be used in line with the operation of "ignition before gas supply". Otherwise, the combustible gas will be fully mixed with air to trigger the mixed combustion described above, manifested as "deflagration". In case of any flammables nearby, such as liquor, it will immediately cause combustion.

(2) Natural gas or coal gas itself is colorless and odorless. Natural gas and coal gas companies add a smelly gas called tetrahydrothiophene, so that we can detect its leakage. When you smell natural gas or coal gas leaks at home, please open the doors and windows for ventilation. Remember not to use open flame, nor turn on the light, otherwise it may cause mixed combustion or even explosion. Then close the valve.

(3) If you can smell tetrahydrothiophene when the valve is closed, the pipe has leaked. Open the door and window, shut off the master valve, and then call the natural gas company to check.

(4) No one is allowed to change, disassemble or transfer valves, meters and other devices in natural gas and coal gas piping.

(5) Natural gas and coal gas hoses are special ones for fuel gas, which cannot be replaced by water pipes or other plastic pipes. Instead, we should purchase and replace the gas hoses by following the instructions of those from the natural gas and coal gas companies.

(6) If the liquefied gas tank is used, the valve is damaged and cannot be closed. This is divided into two situations: no ignition, place the gas tank outside in an open area, keep it off the ignition source, and call 119; already ignited, if the flame is not large enough to ignite other objects, leave it alone and never extinguish it (otherwise, there will be a mixture of gas and air, causing mixed combustion). Keep other combustibles away from the flame, and call 119.

(7) If the liquefied gas tank itself catches fire, do not panic. There is only combustible gas (no oxidizing agent) in the gas tank. As long as it is not overheated, no explosion will happen. At this time, you should close the cylinder corner valve with wet gloves to cut off the air source, and then cover the cylinder with a wet rag or use a dry powder extinguisher to put out the fire.

(8) Do not leave the cooking range unattended while cooking with natural gas and coal gas, otherwise it may cause a kitchen fire as a result of forgetfulness.

(9) If the frying pan catches fire, do not use water to put it out in a panic, which will only increase the fire, as shown in Fig. 9-26. The most direct way is to cover the pot and suffocate the fire.

Fig. 9-26 Frying pan on fire

(Ⅲ) Fire protection for electric appliances

As electricity becomes increasingly common, the proportion of fires caused by electricity is increasing year by year. The improper use of household appliances and other live items will bring greater fire hazards. If any home appliance catches fire, remember to power off and then extinguish the fire.

(1) Washing machine: Washing machines are generally less of a fire risk, but be aware of some special circumstances: If stained with special stains such as paint, clothes should be cleaned with ethanol, gasoline and other flammable liquids, and then cleansed with water before being put into the washing machine. Do not put too much clothes into the vintage washing machine at a time, otherwise it will easily lead to excessive motor load and even locked rotor. Besides, the vintage washing machine does not have overheating protection, which will cause the motor to burn or even cause a fire.

(2) Air conditioning: Air conditioning is generally divided into internal and external units, and the internal unit is very unlikely to catch fire. If the external unit is facing the sun and free from any shading measures, turning on the air conditioner for a long time in summer may cause the external unit to overheat and cause fire hazards.

(3) Electric iron: When in use, the electric iron without overheating protection must not be left unattended, otherwise it is prone to a fire once forgotten.

(4) Electric blanket: Do not buy fake and shoddy goods without safety protection measures. The fake and shoddy goods are prone to inter-turn short circuit, leading to partial overheating and even burning the bedding. What's worse, the electric leakage may give rise to electric shock casualties, as shown in Fig. 9-27. The electric blanket must be turned off when the user leaves.

(5) Electric kettle: The electric kettle may overflow when boiling water. It cannot be placed next to the socket or other live equipment, so that there will be no leakage or short circuit even if water overflows.

Fig. 9-27 Spontaneous combustion of inferior electric blanket

(6) Electric hair dryer: When used, the electric hair dryer cannot be left unattended, nor be placed on the sofa, bedding and other combustibles. Moreover, you must not block the air outlet or intake with towels and other items, as shown in Fig. 9-28. This is particularly important for the vintage hair dryer without overheating cutout. The electric heating wire is insulated by air, no other insulation, included in the scope of open flame management. The electric hair dryer cannot be used in case of a lot of ethanol gas, suspended flour and other combustible gases or combustible solid powders in the space.

(7) Refrigerator: For the new refrigerator, the plastic plate and foam for packaging must be removed to ensure the normal heat dissipation. The back of the refrigerator is the compressor radiating tube, and cannot be placed flammables and explosives with low fire points, such as liquor, gasoline, and banana oil in summer.

(8) Electric heater: There are many kinds of electric heaters, including oil tincture, heat radiation, and hot air. Despite the kind, the heating part cannot be close to flammables and explosives, nor can clothes block heat dissipation for quick drying. This is prone to a fire. See Fig. 9-29.

Fig. 9-28 Fire caused by the blocking of air intake and outlet of the electric hair dryer

Fig. 9-29　Towel scorched by an electric heater

　　(9) Socket, plug board, switch, etc.: Regular products are preferred. Most of the regular products use copper sheets, with high clamping force. Many fake and shoddy goods use iron sheets, whose clamping force is not enough, resulting in high contact resistance, serious heating, overload, and even fire. Pay attention to its rated power when in use. The power of electrical appliances should not be greater than the rated power of sockets, plugboards and switches, otherwise long-term overloading will cause fire due to overheating. In particular, the plugboard should not be connected with other external plugboards, which can easily cause overload.

　　(10) Range hood: The fire caused by the range hood is usually not caused by electricity, but an open flame. The surface of the range hood is easy to be contaminated with greasy dirt, which is easily ignited by open flames. It works right above the cooking range, a very harsh environment; especially when the pot is on fire, the flame may ignite the surface of the range hood, resulting in a fire. Therefore, we should clean the surface of the range hood regularly.

(Ⅳ) Fire prevention for smoking at home

　　The burning cigarette butt is generally 200–300 °C, and the central temperature reaches 700–800 °C. General paper, cotton, cloth and other combustibles have a fire point of mostly below 300 °C, and something like toilet paper has a lower fire point. This shows that the cigarette butt has actually exceeded the fire point of many combustibles; thus we must pay attention to fire prevention when smoking at home.

　　(1) Do not smoke on the sofa and bed, as the falling hot soot may ignite the sofa, bedding, etc.

　　(2) Do not smoke in the room when you are drunk or overtired. You might have fallen asleep before a cigarette is finished, and cigarette butts fall on combustibles, resulting in a fire.

　　(3) Don't leave lit cigarette butts lying around, such as on the writing desk, windowsill, dining table, etc. When leaving to do other things, you may forget the burning cigarette butt, thus igniting other combustibles.

　　(4) Dispose of the cigarette butt after smoking. Be sure to extinguish the fire completely before throwing it in the trash, otherwise it may ignite other garbage in the trash can.

任务四　灭火器选择与检查实训

一、作业任务

（1）教师模拟火灾场景，学生根据模拟火灾场景的可燃物与氧化剂类别、燃烧状况、现场各种环境信息，选择正确的灭火器、灭火方式和逃生方式。

（2）正确检查手提式干粉灭火器。

二、引用标准及文件

（1）《安全生产技术》。

（2）《建筑设计防火规范》（GB 50016—2014）。

三、作业条件：室内模拟现场

（1）消防设施：见表9-6。

表9-6　虚拟现场消防设施

序号	名称
1	室内手动防排烟系统
2	室内手动警报装置
3	室内手动喷淋系统
4	室内（室外）消防栓

（2）消防器材：见表9-7。

表9-7　虚拟现场消防器材

序号	名称	灭火剂类别
1	手提式灭火器	二氧化碳
2		清水
3		化学泡沫
4		机械泡沫（水基式）
5		ABC干粉
6		1211卤代烷
7		7150灭火剂
8		酸碱
9	消防沙	石英砂

四、作业前准备

室内笔试，准备纸笔。

五、作业规范及要求

（1）根据灭火器配置场所的使用性质与可燃物种类，可初步判断该场所可能发生哪种类别的火灾。

（2）在灭火机理相同的情况下，有几种灭火器均适用于扑救同一种类的火灾。但它们在灭火程度上有一定的差别，因此，对于不同程度、不同范围的同类火灾，选择灭火器时应当充分考虑该因素。

（3）为了保护贵重物资和设备，免受不必要的污染和损失，灭火器的选择应考虑其对保护物品的污染程度。

（4）要选择适用的灭火器，应先对使用人员的年龄、性别和身手是否敏捷等进行大概分析估计，然后正确选择合适的灭火器。

（5）环境温度过低，某些灭火器灭火性能会显著降低，温度过高，灭火器有爆炸伤人危险，在不同温度环境采用不同的灭火器。

（6）在选择灭火器时，应考虑不同灭火剂之间可能产生的相互反应、污染及其对灭火的影响，部分灭火剂之间可能会发生相互作用。

（7）灭火器的检查要点。

① 检查手提式干粉灭火器瓶体无变形，无破损，见图9-30。

② 把手无变形，见图9-31。

图9-30　手提式干粉灭火器瓶体

图9-31　手提式干粉灭火器把手

③ 保险销、铅封无损坏、丢失，见图9-32。

④ 皮管无老化、龟裂，见图9-33。

图9-32　手提式干粉灭火器保险销

图9-33　手提式干粉灭火器皮管

⑤ 压力表指示在绿色范围，见图9-34。

⑥ 喷嘴通畅，见图9-35。

图9-34 手提式干粉灭火器压力表　　图9-35 手提式干粉灭火器喷嘴

⑦ 合格证处于有效期内，见图9-36。

图9-36 手提式干粉灭火器合格证

⑧ 瓶底钢印可见，见图9-37。

图9-37 手提式干粉灭火器钢印

⑨ 底部无锈蚀，见图9-38。

图9-38 手提式干粉灭火器底部

六、作业流程及标准（表 9-8）

表 9-8 灭火器选择与检查评分标准

班级		姓名		学号		考评员		成绩		
序号	作业名称	质量标准			分值/分	扣分标准			扣分	得分
1	灭火方式选择	根据老师模拟的现场，选择合适的灭火方式灭火			40	灭火方式选择错误造成人身伤亡后果扣 40 分；灭火方式选择错误造成重要设备损坏、无法扑灭等后果扣 20 分；灭火方式选择不准确导致延误扑灭时间等扣 10 分				
2	逃生方式选择	根据老师模拟的现场，选择合适的逃生方式			33	逃生方式选择错误直接造成人身伤亡后果扣 33 分；逃生方式错误导致增加自身、他人伤亡风险等扣 20 分；逃生方式不准确导致延误救援等扣 10 分				
3	灭火器检查	检查手提式干粉灭火器瓶体无变形，无破损；把手无变形；保险销、铅封无损坏、丢失；皮管无老化、龟裂；压力表指示在绿色范围；喷嘴通畅；合格证处于有效期内；瓶底钢印可见；底部无锈蚀			27	灭火器检查漏一项或错一项扣 3 分				
合计					100					

Task Ⅳ Training in the selection and inspection of fire extinguishers

Ⅰ. Operating tasks

(1) The teacher simulates the fire scenes, and students are expected to select the correct fire extinguishers, extinguishing methods and escape routes according to the combustibles and oxidizing agents, burning conditions and environmental information of the simulated fire scenes.

(2) Properly check portable dry powder fire extinguishers.

Ⅱ. Referenced standards and documents

(1) *Work Safety Technology*;
(2) *Code for Fire Protection Design of Buildings* (GB 50016—2014).

Ⅲ. Operating conditions: Indoor simulated scenes

(1) Fire-fighting equipment: See Table 9-6.

Table 9-6 Fire-fighting equipment on virtual scenes

S/N	Name
1	Indoor manual smoke control and exhaust system
2	Indoor manual alarm device
3	Indoor manual sprinkler system
4	Indoor (outdoor) fire hydrants

(2) Fire-fighting facilities: See Table 9-7.

Table 9-7 Fire-fighting facilities on virtual scenes

S/N	Name	Category of extinguishing agent
1	Portable fire extinguisher	Carbon dioxide
2		Plain water
3		Chemical foam
4		Mechanical foam (water-based)
5		ABC dry powder
6		Halon 1211
7		7150 extinguishing agent
8		Soda-acid
9	Fire sand	Quartz sand

Ⅳ. Preparation before operation

Prepare paper and pen for indoor written examination.

Ⅴ. Operating specifications and requirements

(1) According to the nature of fire extinguishers and the type of combustibles, we can preliminarily judge which type of fire may occur at the scene.

(2) Under the same extinguishing mechanism, several fire extinguishers are applicable to fighting the same kind of fire. But they differ somewhat in the degree of fire suppression. Therefore, for different degrees and ranges of similar fires, fire extinguishers should be selected by taking this factor into full account.

(3) In order to protect valuable materials and equipment from unnecessary contamination and loss, fire extinguishers should be selected by considering their degree of contamination to the protected items.

(4) We should first estimate the age, gender and agility of users, and then choose the appropriate fire extinguishers.

(5) If the ambient temperature is too low, the fire extinguishing performance of some fire extinguishers will be significantly reduced; if the temperature is too high, the fire extinguisher may explode and injure people. Hence, different fire extinguishers are used in different temperature environments.

(6) When selecting a fire extinguisher, we must consider the possible interaction between different extinguishing agents, contamination and their effects on fire extinguishing. Some extinguishing agents may interact with each other.

(7) Key inspection points for fire extinguishers.

① Check the body of the portable dry powder fire extinguisher, free from deformation and damage, as shown in Fig. 9-30.

② The handle is intact, as shown in Fig. 9-31.

Fig. 9-30　Body of the portable dry powder fire extinguisher

Fig. 9-31　Handle of the portable dry powder fire extinguisher

③ The safety pin and lead seal are neither damaged nor lost, as shown in Fig. 9-32.

④ The hose is free from ageing and crack, as shown in Fig. 9-33.

Fig. 9-32　Safety pin for the portable dry powder fire extinguisher

Fig. 9-33　Hose for the portable dry powder fire extinguisher

⑤ Pressure gauge indications are in the green range, as shown in Fig. 9-34.

⑥ The nozzle is unobstructed, as shown in Fig. 9-35.

Fig. 9-34　Pressure gauge of the portable dry powder fire extinguisher

Fig. 9-35　Nozzle of the portable dry powder fire extinguisher

⑦ The certificate is within the validity period, as shown in Fig. 9-36.

⑧ There is a visible steel stamp on the bottom of the extinguisher, as shown in Fig. 9-37.

Fig. 9-36　Certificate of the portable dry powder fire extinguisher

Fig. 9-37　Steel stamp of the portable dry powder fire extinguisher

⑨ The bottom is not corroded, as shown in Fig. 9-38.

Fig. 9-38　Bottom of the portable dry powder fire extinguisher

VI. Operation procedures and standards (see Tab. 9-8)

Tab. 9-8 Grading criteria for the selection and inspection of fire extinguishers

Class		Name		Student ID		Examiner		Score	
S/N	Operation name	Quality standard			Points	Deduction criteria		Deduction	Score
1	Choice of extinguishing methods	Choose the appropriate extinguishing method according to the simulated scene			40	Deduct 40 points if the wrong selection of extinguishing method causes the casualty of life; Deduct 20 points if the wrong selection of extinguishing method causes critical equipment damage and failure to put out a fire; Deduct 10 points if the inaccurate selection of extinguishing method leads to the delay in extinguishing time			
2	Choice of escape mode	Choose the right way to escape according to the simulated scene			33	Deduct 33 points if the wrong choice of escape mode directly results in personal injury and death; Deduct 20 points if the wrong way of escape increases the risk of injury and death to oneself and others; Deduct 10 points if the inaccurate means of escape delays the rescue, etc.			
3	Inspection of fire extinguisher	Check the body of the portable dry powder fire extinguisher, which is free from deformation and damage; The handle is intact; The safety pin and lead seal are neither damaged nor lost; The hose is free from ageing and crack; Pressure gauge indications are in the green range; The nozzle is unobstructed; The certificate is within the validity period; There is a visible steel stamp on the bottom of the extinguisher; There is no rust on bottom			27	Deduct 3 points if any item is missing or wrong in the inspection of fire extinguishers			
Total					100				

参考文献

[1] 顾飚. 电力安全知识及案例[M]. 北京：中国电力出版社，2015.
[2] 许庆海. 电力安全基本技能[M]. 北京：中国电力出版社，2008.
[3] 张良瑜. 电业安全[M]. 北京：中国电力出版社，2010.
[4] 中国安全生产协会注册安全工程工作委员会. 安全生产技术[M]. 北京：中国大百科全书出版社，2011.
[5] 国家电网公司. 国家电网公司电力安全工作规程（变电部分）[M]. 北京：中国电力出版社，2009.